# Materials for Civil and Construction Engineers

## Michael S. Mamlouk
*Arizona State University*

■

## John P. Zaniewski
*West Virginia University*

 **ADDISON-WESLEY**

An imprint of Addison Wesley Longman, Inc.

Menlo Park, California • Reading, Massachusetts • Harlow, England
Berkeley, California • Don Mills, Ontario • Sydney • Bonn • Amsterdam • Tokyo • Mexico City

Acquisitions Editor: Michael Slaughter
Associate Editor: Susan Slater
Production Manager: Pattie Myers
Production Coordinator: Kamila Storr
Art and Design Supervisor: Kevin Berry
Production: The Book Company
Composition: Proctor-Willenbacher

Cover Design: Vargas Williams Design
    (Juan Vargas)
Cover Image: Image # BU05508
    © 1998 PhotoDisc, Inc.
Text Printer and Binder: Courier Westford
Cover Printer: Phoenix Color Corp.

Many of the designations used by manufacturers and sellers to distinguish their products are claimed as trademarks. Where those designations appear in this book, and Addison-Wesley was aware of a trademark claim, the designations have been printed in initial caps or in all caps.

Library of Congress Cataloging-in-Publication Data
Mamlouk, Michael S.
    Materials for civil and construction engineers / Michael S. Mamlouk, John P.
Zaniewski.
        p.        cm.
    Includes bibliographical references and index.
    ISBN 0-673-98040-5
    1. Materials.    I. Zaniewski, John P.    II. Title.
TA403.2.M35 1999                                                    98-9934
624.1'8dc21                                                          CIP

**Instructional Material Disclaimer**
The programs presented in this book have been included for their instructional value. They have been tested with care but are not guaranteed for any particular purpose. Neither the publisher nor the author offers any warranties or representations, nor do they accept any liabilities with respect to the programs.

The full complement of supplemental teaching materials is available to qualified instructors.

ISBN 0-673-98187-8

    4 5 6 7 8 9 10—CW—02

**Photo and Art Credits:** p. 24, 28: Burati, J.L. and Hughes, C.S. (199) "Highway Materials Engineering. Module I: Materials Control and Acceptance—Quality Assurance." Publication No. FHWA-HI-90-004, Federal Highway Administration; p. 43, 44, 46, 47: Van Vlack, L.H., *Elements of Materials Science and Engineering*, 6th edition. Copyright © 1989 by Addison Wesley Publishing Company, Reprinted with permission. Continued on page 388 which is a continuation of this copyright page.

**Addison Wesley Longman, Inc.**
2725 Sand Hill Road
Menlo Park, California 94025

# CONTENTS

# CHAPTER 6    Portland Cement    140

# CHAPTER 7    Portland Cement Concrete    165

# CHAPTER 8　Masonry　207

# CHAPTER 9　Asphalt and Asphalt Mixture　217

# CHAPTER 10   Wood   269

# CHAPTER 11   Composites   299

# APPENDIX: Laboratory Manual    314

# Index    381

# PREFACE

Civil and construction engineering provides and maintains society's infrastructure needs, such as buildings, water treatment and distribution systems, waste-water removal and processing, dams, and highway and airport bridges and pavements. Although some civil and construction engineers are involved in the planning process, most are concerned with the design, construction, and maintenance of facilities. The common denominator between these responsibilities is the need to understand how materials behave and perform. Although not all civil and construction engineers need to be material specialists, a basic understanding of the material selection process, is a fundamental requirement for all civil and construction engineers performing design, construction, and maintenance tasks.

Material requirements in civil engineering and construction facilities are different from material requirements in other engineering disciplines. Frequently, civil engineering structures require tons of materials with relatively low replications of specific designs. Generally, the materials used in civil engineering have relatively low unit costs. In many cases, civil engineering structures are formed or fabricated in the field under adverse conditions. Finally, many civil engineering structures are directly exposed to detrimental effects of the environment.

Engineering materials have advanced greatly in the last few decades. As a result, many of the conventional materials have either been replaced by more efficient materials or modified to improve their performance. Civil and construction engineers must be aware of these advances and be able to select the most cost-effective material or use the appropriate modifier to make the material appropriate for the specific application at hand.

This text is organized into three parts: 1) introduction to the materials engineering, 2) characteristics of materials used in civil and construction engineering, and 3) laboratory methods for the evaluation of materials.

The introduction to materials engineering includes information on the basic mechanistic properties of materials, environmental influences, and basic material classes. Because one of the responsibilities of civil and construction engineers is the inspection and quality control of materials in construction process, an understanding of material variability and testing procedures is required. The atomic structure of materials is discussed to provide basic understanding of material behavior and to relate the molecular structure to the engineering response.

The second section, which represents a large portion of the book, presents the characteristics of the primary material types used in civil and construction engineering: steel, aluminum, concrete, masonry, asphalt, and wood. Since the discussion of concrete and asphalt materials requires a basic knowledge of aggregates, there is a chapter on aggregates. Moreover, since composites are gaining wide acceptance among engineers and are replacing many of the conventional materials, there is a chapter introducing composites.

The discussion of each material includes information on the

- basic structure,
- material production process,
- mechanistic behavior and other properties,
- environmental influences,
- construction considerations, and
- special topics.

Finally, each chapter includes an overview of various test procedures to introduce the test methods used with each material. However, the detailed description of the test procedures is left to the appropriate standards organizations such as the American Society for Testing and Materials (ASTM) and the American Association of State Highway and Transportation Officials (AASHTO). These ASTM and AASHTO standards are usually available in college libraries, and students are encouraged to use them. Also, there are sample problems in most chapters, as well as selected questions and problems at the end of each chapter. Answering these questions and problems will lead to a better understanding of the subject matter.

There are volumes of information available for each of these materials. It is not possible, or desirable, to cover these materials exhaustively in a single introductory text. Instead, this book limits the information to an introductory level, concentrates on current practices, and extracts information that is relevant to the general education of civil and construction engineers.

The content of the book is intended to be covered in one academic semester, although quarter system courses can use it, too. The instructor of the course can also change the emphasis of some topics to match the specific curriculum of the department. Furthermore, since such a course usually includes a laboratory portion, a number of laboratory test methods are described. In the appendix there are more laboratory tests provided than needed for a typical semester so that the instructor has the flexibility to use available equipment. Laboratory tests should be coordinated with the topics covered in the lectures so that the students get the most benefit from the laboratory experience.

The authors would like to acknowledge the contributions of Drs. Barzin Mobasher and Les Hendrickson of Arizona State University for their advice for and providing some photos. Appreciation also goes to Mr. Mofreh Saleh and Mrs. Alejandra Medina of Arizona State University for their contributions in the preparation of the book materials.

# About the Authors

Michael S. Mamlouk is a Professor of Civil and Environmental Engineering at Arizona State University. He has many years of experience in teaching courses of civil engineering materials and other related subjects at both undergraduate and graduate levels. Dr. Mamlouk is the author of many publications in the fields of pavement and materials, and is a member of several professional societies, including the American Society of Civil Engineers, the American Society for Testing and Materials, the Transportation Research Board, and the Association of Asphalt Paving Technologies. As a registered professional engineer, Dr. Mamlouk has served many agencies as a consultant and expert witness in various aspects of pavement, asphalt, and concrete, and has been the principal investigator of numerous research projects funded by international, federal, state, and local agencies. His interests include civil engineering and highway materials, pavement analysis and design, pavement management systems, pavement evaluation and maintenance, vehicle-pavement interaction, and computer applications.

John P. Zaniewski is the Asphalt Technology Professor in the Department of Civil and Environmental Engineering at West Virginia University. While teaching at Arizona State University, Dr. Zaniewski was awarded Chi Epsilon's James M. Robbins Excellence-in-Teaching Award. He coauthored the textbook *Modern Pavement Management*. Dr. Zaniewski has been active in the field of technology transfer, and has presented more than 40 seminars in support of the $T^2$ Program in several states. He also has taught a number of Federal Highway Administration short courses. Among others, his fields of research include pavement management and performance, materials for pavements, and vehicle operating costs.

# 1 Materials Engineering Concepts

Materials engineers are responsible for the selection, specification, and quality control of materials to be used in a job. These materials must meet certain classes of criteria or materials properties (Ashby and Jones 1980). These classes of criteria include

- economic factors
- mechanical properties
- nonmechanical properties
- production/construction considerations
- aesthetic properties

When engineers select the material for a specific application they must consider the various criteria and make compromises. Both the client and purpose of the facility or structure dictate, to a certain extent, the emphasis that will be placed on the different criteria.

Civil and construction engineers must be familiar with materials used in the construction of a wide range of structures. Materials most frequently used include steel, aggregate, concrete, masonry, asphalt, and wood. Materials used to a lesser extent include aluminum, glass, plastics, and fiber-reinforced composites. Geotechnical engineers make a reasonable case for including soil as the most widely used engineering material, since it provides the basic support for all civil engineering structures. However, the properties of soils will not be discussed in this text since this is generally the topic of a separate course.

Recent advances in the technology of civil engineering materials have resulted in the development of better quality, more economical, and safer materials. These materials are commonly referred to as high-performance materials. Because more is known about the molecular structure of materials and because of the continuous research efforts by scientists and engineers, new materials such as polymers, adhesives, composites, geotextiles, coatings, cold-formed metals, and various synthetic products are competing with traditional civil engineering materials. In addition, improvements have been made to existing materials through changing their molecular structures or including additives to improve quality, economy, and performance. For example, superplasticizers have made a breakthrough in the concrete industry, allowing the production of much stronger concrete. Joints made of elastomeric materials have improved the safety of high-rise structures in earthquake-active areas. Lightweight synthetic aggregates have decreased the weight of concrete structures, allowing small cross-sectional areas of components. Polymers have been mixed with asphalt, allowing pavements to last longer under the effect of vehicle loads and environmental conditions.

The field of fiber composite materials has developed rapidly in the last 30 years. Many recent civil engineering projects have used fiber-reinforced composites. These advanced composites compete with traditional materials due to their higher strength to weight ratio and their ability to overcome such shortcomings as corrosion. For example, fiber-reinforced concrete has much greater toughness than conventional portland cement concrete. Composites can replace reinforcing steel in concrete structures. In fact, composites have allowed the construction of structures that could not have been built in the past.

The nature and behavior of civil engineering materials are as complicated as those of materials used in any other field of engineering. Due to the high quantity of materials used in civil engineering projects, the civil engineer frequently works with locally available materials that are not as highly refined as the materials used in other engineering fields. As a result, civil engineering materials frequently have highly variable properties and characteristics.

This chapter reviews how the properties of materials affect their selection and performance in civil engineering applications. In addition, the chapter reviews some basic definitions and concepts of engineering mechanics required for understanding material behavior. The variable nature of material properties is also discussed so that the engineer will understand the concepts of precision and accuracy, sampling, quality assurance, and quality control. Finally, instruments used for measuring material response are described.

## Economic Factors

The economics of the material selection process are affected by much more than just the cost of the material. Factors that should be considered in the selection of the material include

- availability and cost of raw materials
- manufacturing costs
- transportation
- placing
- maintenance

The materials used for civil engineering structures have changed over time. Early structures were constructed of stone and wood. These were in ready supply and could be cut and shaped with available tools. Later, cast iron was used because mills were capable of crudely refining iron ore. As the industrial revolution took hold, quality steel could be produced in the quantities required for large structures. In addition, portland cement, developed in the mid 1800s, provided civil engineers with a durable inexpensive material with broad applications.

Due to the efficient transportation system in the United States, availability is not as much of an issue as it once was in the selection of a material. However, transportation can significantly add to the cost of the materials at the job site. For example, in many locations in the United States quality aggregates for concrete and asphalt are in short supply. The closest aggregate source to Houston, Texas, is 150 km (90 miles) from the city. This haul distance approximately doubles the cost of the aggregates in the city and hence puts concrete at a disadvantage when compared to steel.

The type of material selected for a job can greatly affect the ease of construction and the construction costs and time. For example, the structural members of a steel-frame building can be fabricated in a shop, transported to the job site, lifted into place with a crane, and bolted or welded together. In contrast, for a reinforced concrete building, the forms must be built; reinforcing steel placed; concrete mixed, placed, and allowed to cure; and the forms removed. Constructing the concrete frame building can be more complicated and time consuming than constructing steel structures. To overcome this shortcoming, precast concrete units commonly have been used, especially for bridge construction.

All materials deteriorate over time and with use. This deterioration affects both the maintenance cost and the useful life of the structure. The rate of deterioration varies between materials. Thus, in analyzing the economic selection of a material, the life cycle cost should be evaluated in addition to the initial costs of the structure.

## Mechanical Properties

The mechanical behavior of materials is the response of the material to external loads. All materials deform in response to loads; however, the specific response of a material depends on its properties, the magnitude and type of load, and the geometry of the element. Whether or not the material "fails" under the load conditions depends on the failure criterion. Catastrophic failure of a structural member, resulting in the collapse of the structure, is an obvious material failure. However, in some cases the failure is more subtle, but with equally severe consequences. For example, pavement may fail due to excessive roughness at the surface even though the stress levels are well within the capabilities of the material. A building may have to be closed due to excessive vibrations by wind or other live loads, although it could be structurally sound. These are examples of *functional* failures.

### Loading Conditions

One of the considerations in the design of a project is the type of loading the structure will be subjected to during its design life. The two basic types of loads are static and dynamic. Each type affects the material differently, and frequently the interactions between the load types are important. Civil engineers encounter both when designing a structure.

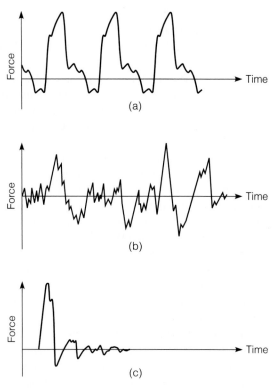

**FIGURE 1.1** Types of dynamic loads: (a) periodic, (b) random, and (c) transient.

    *Static* loading implies a sustained loading of the structure over a period of time. Generally, static loads are slowly applied such that no shock or vibration is generated in the structure. Once applied, the static load may remain in place or be slowly removed. Loads that remain in place for an extended period of time are called *sustained* (dead) loads. In civil engineering, much of the load the materials must carry is due to the weight of the structure and equipment in the structure.

    Loads that generate a shock or vibration in the structure are *dynamic* loads. Dynamic loads can be classified as *periodic, random, or transient,* as shown in Figure 1.1 (Richart et al. 1970). A periodic load, such as a harmonic or sinusoidal load, repeats itself with time. For example, rotating equipment in a building can produce a vibratory load. In a random load, the load pattern never repeats, such as that produced by earthquakes. Transient load, on the other hand, is an impulse load that is applied over a short time interval, after which the vibrations decay until the system returns to a rest condition. For example, bridges must be designed to withstand the transient loads of trucks.

## Stress-Strain Relations

Materials deform in response to loads or forces. In 1678 Robert Hooke published the first findings that documented a linear relationship between the amount of force applied to a member and its deformation. The amount of deformation is proportional

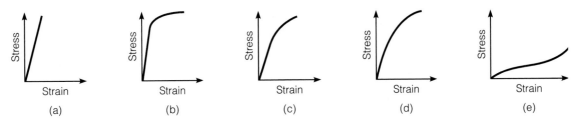

**FIGURE 1.2** Typical uniaxial stress-strain diagrams for some engineering materials: (a) glass and chalk, (b) steel, (c) aluminum alloys, (d) concrete, and (e) soft rubber.

to the properties of the material and its dimensions. The effect of the dimensions can be normalized. Dividing the force by the cross-sectional area of the specimen normalizes the effect of the loaded area. The force per unit area is defined as the stress $\sigma$ in the specimen (i.e., $\sigma$ = Force / Area). Dividing the deformation by the original length is defined as strain $\epsilon$ of the specimen (i.e., $\epsilon$ = Change in length / Original length). Much useful information about the material can be determined by plotting the stress-strain diagram.

Figure 1.2 shows typical uniaxial tensile or compressive stress-strain curves for several engineering materials. Figure 1.2(a) shows a linear stress-strain relationship up to the point where the material fails. Glass and chalk are typical of materials exhibiting this tensile behavior. Figure 1.2(b) shows the behavior of steel in tension, here, a linear relationship is obtained up to a certain point (proportional limit), after which the material deforms without much increase in stress. On the other hand, aluminum alloys in tension exhibit a linear stress-strain relation up to the proportional limit, after which a nonlinear relation follows, as illustrated in Figure 1.2(c). Figure 1.2(d) shows a nonlinear relation throughout the whole range. Concrete and other materials exhibit this relationship, although the first portion of the curve for concrete is very close to being linear. Soft rubber in tension differs from most materials in such a way that it shows an almost linear stress-strain relationship followed by a reverse curve, as shown in Figure 1.2(e).

## Elastic Behavior

If a material exhibits true elastic behavior, it must have an instantaneous response (deformation) to load, and the material must return to its original shape when the load is removed. Many materials, including most metals, exhibit elastic behavior, at least at low stress levels. As discussed in Chapter 2, elastic deformation does not change the arrangement of atoms within the material, but rather it stretches the bonds between atoms. When the load is removed the atomic bonds return to their original position.

Young observed that different elastic materials have different proportional constants between stress and strain. For a homogeneous, isotropic, and linear elastic material, the proportional constant between normal stress and normal strain of an axially loaded member is the *modulus of elasticity* or *Young's modulus, E*, and is equal to

$$E = \frac{\sigma}{\epsilon} \tag{1.1}$$

where $\sigma$ is the normal stress and $\epsilon$ is the normal strain.

In the axial tension test, as the material is elongated, there is a reduction of the cross section in the lateral direction. In the axial compression test, the opposite is true. The ratio of the lateral strain, $\epsilon_l$, to the axial strain, $\epsilon_a$, is *Poisson's ratio, $\nu$*.

$$\nu = \frac{-\epsilon_l}{\epsilon_a} \tag{1.2}$$

Since the axial and lateral strains will always have different signs, the negative sign is used in Equation 1.2 to make the ratio positive. Poisson's ratio has a theoretical range of 0.0 to 0.5, where 0.0 is for a compressible material where the axial and lateral directions are not affected by each other and where the 0.5 value is for a material that does not change its volume when the load is applied. Most solids have Poisson's ratios between 0.10 and 0.45.

Although Young's modulus and Poisson's ratio were defined for the uniaxial stress condition, they are important when describing the three-dimensional stress-strain relationships, as well. If a homogeneous, isotropic cubical element with linear elastic response is subjected to normal stresses $\sigma_x$, $\sigma_y$, and $\sigma_z$ in the three orthogonal directions (as shown in Figure 1.3), the normal strains $\epsilon_x$, $\epsilon_y$, and $\epsilon_z$ can be computed by the *generalized Hooke's law*, that is,

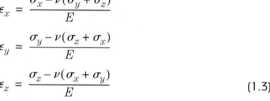

**FIGURE 1.3** Normal stresses applied on a cubical element.

$$\epsilon_x = \frac{\sigma_x - \nu(\sigma_y + \sigma_z)}{E}$$

$$\epsilon_y = \frac{\sigma_y - \nu(\sigma_z + \sigma_x)}{E}$$

$$\epsilon_z = \frac{\sigma_z - \nu(\sigma_x + \sigma_y)}{E} \tag{1.3}$$

**SAMPLE PROBLEM 1.1**

A cube made of an alloy with dimensions of 50 mm × 50 mm × 50 mm is placed into a pressure chamber and subjected to a pressure of 90 MPa. If the modulus of elasticity of the alloy is 100 GPa and Poisson's ratio is 0.28, what will be the length of each side of the cube, assuming that the material remains within the elastic region?

*Solution:*

$$\epsilon_x = \frac{\sigma_x - \nu(\sigma_y + \sigma_z)}{E} = \frac{-90 - 0.28 \times (-90 - 90)}{100000} = -0.000396 \text{ m/m}$$

$$\epsilon_y = \epsilon_z = -0.000396 \text{ m/m}$$

$$\Delta x = \Delta y = \Delta z = -0.000396 \times 50 = -0.0198 \text{ mm}$$

$$L_{new} = 50 - 0.0198 = 49.9802 \text{ mm} \qquad \blacklozenge$$

Linearity and elasticity should not be confused. A *linear material's* stress-strain relation follows a straight line. An *elastic material* returns to its original shape when the load is removed and reacts instantaneously to changes in load. For example, Figure 1.4(a) represents a linear elastic behavior, while Figure 1.4(b) represents a nonlinear elastic behavior.

For materials that do not display any linear behavior, such as concrete and soils, determining a Young's modulus or elastic modulus can be problematical. There are

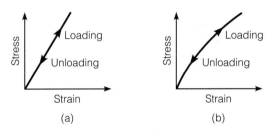

**FIGURE 1.4** Elastic behavior: (a) linear and (b) nonlinear.

several options for arbitrarily defining the modulus for these materials. Figure 1.5 shows four options: the initial tangent, tangent, secant, and chord moduli. The *initial tangent modulus* is the slope of the tangent of the stress-strain curve at the origin. The *tangent modulus* is the slope of the tangent at a point on the stress-strain curve. The *secant modulus* is the slope of a chord drawn between the origin and an arbitrary point on the stress-strain curve. The *chord modulus* is the slope of a chord drawn between two points on the stress-strain curve. The selection of which modulus to use for a nonlinear material depends on the stress or strain level at which the material typically is used. Also, when determining the tangent, secant, or chord modulus, the stress or strain levels must be defined.

Table 1.1 shows typical modulus and Poisson's ratio values for some materials at room temperature. Note that some materials have a range of modulus values rather than a distinct value. Several factors affect the modulus, such as curing level and proportions of components of concrete or the direction of loading relative to the grain of wood.

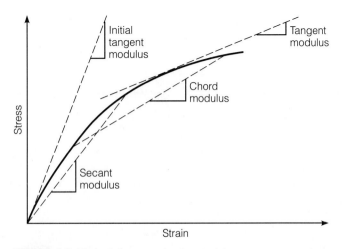

**FIGURE 1.5** Methods for approximating modulus.

**TABLE 1.1**   Typical Modulus and Poisson's Ratio Values (room temperature)

| Material | Modulus, GPa (psi × 10⁶) | Poisson's Ratio |
|---|---|---|
| Aluminum | 69–75 (10–11) | 0.33 |
| Brick | 10–17 (1.5–2.5) | 0.23–0.40 |
| Cast iron | 75–169 (11–23) | 0.17 |
| Concrete | 24–34 (3.5–5) | 0.11–0.21 |
| Copper | 110 (16) | 0.35 |
| Epoxy | 3–140 (0.4–20) | |
| Glass | 62–70 (9–10) | 0.25 |
| Limestone | 58 (8.4) | |
| Rubber (soft) | 0.001–0.014 (0.00015–0.002) | 0.49 |
| Steel | 207 (30) | 0.27 |
| Tungsten | 407 (59) | 0.28 |
| Wood | 6–15 (0.9–2.2) | |

## Elasto-Plastic Behavior

For some materials, as the stress applied on the specimen is increased, the strain will proportionally increase up to a point; after this point the strain will increase with little additional stress. In this case, the material exhibits linear elastic behavior followed by plastic response. The stress level where the behavior changes from elastic to plastic is the *elastic limit*. When the load is removed from the specimen, some of the deformation will be recovered and some of the deformation will remain as seen in Figure 1.6(a). As discussed in Chapter 2, plastic behavior indicates permanent deformation of the specimen so that it does not return to its original shape when the load is removed. This indicates that when the load is applied the atomic bonds stretch, creating elastic response; then, the atoms actually slip relative to each other. When

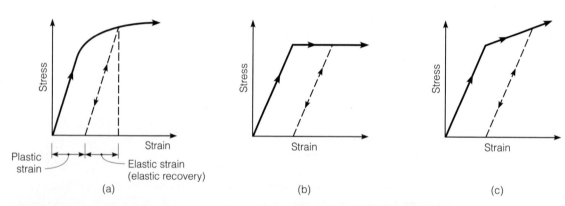

**FIGURE 1.6** Stress-strain behavior of plastic materials: (a) example of loading and unloading, (b) elastic–perfectly plastic, and (c) elasto-plastic with strain hardening.

the load is removed, the atomic slip does not recover; only the atomic stretch is recovered (Callister 1985).

Several models are used to represent the behavior of materials that exhibit both elastic and plastic responses. Figure 1.6(b) shows a linear elastic–perfectly plastic response in which the material exhibits a linear elastic response upon loading, followed by completely plastic response. If such material is unloaded after it has plasticly deformed, it will rebound in a linear elastic manner and follow a straight line parallel to the elastic portion, while some permanent deformation will remain. If the material is loaded again, it will have a linear elastic response followed by plastic response at the same level of stress at which the material was unloaded (Popov 1968).

Figure 1.6(c) shows an elasto-plastic response where the first portion is an elastic response followed by a combined elastic and plastic response. If the load is removed after the plastic deformation, the stress-strain relation will follow a straight line parallel to the elastic portion; consequently, some of the strain in the material will be removed, and the remainder of the strain will be permanent. Upon reloading, the material again behaves in a linear elastic manner up to the stress level that was attained in the previous stress cycle. After that point the material will follow the original stress-strain curve. Thus the stress required to cause plastic deformation actually increases. This process is called *strain hardening* or *work hardening*. Strain hardening is beneficial in some cases since it allows more stress to be applied without permanent deformation. Mild steel is an example of material that experiences strain hardening during plastic deformation.

Some materials exhibit *strain softening*, in which plastic deformation causes weakening of the material. Portland cement concrete is a good example of material of this kind. In this case, plastic deformation causes microcracks at the interface between aggregate and cement paste.

**◄SAMPLE PROBLEM 1.2**

An elasto-plastic material with strain hardening has the stress-strain relation shown in Figure 1.6(c). The modulus of elasticity is $25 \times 10^6$ psi, yield strength is 70 ksi, and the slope of the strain hardening portion of the stress-strain diagram is $3 \times 10^6$ psi.

   **a.** Calculate the strain that corresponds to a stress of 80 ksi.
   **b.** If the 80 ksi stress is removed, calculate the permanent strain.

*Solution:*

   **a.** $\epsilon = \dfrac{70000}{25 \times 10^6} + \dfrac{80000 - 70000}{3 \times 10^6} = 0.0028 + 0.0033 = 0.0061$ in./in.

   **b.** $\epsilon_{\text{permanent}} = 0.0061 - \dfrac{80000}{25 \times 10^6} = 0.0061 - 0.0032 = 0.0029$ in./in.   ◆

Materials that do not undergo plastic deformation prior to failure, such as concrete, are said to be *brittle*, whereas materials that display appreciable plastic deformation, such as mild steel, are *ductile*. Generally, ductile materials are preferred for construction. When a brittle material fails, the structure can collapse in a catastrophic manner. On the other hand, overloading a ductile material will result in distortions of the structure, but the structure will not necessarily collapse. Thus, the ductile material provides the designer with a margin of safety.

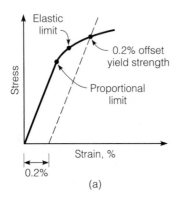

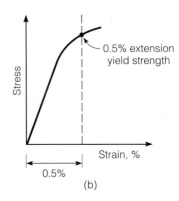

**FIGURE 1.7** Methods for estimating yield stress: (a) offset method and (b) extension method.

**TABLE 1.2**   Offset Values Typically Used to Determine Yield Stress

| Material | Stress Condition | Offset, % | Corresponding Strain |
|---|---|---|---|
| Steel | Tension | 0.20 | 0.0020 |
| Wood | Compression parallel to grain | 0.05 | 0.0005 |
| Gray cast iron | Tension | 0.05 | 0.0005 |
| Concrete | Compression | 0.02 | 0.0002 |
| Aluminum alloys | Tension | 0.20 | 0.0020 |
| Brass and bronze | Tension | 0.35 | 0.0035 |

Figure 1.7(a) demonstrates three concepts of the stress-strain behavior of elasto-plastic materials. The lowest point shown on the diagram is the *proportional limit,* defined as the transition point between linear and nonlinear behavior. The second point is the *elastic limit,* which is the transition between elastic and plastic behavior. However, most materials do not display an abrupt change in behavior from elastic to plastic. Rather there is a gradual, almost imperceptible transition between the behaviors, making it difficult to locate an exact transition point (Polowski and Ripling 1966). For this reason, arbitrary methods are used to identify the elastic limit, thereby defining the *yield stress (yield strength),* such as the *offset* and the *extension* methods. In the offset method a specified offset is measured on the abscissa, and a line with a slope equal to the initial tangent modulus is drawn through this point. The point where this line intersects the stress-strain curve is the *offset yield stress* of the material as seen in Figure 1.7(a). Different offsets are used for different materials (Table 1.2). The *extension yield stress* is located where a vertical projection, at a specified strain level, intersects the stress-strain curve. Figure 1.7(b) shows the yield stress corresponding to 0.5% extension.

**SAMPLE PROBLEM 1.3**

A rod made of aluminum alloy, with a gauge length of 100 mm, diameter of 10 mm, and yield strength of 150 MPa, was subjected to a tensile load of 5.85 kN. If the gauge length was changed to 100.1 mm and the diameter was changed to 9.9967 mm, calculate the modulus of elasticity and Poisson's ratio.

*Solution:*

$$\sigma = \frac{P}{A} = \frac{5850 \text{ N}}{\pi(5 \times 10^{-3} \text{ m})^2} = 74.5 \times 10^6 \text{ Pa} = 74.5 \text{ MPa}$$

Since the applied stress is well below the yield strength, the material is within the elastic region.

$$\epsilon_a = \frac{\Delta L}{L} = \frac{100.1 - 100}{100} = 0.001$$

$$E = \frac{\sigma}{\epsilon_a} = \frac{74.5}{0.001} = 74500 \text{ MPa} = 74.5 \text{ GPa}$$

$$\epsilon_l = \frac{\text{change in diameter}}{\text{diameter}} = \frac{9.9967 - 10}{10} = -0.00033$$

$$\nu = \frac{-\epsilon_l}{\epsilon_a} = \frac{0.00033}{0.001} = 0.33$$

## Work and Energy

When a material is tested, the testing machine is actually generating a force in order to move or deform the specimen. Since work is force times distance, on a diagram the area under a force-displacement curve is the work done on the specimen. When the force is divided by the cross-sectional area of the specimen to compute the stress, and the deformation is divided by the length of the specimen to compute the strain, the force-displacement diagram becomes a stress-strain diagram. However, the area under the stress-strain diagram no longer has the units of work. By manipulating the units of the stress-strain diagram, we can see that the area under the stress-strain diagram equals the work per unit volume of material required to deform or fracture the material. This is a useful concept, for it tells us the energy that is required to deform or fracture the material. This information is used for selecting materials to use where energy must be absorbed by the member. The area under the elastic portion of the curve is the *modulus of resilience,* Figure 1.8(a). The amount of energy required to fracture a specimen is a measure of the *toughness* of the material, as in Figure 1.8(b).

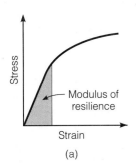

(a)

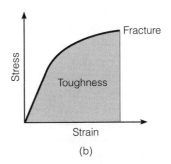

(b)

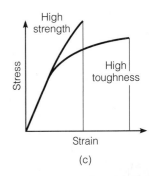

(c)

**FIGURE 1.8** Areas under stress-strain curves: (a) modulus of resilience, (b) toughness, and (c) high-strength and high-toughness materials.

As shown in Figure 1.8(c), a high-strength material is not necessarily a tough material. For instance, as discussed in Chapter 3, increasing the carbon content of steel increases the yield strength, but reduces ductility. Therefore, the strength is increased but the toughness may be reduced.

## Time-Dependent Response

The previous discussion assumed that the strain was an immediate response to stress. This is an assumption for elastic materials, however, no material has this property under all conditions. In some cases, materials have a delayed response. The amount of deformation depends on the duration of the load, the temperature, and the material characteristics. There are several mechanisms associated with time-dependent deformation, such as *creep* and *viscous flow.* There is no clear distinction between these terms. Creep is generally associated with long-term deformations and can occur in metals, ionic and covalent crystals, and amorphous materials. On the other hand, viscous flow is associated only with amorphous materials and can occur under short-term load duration. For example, concrete, a material with predominantly covalent crystals, can creep over a period of decades. Asphalt concrete pavements, an amorphous-binder material, can have ruts caused by the accumulated effect of viscous flows the result of traffic loads that have a load duration of only a fraction of a second.

Creep of metals occurs in three phases. The first phase is the result of dislocation movements in the molecular structure of the metal. The second phase is associated with slip at the grain boundaries, similar to plastic deformation but accelerated due to the high temperature. The third phase is associated with an increase in the strain due to a reduction of the cross section of the specimen. In steel, creep can occur at temperatures greater than 30% of the melting point on the absolute scale. This may be a concern in the design of boilers and nuclear reactor containment vessels. Creep is also considered in the design of wood and advanced composite structural members. Wood elements loaded for a few days can carry higher stresses than elements designed to carry "permanent" loads. On the other hand, creep of concrete is associated with microcracking at the interface of the cement paste and the aggregate particles (Mehta and Monteiro 1993).

The viscous flow models are similar in nature to Hooke's law. In linearly viscous materials, the rate of deformation is proportional to the stress level. These materials are not compressible and do not recover when the load is removed. Materials with these characteristics are *Newtonian fluids.*

Figure 1.9(a) shows a typical creep test where a constant compressive stress is applied to an asphalt concrete specimen. In this case an elastic strain will develop, followed by time-dependent strain or creep. If the specimen is unloaded, a part of the strain will recover instantaneously, while the remaining strain will recover, either completely or partially, over a period of time. Another phenomenon typical of time-dependent materials is *relaxation,* or dissipation of stresses with time. For example, if an asphalt concrete specimen is placed in a loading machine and subjected to a constant strain, the stress within the specimen will initially be high, then gradually dissipate due to relaxation as shown in Figure 1.9(b). Relaxation is an important concern in the selection of steel for a prestressed concrete design.

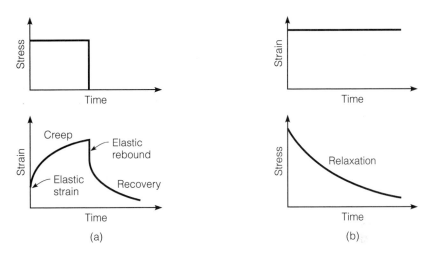

**FIGURE 1.9** Creep and relaxation of time-dependent materials: (a) creep and (b) relaxation.

Materials exhibiting both viscous and elastic responses are known as *viscoelastic*. In viscoelasticity, there are two approaches used to describe how stresses, strains, and time are interrelated. One approach is to postulate mathematical relations between these parameters based on material functions obtained from laboratory tests. The other approach is based on combining a number of discrete *rheological elements* to form *rheological models*, which describe the material response.

## Rheological Models

Rheological models are used to model mechanically the time-dependent behavior of materials. There are many different modes of material deformation, particularly in polymer materials. These materials cannot be described as simply elastic, viscous, etc. However, these materials can be modeled by a combination of simple physical elements. The simple physical elements have characteristics that can be easily visualized. Rheology uses three basic elements, combined in either series or parallel to form models that define complex material behaviors. The three basic rheological elements, Hookean, Newtonian, and St. Venant, are shown in Figure 1.10 (Polowski and Ripling 1966).

The *Hookean* element, as in Figure 1.10(a), has the characteristics of a linear *spring*. Deformation $\delta$ is proportional to force $F$ by a constant $M$

$$F = M\delta \tag{1.4}$$

This represents a perfectly linear elastic material. The response to a force is instantaneous and the deformation is completely recovered when the force is removed. Thus, the Hookean element represents a perfectly linear elastic material.

A *Newtonian* element models a perfectly viscous material and is modeled as a *dashpot* or shock absorber as seen in Figure 1.10(b). Deformation for a given level of force is proportional to the amount of time the force is applied. Hence, the rate of deformation, for a constant force, is a constant $\beta$.

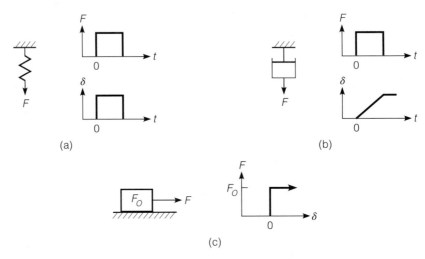

**FIGURE 1.10** Basic elements used in rheology: (a) Hookean, (b) Newtonian, and (c) St. Venant.

$$F = \beta \dot{\delta} \qquad (1.5)$$

The dot above the $\delta$ defines this as the rate of deformation with respect to time. If $\delta = 0$ at time $t = 0$ when a constant force $F$ is applied, the deformation at time $t$ is

$$\delta = \frac{Ft}{\beta} \qquad (1.6)$$

When the force is removed, the specimen retains the deformed shape. There is no recovery of any of the deformation.

The *St. Venant* element, as seen in Figure 1.10(c), has the characteristics of a sliding block that resists movement by friction. When the force $F$ applied to the element is less than the critical force $F_O$, there is no movement. If the force is increased to overcome the static friction, the element will slide and continue to slide as long as the force is applied. This element is unrealistic since any sustained force sufficient to cause movement would cause the block to accelerate. Hence the St. Venant element is always used in combination with the other basic elements.

The basic elements are usually combined in parallel or series to model material response. Figure 1.11 shows the three primary two-component models: the Maxwell, Kelvin, and Prandtl. The Maxwell and Kelvin models have a spring and dashpot in series and parallel, respectively. The Prandtl model uses spring and St. Venant elements in series.

In the *Maxwell* model [Figure 1.11(a)], the total deformation is the sum of the deformations of the individual elements. The force in each of the elements must be equal to the total force ($F = F_1 = F_2$). Thus the equation for the total deformation at any time after a constant load is applied is simply

$$\delta = \delta_1 + \delta_2 = \frac{F}{M} + \frac{Ft}{\beta} \qquad (1.7)$$

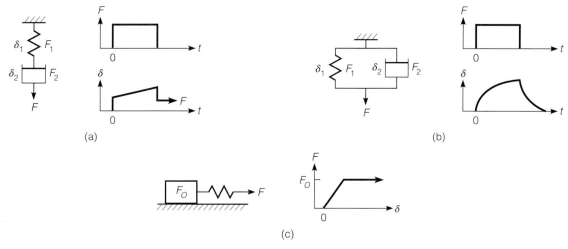

**FIGURE 1.11** Two-element rheological models: (a) Maxwell, (b) Kelvin, and (c) Prandtl.

In the *Kelvin* model, see Figure 1.11(b), the deformation of each of the elements must be equal at all times due to the way the model is formed. Thus the total deformation is equal to the deformation of each element ($\delta = \delta_1 = \delta_2$). Since the elements are in parallel, they will share the force such that the total force is equal to the sum of the force in each element. If $\delta = 0$ at time $t = 0$ when a constant force $F$ is applied, Equation 1.4 then requires zero force in the spring. Hence, when the load is initially applied, before any deformation takes place, all of the force must be in the dashpot. Under constant force the deformation of the dashpot must increase since there is force on the element. However, this also requires deformation of the spring, indicating that some of the force is carried by the spring. In fact, with time the amount of force in the dashpot decreases and the force in the spring increases. The proportion is fixed by the fact that the sum on the forces in the two elements must be equal to the total force. After a sufficient amount of time, all of the force will be transferred to the spring and the model will stop deforming. Thus the maximum deformation of the Kelvin model is $\delta = F/M$. Mathematically, the equation for the deformation in a Kelvin model is derived as

$$F = F_1 + F_2 = M\delta + \beta\dot{\delta} \tag{1.8}$$

Integrating Equation 1.8, using the limits that $\delta = 0$ at $t = 0$, and solving for the deformation $\delta$ at time $t$ results in

$$\delta = \frac{F}{M}(1 - e^{-Mt/\beta}) \tag{1.9}$$

The *Prandtl* model [Figure 1.11(c)] consists of St. Venant and Hookean bodies in series. The Prandtl model represents a material with an elastic–perfectly plastic response. If a small load is applied, the material responds elasticly until it reaches the yield point, after which the material exhibits plastic deformation.

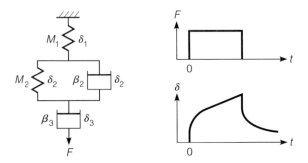

**FIGURE 1.12** Burgers model of viscoelastic materials.

Neither the Maxwell nor Kelvin model adequately describes the behavior of some common engineering materials, such as asphalt concrete. However, the Maxwell and the Kelvin models can be put together in series, producing the *Burgers* model, which can be used to describe simplistically the behavior of asphalt concrete. As shown in Figure 1.12, the Burgers model is generally drawn as a spring in series with a Kelvin model in series with a dashpot. The total deformation, when $\delta = 0$ at time $t = 0$, is then the sum of the deformations of these three elements.

$$\delta = \delta_1 + \delta_2 + \delta_3 = \frac{F}{M_1} + \frac{F}{M_2}(1 - e^{-M_2 t / \beta_2}) + \frac{Ft}{\beta_3} \qquad (1.10)$$

The deformation-time diagram for the loading part of the Burgers model demonstrates three distinct phases of behavior. First is the instantaneous deformation of the spring when the load is applied. Second is the combined deformation of the Kelvin model and the dashpot. Third, after the Kelvin model reaches maximum deformation, there is a continued deformation of the dashpot at a constant rate of deformation. The unloading part of the Burgers model follows similar behavior.

Some materials require more complicated rheological models to represent their response. In such cases a number of Maxwell models can be combined in parallel to form the generalized Maxwell model or a number of Kelvin models in series to form the generalized Kelvin model.

The use of rheological models requires quantifying material parameters associated with each model. Laboratory tests, such as creep tests, can be used to obtain time-deformation curves from which material parameters can be determined.

Although the rheological models are useful in describing the time-dependent response of materials, they can only be used to represent uniaxial responses. The three-dimensional behavior of materials and the Poisson's effect cannot be represented by these models.

**SAMPLE PROBLEM 1.4**    Derive the response relation for the following model assuming that the force $F$ is constant and instantaneously applied.

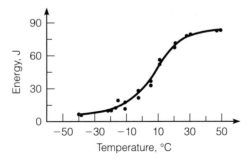

**Solution:**

$$\text{For } F \leq F_O : \delta = \frac{F}{M}$$

$$\text{For } F > F_O : \text{movement}$$

◆

## Temperature and Time Effects

The mechanical behavior of all materials is affected by temperature. Some materials, however, are more temperature-susceptible than others. For example, viscoelastic materials such as plastics and asphalt are greatly affected by temperature, even if the temperature is changed by a few degrees. Other materials, such as metals or concrete, are less affected by temperatures, especially when they are near room temperature.

Ferrous metals, including steel, demonstrate a change from ductile to brittle behavior as the temperature drops below the *transition temperature*. This change from ductile to brittle behavior greatly reduces the toughness of the material. While this could be determined by evaluating the stress-strain diagram at different temperatures, it is more common to evaluate the toughness of a material with an impact test that measures the energy required to fracture a specimen. Figure 1.13 shows how the energy required to fracture a mild steel changes with temperature (Flinn and Trojan 1986). The test results seen in Figure 1.13 were achieved by applying impact forces on bar specimens with a "defect" (a simple V notch) machined into the specimens (ASTM E23). During World War II many Liberty ships sank because the steel used in the ships met specifications at room temperature but became brittle in the cold waters of the North Atlantic.

In addition to temperature, some materials, such as viscoelastic materials, are affected by the load duration. The longer the load is applied, the larger the amount of

**FIGURE 1.13** Fracture toughness of steel under impact testing.

deformation or creep. In fact, increasing the load duration and increasing the temperature cause similar material responses. Therefore, temperature and time can be interchanged. This concept is very useful in running some tests. For example, a creep test on an asphalt concrete specimen can be performed with short load durations by increasing the temperature of the material. A time-temperature shift factor is then used to adjust the results for lower temperatures.

Viscoelastic materials are not only affected by the duration of the load, but also by the rate of load application. If the load is applied at a fast rate, the material is stiffer than if the load is applied at a slow rate. For example, if a heavy truck moves at a high speed on an asphalt pavement, no permanent deformation may be observed. However, if the same truck is parked on an asphalt pavement on a hot day, some permanent deformations on the pavement surface may be observed.

## Failure and Safety

Failure occurs when a member or structure ceases to perform the function for which it was designed. Failure of a structure can take several modes including fracture fatigue, general yielding, buckling, and excessive deformation. *Fracture* is a common failure mode. A brittle material typically fractures suddenly when the static stress reaches the strength of the material, where the strength is defined as the maximum stress the material can carry. On the other hand, a ductile material may fracture due to excessive plastic deformation.

Many structures, such as bridges, are subjected to repeated loadings, creating stresses that are less than the strength of the material. Repeated stresses can cause a material to fail or *fatigue*, at a stress well below the strength of the material. The number of applications a material can withstand depends on the stress level relative to the strength of the material. As shown in Figure 1.14, as the stress level decreases, the number of applications before failure increases. Ferrous metals have an apparent

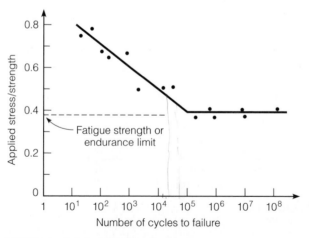

**FIGURE 1.14** An example of endurance limit under repeated loading.

*endurance limit*, or stress level, below which fatigue does not occur. The endurance limit for steels is generally in the range of one-quarter to one-half the ultimate strength (Flinn and Trojan 1986).

Another mode of failure is *general yielding.* This failure happens in ductile materials and it spreads throughout the whole structure, which results in a total collapse.

Long and slender members subjected to axial compression may fail due to *buckling.* Although the member is intended to carry axial compressive loads, a small lateral force might be applied, which causes deflection and eventually might cause failure.

Sometimes *excessive deformation* (elastic or plastic) could be defined as failure, depending on the function of the member. For example, excessive deflections of floors make people uncomfortable and in an extreme case may render the building unusable even though it is structurally sound.

To minimize the chance of failure, structures are designed to carry a load greater than the maximum anticipated load. The *factor of safety* is defined as the ratio of the load at failure to the maximum anticipated load. Typically, the larger the factor of safety, the larger the required cross section of the structure and, consequently, the higher the cost. The proper value of the factor of safety varies from one structure to another and depends on several factors, including the

- cost of unpredictable failure in lives, dollars, and time,
- variability in material properties,
- degree of accuracy in considering all possible loads applied to the structure, such as earthquakes,
- possible misuse of the structure, such as improperly hanging an object from a truss roof,
- degree of accuracy of considering the proper response of materials during design, such as assuming elastic response although the material might not be perfectly elastic.

## Nonmechanical Properties

Nonmechanical properties refer to characteristics of the material, other than load-response, that affect selection, use, and performance. There are several types of properties that are of interest to engineers, but those that are of the greatest concern to civil engineers are density, thermal properties, and surface characteristics.

### Density and Unit Weight

In many structures the dead weight of the materials in the structure significantly contributes to the total design stress. If the weight of the materials can be reduced, the size of the structural members can be also reduced. Thus the weight of the materials is an important design consideration. In addition, in the design of asphalt and concrete mixes, the weight–volume relationship of the aggregates and binders must be used to select the mix proportions.

There are three general terms used to describe the mass, weight, and volume relationship of materials. *Density* is the mass per unit volume of material. *Unit weight* is the weight per unit volume of material. By manipulation of units, it can be shown that

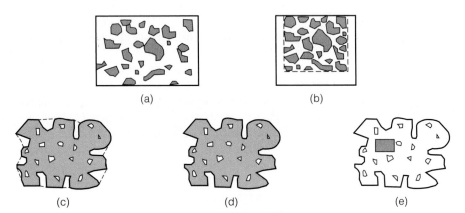

**FIGURE 1.15** Definitions of volume used for determining density: (a) loose, (b) compacted, (c) total particle volume, (d) volume not accessible to water, and (e) volume of solids.

$$\gamma = \rho g \qquad\qquad (1.11)$$

where

$\gamma$ = unit weight

$\rho$ = density

$g$ = acceleration of gravity.

*Specific gravity* is the ratio of the mass of a substance relative to the mass of an equal volume of water at a specified temperature. The density of water in SI units is 1 Mg/m$^3$ and 62.4 lb/ft$^3$ in English units at 4°C (39.2°F).

For solid materials, such as metals, the unit weight, density, and specific gravity have definite numerical values. For other materials such as wood and aggregates, voids in the materials require definitions for a variety of densities and specific gravities. As shown in Figure 1.15(a) and (b), the bulk volume aggregates will occupy depends on the compaction state of the material. In addition, the density of the material will change depending on how the volume of the particles is measured. Several types of particle volume can be used, such as the total volume enclosed within the boundaries of the individual particles, volume not accessible to water or asphalt, and volume of solids, as seen in Figure 1.15(c), (d), and (e). These are important factors in the mix designs of portland cement concrete and asphalt concrete.

## Thermal Expansion

Practically all materials expand as temperature increases and contract as temperature falls. The amount of expansion per unit length due to one unit of temperature increase is a material constant and is expressed as the coefficient of thermal expansion

$$\alpha_L = \frac{\delta L / \delta T}{L} \qquad\qquad (1.12)$$

$$\alpha_V = \frac{\delta V / \delta T}{V} \qquad\qquad (1.13)$$

Where

$\alpha_L$ = linear coefficient of thermal expansion
$\alpha_V$ = volumetric coefficient of thermal expansion
$\delta L$ = change in the length of the specimen
$\delta T$ = change in temperature
$L$ = original length of the specimen
$\delta V$ = change in the volume of the specimen
$V$ = original volume of the specimen

For isotropic materials $\alpha_V = 3\alpha_L$.

The coefficient of thermal expansion is very important in the design of structures. Generally, structures are composed of many materials that are bound together. If the coefficients of thermal expansion are different, the materials will strain at different rates. The material with the lesser expansion will restrict the straining of other materials. This constraining effect will cause stresses in the materials that can lead directly to fracture.

Stresses can also be developed as a result of a thermal gradient in the structure. As the temperature outside the structure changes and the temperature inside remains constant, a thermal gradient develops. When the structure is restrained from straining, stress develops in the material. This mechanism has caused brick facades on buildings to fracture and, in some cases, fall off the structure. Also, since concrete pavements are restrained from movement, they may crack in the winter due to a drop in temperature and may "blow up" in the summer due to an increase in temperature. Joints are, therefore, used in buildings, bridges, concrete pavements, and various structures to accommodate this thermal effect.

**SAMPLE PROBLEM 1.5**

A steel bar with a length of 3 m, diameter of 25 mm, modulus of elasticity of 207 GPa, and a linear coefficient of thermal expansion of 0.000009 m/m/°C, is fixed at both ends when the ambient temperature was 40°C. If the ambient temperature is decreased to 15°C, what internal stress will develop due to this temperature change? Is this stress tension or compression? Why?

**Solution:** If the bar was fixed at one end and free at the other end, the bar would have contracted and no stresses would have developed. In that case, the change in length can be calculated using Equation 1.12 as follows.

$$\delta L = \alpha_L \times \delta T \times L = 0.000009 \times (-25) \times 3 = -0.000675 \text{ m}$$

$$\epsilon = \frac{\delta L}{L} = \frac{-0.000675}{3} = -0.000225 \text{ m/m}$$

Since the bar is fixed at both ends, the length of the bar will not change. Therefore, a tensile stress will develop in the bar as follows.

$$\sigma = \epsilon E = 0.000225 \times 207000 = 46.575 \text{ MPa}$$

The stress will be tension; in effect, the length of the bar at 15°C without restraint would be 2.999325 m and the stress would be zero. Restraining the bar into a longer condition requires a tensile force.

## Surface Characteristics

The surface properties of materials of interest to civil engineers include corrosion and degradation, the ability of the material to resist abrasion and wear, and surface texture.

**Corrosion and Degradation** Nearly all materials deteriorate over their service lives. The mechanisms contributing to the deterioration of a material differ depending on the characteristics of the material and the environment. Crystalline materials, such as metals and ceramics, deteriorate through a *corrosion* process where there is a loss of material either by dissolution or by the formation of nonmetallic scale or film. Polymers, such as asphalt, deteriorate by *degradation* of the material, including the effects of solvent and ultraviolet radiation on the material.

The protection of materials from environmental degradation is an important design concern, especially when the implications of deterioration and degradation on the life and maintenance costs of the structure are considered. The selection of a material should consider both how the material will react with the environment and the cost of preventing the resulting degradation.

**Abrasion and Wear Resistance** Since most structures in civil engineering are static, the abrasion or wear resistance is of less importance than in other fields of engineering. For example, mechanical engineers must be concerned with the wear of parts in the design of machinery. This is not to say that wear resistance can be totally ignored in civil engineering. Pavements must be designed to resist the wear and polishing from vehicle tires in order to provide adequate skid resistance for braking and turning. Resistance to abrasion and wear is, therefore, an important property of aggregates used in pavements.

**Surface Texture** The surface texture of some materials and structures is of importance to civil engineers. For example, smooth texture of aggregate particles is needed in portland cement concrete to improve workability during mixing and placing. In contrast, rough texture of aggregate particles is needed in asphalt concrete mixtures to provide a stable pavement layer that resists deformation under the action of load. Also, a certain level of surface texture is needed in the pavement surface to provide adequate friction resistance and prevent skidding of vehicles.

# Production and Construction

Even if a material is well suited to a specific application, production and construction considerations may block the selection of the material. Production considerations include the availability of the material and the ability to fabricate the material into the desired shapes and required specifications. Construction considerations address all the factors that relate to the ability to fabricate and erect the structure on site. One of the primary factors is the availability of a trained work force. For example, in some cities high-strength concrete is used for skyscrapers, whereas in other cities steel is the material of choice. Clearly, either concrete or steel can be used for high-rise buildings. Regional preferences for one material develop as engineers in the region become comfortable and confident in designing with one of the materials and constructors respond with a trained work force and specialized equipment.

**FIGURE 1.16** An example of artist-engineer collaboration in an
engineering project: Air Force Academy, Colorado Springs, Colorado.

## Aesthetic Characteristics

The aesthetic characteristics of a material refer to the appearance of the material.
Generally, this characteristic is the responsibility of the architect. However, the civil
engineer is responsible for working with the architect to ensure that the aesthetic
characteristics of the facility are compatible with the structural requirements. During
the construction of many public projects, a certain percentage of the capital budget
typically goes toward artistic input. The collaboration between the civil engineer and
the architect is greatly encouraged, and the result can increase the value of the structure (see Figure 1.16).

In many cases, the mix of artistic and technical design skills make the project
acceptable to the community. In fact, political views are often more difficult to deal
with than technical design problems. Thus engineers should understand that there are
many factors beyond the technical needs that must be considered when selecting
materials and designing public projects.

## Material Variability

It is essential to understand that engineering materials are inherently variable. For
example, steel properties vary depending on chemical composition and method of manufacture. Concrete properties change depending on type and amount of cement, type of
aggregate, air content, slump, method of curing, etc. The properties of asphalt concrete

vary depending on the binder amount and type, aggregate properties and gradation, amount of compaction, and age. Wood properties vary depending on the tree species, method of cut, and moisture content. Some materials are more homogeneous than others, depending on the nature of the material and the method of manufacturing. For example, the variability of the yield strength of one type of steel is less than the variability of the compressive strength of one batch of concrete. Therefore, variability is an important parameter in defining the quality of civil engineering materials.

When materials from a particular lot are tested, the observed variability is the cumulative effect of three types of variance: the inherent variability of the material, variance caused by the sampling method, and variance associated with the way the tests are conducted. Just as materials have an inherent variability, sampling procedures and test methods can produce variable results. Frequently, statisticians call variance associated with sampling and testing *error*. However, this does not imply the sampling or testing was performed incorrectly. When an incorrect procedure is identified, it is called a *blunder*. The goal of a sampling and testing program is to minimize sampling and testing variance so the true statistical features of the material can be identified.

The concepts of precision and accuracy are fundamental to the understanding of variability. *Precision* refers to the variability of repeat measurements under carefully controlled conditions. *Accuracy* is the conformity of results to the true value or the absence of bias. *Bias* is a tendency of an estimate to deviate in one direction from the true value. In other words, bias is a systematic error between a test value and the true value. A simple analogy to understand the relationship between precision and accuracy is the target shown in Figure 1.17. When all shots are concentrated at one location away from the center, that indicates good precision and poor accuracy (biased) [Figure 1.17(a)]. When shots are scattered around the center, that indicates poor precision and good accuracy [Figure 1.17(b)]. Finally, good precision and good accuracy are obtained if all shots are concentrated close to the center [Figure 1.17(c)] (Burati and Hughes 1990). Many standardized test methods, such as those of the American Society for Testing and Materials (ASTM) and the American Association of State Highway and Transportation Officials (AASHTO), contain precision and bias

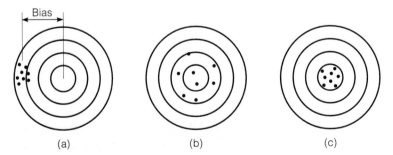

**FIGURE 1.17** Exactness of measurements: (a) precise but not accurate, (b) accurate but not precise, and (c) precise and accurate.

statements. These statements provide the limits of acceptable test results variability. Laboratories are usually required to demonstrate testing competence and can be certified by the American Material Reference Laboratory (AMRL).

## Sampling

Typically, *samples* are taken from a *lot* or *population* since it is not practical or possible to test the entire lot. By testing sufficient samples, it is possible to estimate the properties of the entire lot. In order for the samples to be valid they must be *randomly* selected. Random sampling requires that all elements of the population have an equal chance for selection. Another important concept in sampling is that the sample must be *representative* of the entire lot. For example, when sampling a stockpile of aggregate, it is important to collect samples from the top, middle, and bottom of the pile and to combine them, since different locations within the pile are likely to have different aggregate sizes. The sample size needed to quantify the characteristics of a population depends on the variability of the material properties and the confidence level required in the evaluation.

Statistical parameters describe the material properties. The mean and the standard deviation are two commonly used statistics. The *arithmetic mean* is simply the average of test results of all specimens tested. It is a measure of the central tendency of the population. The *standard deviation* is a measure of the dispersion or spread of the results. The equations for the mean $\bar{x}$ and standard deviation $s$ of a sample are

$$\bar{x} = \frac{\sum\limits_{i=1}^{n} x_i}{n} \tag{1.14}$$

$$s = \left( \frac{\sum\limits_{i=1}^{n} (x_i - \bar{x})^2}{n-1} \right)^{1/2} \tag{1.15}$$

where $n$ is the sample size. The mean and standard deviation of random samples are estimates of the mean and standard deviation, respectively, of the population.

## Normal Distribution

The normal distribution is a symmetrical function around the mean, as shown in Figure 1.18. The normal distribution describes many populations that occur in nature, research, and industry, including material properties. The area under the curve between any two values represents the probability of occurrence of an event of interest. Expressing the results in terms of mean and standard deviation, it is possible to determine the probabilities of an occurrence of an event. For example, the probability of occurrence of an event between the mean and ±1 standard deviation is 68.3%, between the mean and ±2 standard deviations is 95.5%, and between the mean and ±3 standard deviations is 99.7%. If a materials engineer tests 20 specimens of concrete

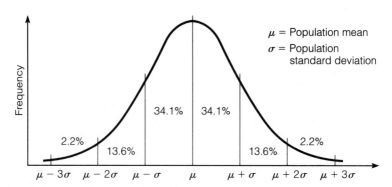

**FIGURE 1.18** A normal distribution.

and determines the average as 22 MPa and the standard deviation as 3 MPa, the statistics will show that 95.5% of the time the true mean of the population will be in the range of $22 \pm (2 \times 3)$ or 16 to 28 MPa.

## Control Charts

Control charts have been used in manufacturing industry and construction applications to verify that a process is in control. It is important to note that control charts do not get or keep a process under control; they provide only a visual warning mechanism to identify when a contractor or material supplier should look for possible problems with the process. Control charts have many benefits (Burati and Hughes 1990), such as

- detect trouble early
- decrease variability
- establish process capability
- reduce price adjustment cost
- decrease inspection frequency
- provide a basis for altering specification limits
- provide a permanent record of quality
- provide basis for acceptance
- instill quality awareness

There are many types of control charts, the simplest of which plots individual results in chronological order. For example, Figure 1.19 shows a control chart of the compressive strength of concrete specimens tested at a ready-mix plant. The control chart can also show the specification tolerance limits so that the operator can identify when the test results are out of the specification requirements. Although this type of control chart is useful, it is based on a sample size of one, and it therefore fails to consider variability within the sample.

Statistical control charts can be developed such as the control chart for means (X-bar chart) and the control chart for the ranges (R chart) in which the means or the ranges of the test results are chronologically plotted. Figure 1.20(a) shows a control

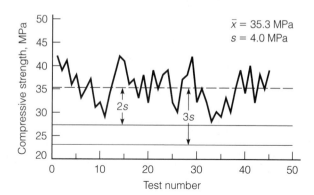

**FIGURE 1.19** Control chart of compressive strength of concrete specimens.

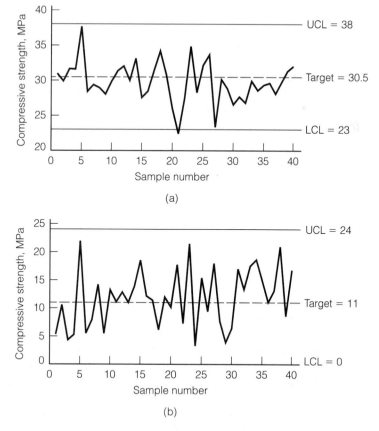

**FIGURE 1.20** Statistical control charts: (a) X-bar chart and (b) R chart. UCL indicates upper control limit; LCL indicates lower control limit.

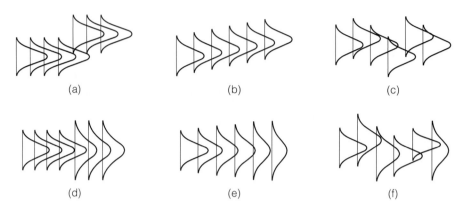

**FIGURE 1.21** Possible trends of means and ranges in statistical control charts: (a) sudden change in mean, (b) gradual changes in mean, (c) irregular change in mean, (d) sudden change in range, (e) gradual change in range, and (f) irregular change in mean and range.

chart for the moving average of each three consecutive compressive strength tests. For example, the first point represents the mean of the first three tests, the second point represents the mean of tests two through four, and so on. Figure 1.20(b) shows a control chart for the moving range of each three consecutive compressive strength tests. The key element in the use of statistical control charts is the proper designation of the *control limits* that are set for a given process. These control limits are not necessarily the same as the tolerance or specification limits and can be set using probability functions. For example, the control chart for means relies on the fact that, for a normal distribution, essentially all of the values fall within ±3 standard deviations from the mean. Thus control limits can be set between ±3 standard deviations from the mean. Warning limits to identify potential problems are sometimes set at ±2 standard deviations from the mean.

Observing the trend of means or ranges in statistical control charts can help eliminate production problems and reduce variability. Figure 1.21 shows possible trends of means and ranges in statistical control charts (Burati and Hughes 1990). Figure 1.21(a) shows sustained sudden shift in the mean. This could indicate a change of a material supplier during the project. A gradual change in the mean as illustrated in Figure 1.21(b) could indicate a progressive change brought on by machine wear. An irregular shift in the mean as shown in Figure 1.21(c) may indicate that the operator is making continuous but unnecessary adjustments to the process settings. Figure 1.21(d) shows a sudden change in range, which could also indicate a change of a material supplier during the project. Figure 1.21(e) shows a gradual increase in the range, which may indicate machine wear. Finally, Figure 1.21(f) shows an irregular shift in both mean and range, which indicates a bad process.

### Experimental Error

When specimens are tested in the laboratory, inaccuracy could occur due to machine or human errors. For example, Figure 1.22 shows a stress-strain curve where a toe region (AC) that does not represent a property of the material exists. This toe region is an artifact caused by taking-up slack and alignment or seating of the specimen. In

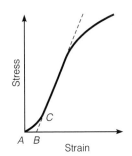

**FIGURE 1.22**
Correction of toe region
in a stress-strain curve.

order to obtain correct values of such parameters as modulus, strain, and offset yield point, this artifact must be compensated for in order to give the corrected zero point on the strain axis. This is accomplished by extending the linear portion of the curve backward until it meets the strain axis at point B. In this case, point B is the corrected zero strain point from which all strains must be measured. In the case of a material that does not exhibit any linear region, a similar correction can be made by constructing a tangent to the maximum slope at the inflection point and extending it until it meets the strain axis.

## Laboratory Measuring Devices

Laboratory tests measure material properties. Frequently, specimens are made of the material in question and tested in the laboratory to measure their response to the applied forces or to certain environmental conditions. These tests require measuring certain parameters such as time, deformation, or force. Some of these parameters are measured directly, while others are measured indirectly through relating parameters to each other. Length and deformation can be measured directly using simple devices such as rulers, dial gauges, and calipers. In other cases, indirect measurements are made by measuring electric voltage and relating it to deformation, force, stress, or strain. Examples of such devices include linear variable differential transformers (LVDTs), strain gauges, and load cells. Noncontact deformation measuring devices using lasers and various optical devices are also available. Electronic measuring devices can be easily connected to chart recorders, digital readout devices, or computers, where the results can be easily displayed and processed.

Each device has a certain *sensitivity*, which is the smallest value that can be read on the device's scale. Sensitivity should not be mistaken for accuracy or precision. Magnification can be designed into a gauge to increase its sensitivity, but wear, friction, noise, drift, and other factors may introduce errors that limit the accuracy and precision.

Measurement accuracy cannot exceed the sensitivity of the measuring device. For example, if a stopwatch with a sensitivity of 0.01 second is used to measure time, the smallest time interval that can be recorded is also 0.01 second. The selection of the measuring device and its sensitivity depends on the required accuracy of measurement. The required accuracy, on the other hand, depends on the significance and use of the measurement. For example, when expressing distance of travel from one city to another, an accuracy of 1 kilometer or even 10 kilometers may be meaningful. In contrast, manufacturing a computer microchip may require an accuracy of one-millionth of a meter or better. In engineering tests the accuracy of measurement must be determined in advance to ensure proper use of such measurements and, at the same time, to avoid unnecessary effort and expense during testing. Many standardized test methods such as those of ASTM and AASHTO state the sensitivity of the measuring devices used in a given experiment. In any case, care must be taken to ensure proper calibration, connections, use, and interpretation of the test results of various measuring devices.

Next we will briefly describe measuring devices commonly used in material testing, such as dial gauges, linear variable differential transformers (LVDTs), strain gauges, proving rings, and load cells.

## Dial Gauge

Dial gauges are used in many laboratory tests to measure deformation. The dial gauge is attached at two points, between which the relative movement is measured. Most of the dial gauges include two scales with two different pointers, as depicted in Figure 1.23. The smallest division of the large scale determines the sensitivity of the device and is usually recorded on the face of the gauge. One division of the small pointer corresponds to one full rotation of the large pointer. The full range of the small pointer determines the range of measurement of the dial gauge. Dial gauges used in civil engineering material testing frequently have sensitivities ranging from 0.1 mm to 0.002 mm. The dial gauge shown in Figure 1.23 has a sensitivity of 0.001 in. and a range of 1 in. The gauge can be "zeroed" by rotating the large scale in order to start the reading at the current pointer position.

Dial gauges can be attached to frames or holders with different configurations to measure the deformation of a certain gauge length or the relative movement between two points. For example, the *extensometer* shown in Figure 1.24 is used to measure the deformation of the gauge length of a metal bar during the tensile test. Note that because of the extensometer configuration shown in the figure, the deformation of the bar is one-half of the reading indicated by the dial gauge.

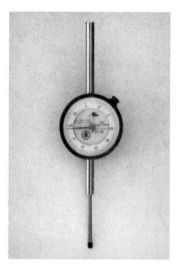

**FIGURE 1.23** Dial gauge.             **FIGURE 1.24** Extensometer with a dial gauge.

## Linear Variable Differential Transformer (LVDT)

The linear variable differential transformer (transducer), or LVDT, is an electronic device commonly used in laboratory experiments to measure small movements or deformations of specimens. The LVDT consists of a nonmagnetic shell and a magnetic core. The shell contains one primary and two secondary electric coils, as illustrated in Figure 1.25. An electric voltage is input to the LVDT and an output voltage is obtained. When the core is in the null position at the center of the shell, the output voltage is zero. When the core is moved slightly in one direction, an output voltage is obtained. Displacing the core in the opposite direction produces an output voltage with the opposite sign. The relationship between the core position and the output voltage is linear within a certain range determined by the manufacturer. If this relation is known, the displacement can be determined by measuring the output voltage using a voltmeter or a readout device. LVDTs can measure both static and dynamic movements.

LVDTs vary widely in sensitivity and range. The sensitivity of commercially available LVDTs ranges from 0.003 to 0.25 V/mm (0.08 to 6.3 V/in.) of displacement per volt of excitation. Normal excitation supplied to the primary coil is 3 Vac with a frequency ranging from 50 Hz to 10 kHz. If 3 V is used, the most sensitive LVDTs provide an output of 18.9 mV/mm (Dally and Riley 1991). In general, very sensitive LVDTs have small linear ranges, whereas LVDTs with greater linear ranges are less sensitive. The sensitivity and the linear range needed depends on the accuracy and the amount of displacement required for the measurement.

Before use, the LVDT must be calibrated to determine the relation between the output voltage and the displacement. A calibration device consisting of a micrometer, a voltmeter, and a holder is used to calibrate the LVDT, as shown in Figure 1.26.

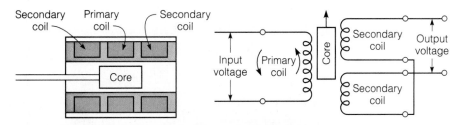

**FIGURE 1.25** LVDT circuit.

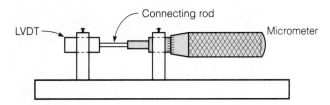

**FIGURE 1.26** LVDT calibration device.

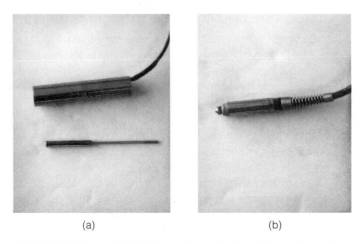

(a)                                            (b)

**FIGURE 1.27** Types of LVDT: (a) nonspring loaded and (b) spring loaded.

The shell and the core of the LVDT can be either separate or attached in a spring-loaded arrangement (see Figure 1.27). When the former type is used, a nonmagnetic threaded connecting rod is attached to the core used to attach the LVDT to the measured object. In either case, to measure the relative movement between two points, the core is attached to one point and the shell is attached to the other point. When the distance between the two points changes, the core position changes relative to the shell proportionally altering the output voltage. Figure 1.28 shows an extensometer with an LVDT that can be used to measure the deformation of a metal rod during the tensile test.

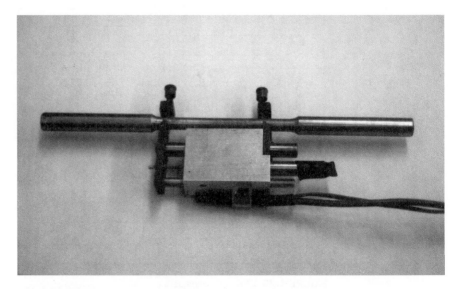

**FIGURE 1.28** Extensometer with an LVDT.

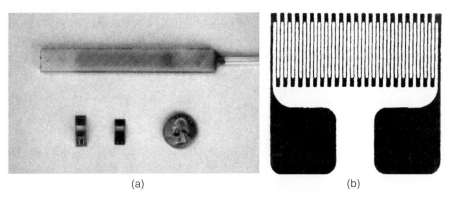

(a)                                    (b)

**FIGURE 1.29** Strain gauges: (a) strain gauges with different sizes and (b) typical foil strain gauge.

## Strain Gauge

Strain gauges are used to measure small deformations within a certain gauge length. There are several types of strain gauge, but the most dominant type is the electrical strain gauge. The electrical strain gauge consists of a foil or wire bonded to a thin base of plastic or paper (Figure 1.29). An electric current is passed through the element (foil or wire). As the element is strained, its electrical resistance changes proportionally. The strain gauge is bonded to the surface on which the strain is desired using an adhesive. As the surface deforms, the strain gauge also deforms and, consequently, the resistance changes. Since the amount of resistance change is very small, an ordinary ohmmeter cannot be used. Therefore, special electric circuits, such as the Wheatstone-bridge, are used to detect the change in resistance (Dally and Riley 1991).

Strain gauges are manufactured with different sizes, but the most convenient strain gauges have a gauge length of about 5 mm to 15 mm (1/4 in. to 1/2 in.). Larger strain gauges can also be made and used in some applications.

A wire gauge consists of a length of very fine wire (about 0.025 mm diameter) that is looped into a pattern. A foil gauge is made by etching a pattern on a very thin metal foil (about 0.0025 mm thick). Foils or wires are made in a great variety of shapes, sizes, and types and are bonded to a plastic or paper base. When the strain gauge is bonded to the object, it is cemented firmly with the foil or wire side out. Foil-plastic gauges are more commonly used than wire gauges.

When using strain gauges, it is very important to have a tight bond between the gauge and the member. The surface must be carefully cleaned and prepared and the adhesive must be properly applied and cured. The adhesive must be compatible with the material being tested.

## Proving Ring

Proving rings are used to measure forces in many laboratory tests. The proving ring consists of a steel ring with a dial gauge attached, as shown in Figure 1.30. When a force is applied on the proving ring, the ring deforms, as measured with the dial gauge. If the relation between the force and dial gauge reading is known, the proving ring can

**FIGURE 1.30** Proving ring.

be used to measure the applied force. Therefore, the proving ring comes with a calibration relationship, either in a form of a linear equation or a table, that allows the user to determine the magnitude of the force based on measuring the deformation of the proving ring. To avoid damage, it is important not to apply a force on the proving ring higher than the capacity reported by the manufacturer. Moreover, periodic calibration of the proving ring is advisable to insure proper measurements.

## Load Cell

The load cell is an electronic force measuring device used for many laboratory tests. In this device strain gauges are attached to a member within the load cell, which is subjected to either axial loading or bending. An electric voltage is input to the load cell and an output voltage is obtained. If the relation between the force and the output voltage is known, the force can be easily determined by measuring the output voltage.

Load cells are manufactured in different shapes and load capacities. Figure 1.31(a) illustrates a tensile load cell fabricated by mounting four strain gauges on the central region of a tension specimen. Figure 1.31(b) shows an S-shaped load cell where strain gauges are bonded to the central portion and calibrated to measure the force applied on top and bottom of the load cell. Figure 1.31(c) illustrates a diaphragm strain gauge bonded to the inside surface of an enclosure that measures the amount of pressure applied on the load cell.

Load cells must be regularly calibrated using either dead loads or a calibrated loading machine. Care must be taken not to overload the load cell. If the load applied on the load cell exceeds the capacity recommended by the manufacturer, permanent deformation can develop ruining the load cell.

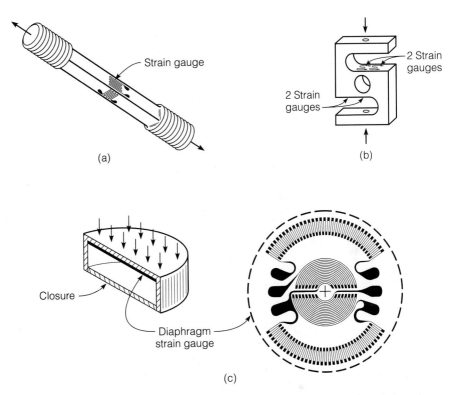

**FIGURE 1.31** Load cells: (a) strain gauges on a tension rod, (b) strain gauges on an S-shaped element, and (c) diaphragm strain gauge on an enclosure.

## SUMMARY

Civil and construction engineers are involved in the selection of construction materials with the mechanical properties needed for the project. In addition, the selection process must weigh other factors beyond the material's ability to carry loads. Economics, production, construction, maintenance and aesthetics must all be considered when selecting a material.

Lately, in all fields of engineering, there has been tremendous growth in the use of new high-performance materials. For example, in the automotive industry, applications of ceramics and plastics are increasing as manufacturers strive for better performance, economy, and safety, while pushing to reduce emissions. Likewise, civil and construction engineers are continuously looking for materials with better quality and higher performance. Advanced composite materials, geotextiles, and various synthetic products are currently competing with traditional civil engineering materials. Although traditional materials such as steel, concrete, wood, and asphalt will continue to be used for some time, improvements of these materials will proceed by changing the molecular structure of such materials and using modifiers to improve their performance. Examples of such improvements include fiber-reinforced concrete, polymer-modified concrete and asphalt, low temperature–susceptible asphalt binder, high-early-strength concrete, superplasticizers, epoxy-coated steel reinforcement, synthetic bar reinforcement, rapid-set concrete patching compounds, prefabricated drainage geocomposites, lightweight aggregates, fire-resistant building materials, and earthquake-resistant joints. Civil engineers are also recycling old materials in an effort to save materials cost, reduce energy, and improve the environment.

**QUESTIONS AND PROBLEMS**

**1.1.** State three examples of a static load application and three examples of a dynamic load application.

**1.2.** A rectangular block of aluminum 30 mm × 60 mm × 90 mm is placed into a pressure chamber and subjected to a pressure of 100 MPa. If the modulus of elasticity is 70 GPa and Poisson's ratio is 0.333, what will be the decrease in the longest side of the block, assuming that the material remains within the elastic region? What will be the decrease in the volume of the block?

**1.3.** A material has a stress-strain relationship that can be approximated by the equation

$$\epsilon = 0.3 \times 10^{-16} \times \sigma^3$$

where the stress is in psi. Find the secant modulus and the tangent modulus for the stress level of 50,000 psi.

**1.4.** The rectangular block shown in figure is subjected to tension within the elastic range. The increase in length of $a$ is $2 \times 10^{-3}$ in. and the contraction of $b$ is $3.25 \times 10^{-4}$ in. If the original lengths of $a$ and $b$ were 2 in. and 1 in., respectively, what is Poisson's ratio for the material of the specimen?

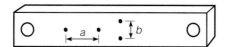

**1.5.** A cylindrical rod with a length of 380 mm and a diameter of 10 mm is to be subjected to a tensile load. The rod must not experience plastic deformation or an increase in length of more than 0.9 mm when a load of 24.5 kN is applied. Which of the four materials listed below are possible candidates? Justify your answer.

| Material | Elastic Modulus, GPa | Yield Strength, MPa | Tensile Strength, MPa |
|----------|---------------------|---------------------|----------------------|
| Copper | 110 | 248 | 289 |
| Aluminum alloy | 70 | 255 | 420 |
| Steel | 207 | 448 | 551 |
| Brass alloy | 101 | 345 | 420 |

**1.6.** The following stress-strain relation was obtained during the tensile test of an aluminum alloy specimen.

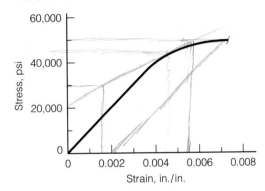

Determine the following:

**a.** Young's modulus within the linear portion

**b.** tangent modulus at a stress of 45,000 psi

**c.** yield stress using an offset of 0.002 strain

**d.** If the yield stress in part c is considered failure stress, what is the maximum working stress to be applied to this material if a factor of safety of 1.5 is used?

**1.7.** A brass alloy has a yield strength of 280 MPa, a tensile strength of 390 MPa, and an elastic modulus of 105 GPa. A cylindrical specimen of this alloy 12.7 mm in diameter and 250 mm long is stressed in tension and found to elongate 7.6 mm. On the basis of the information given, is it possible to compute the magnitude of the load that is necessary to produce this change in length? If so, calculate the load. If not, explain why.

**1.8.** The figure shows (i) elastic–perfectly plastic and (ii) elasto-plastic with strain hardening idealized responses. What stress is needed in each case to have

**a.** a strain of 0.001?

**b.** a strain of 0.004?

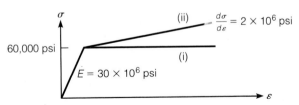

**1.9.** Draw a sketch of the Maxwell model and label all components. Draw a graph showing displacement versus time when the model is subjected to a constant force for a time period $t$ and then released. Comment on why the model responds this way.

**1.10.** Derive the response relation for each of the following models assuming that the force $F$ is constant and instantaneously applied.

           (a)                                         (b)

**1.11.** State four failure modes of materials. Describe typical examples of each mode.

**1.12.** A metal rod having a diameter of 10 mm is subjected to a repeated tensile load. The material of the rod has a tensile strength of 290 MPa and a fatigue failure behavior as shown in Figure 1.14. How many load repetitions can be applied to this rod before it fails if the magnitude of the load is (a) 5 kN, and (b) 11 kN?

**1.13.** What is the factor of safety? On what basis is its value selected?

**1.14.** State the typical values of the densities of three materials commonly used in construction.

**1.15.** Define the coefficient of thermal expansion. What is the relation between the linear and the volumetric coefficients of thermal expansion?

**1.16.** Estimate the tensile strength required to prevent cracking in a concrete-like material in the following situation. The material is cast into a bar, 50 in. long, and is fully restrained at each end against axial movement. The concrete is initially cast and cured at a temperature of 100°F and subsequently cools to a temperature of 0°F. Assume the modulus of elasticity is 5 million psi and the thermal coefficient is $5 \times 10^{-6}$ in./in./°F.

**1.17.** State two examples where corrosion plays an important role in selecting the material to be used in a structure.

**1.18.** Briefly discuss the variability of construction materials. Define the terms accuracy and precision when tests on materials are performed.

**1.19.** In order to evaluate the properties of a material, samples are taken and tested. What are the two most important factors that must be satisfied when the sample is collected? Show how these factors can be satisfied.

**1.20.** A contractor claims that the mean compressive strength for a concrete mix is 32.4 MPa (4700 psi) and that it has a standard deviation of 2.8 MPa (400 psi). If you break 16 cylinders and obtain a mean compressive strength of 30.3 MPa (4400 psi), would you believe the contractor's claim? Why? (Hint: Use statistical $t$-test.)

**1.21.** Briefly discuss the concept behind each of the following measuring devices:

    **a.** LVDT

    **b.** strain gauge

    **c.** proving ring

    **d.** load cell

**REFERENCES**      Ashby, M. F. and D. R. H. Jones. 1980. *Engineering materials, an introduction to their prop-erties and applications.* New York: Pergamon Press.

Burati, J. L. and C. S. Hughes. 1990. *Highway materials engineering, module I: materials control and acceptance—quality assurance.* Publication No. FHWA-HI-90-004. Washington, DC: Federal Highway Administration.

Callister, W. D., Jr. 1985. *Materials science and engineering—an introduction.* New York: John Wiley and Sons.

Dally. J. W. and W. F. Riley. 1991. *Experimental stress analysis.* 3rd ed. New York: McGraw-Hill.

Flinn, R. A. and P. K. Trojan. 1986. *Engineering materials and their applications.* 3rd ed. Boston, MA: Houghton Mifflin.

Mehta, P. K. and P. J. M. Monteiro. 1993. *Concrete structure, properties, and materials.* 2nd ed. Englewood Cliffs, NJ: Prentice-Hall.

Polowski, N. H. and E. J. Ripling. 1966. *Strength and structure of engineering materials.* Englewood Cliffs, NJ: Prentice-Hall.

Popov, E. P. 1968. *Introduction to mechanics of solids.* Englewood Cliffs, NJ: Prentice-Hall.

Richart, F. E., Jr., J. R. Hall, Jr., and R. D. Woods. 1970. *Vibrations of soils and foundations.* Englewood Cliffs, NJ: Prentice-Hall.

# 2 Nature of Materials

To a large extent, the behavior of materials is dictated by the structure and bonding of the atoms that are the building blocks for all matter. Knowledge of the bonding and structure of materials at the molecular level allows us to understand their behavior. This chapter presents a broad overview of concepts essential to our understanding of these behaviors. This chapter reviews the basic types of bonds and then, based on the type of bond, classifies materials as metallic, ceramic, or amorphous. The general nature of each of these classes of materials is presented.

## Basic Materials Concepts

Atoms are the basic building block of all materials. For the purpose of this text, atoms are made up of three subatomic particles: *protons*, *neutrons*, and *electrons*. The protons and neutrons are at the center of the atom, while the electrons travel about the nucleus in paths or shells. The *atomic number* is the number of protons in the nucleus of the atom. The *atomic mass* of an atom is the number of protons plus the number of neutrons in the center of the atom. An *element* is an atom or group of atoms with the same atomic number. *Isotopes* are elements with different numbers of neutrons in the nucleus.

### Electron Configuration

The behavior of an atom's electrons controls, to a large extent, the characteristics of an element. An electrically neutral (or complete) atom has equal numbers of electrons and protons. However, an atom may either release or attract electrons to reach a more stable configuration. Electrons travel around the nucleus in orbital paths, or orbits, around the nucleus. The distance between electrons and nucleus is not fixed; it is better described as a random variable with a distribution, as shown in Figure 2.1.

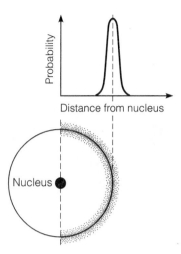

**FIGURE 2.1** Orbital path of model electrons.

Each orbital path, or *shell,* can hold only a fixed number of electrons. The orbital path of electrons is defined by four parameters: the principal quantum number or shell designation, the subshell designation, the number of energy states of the subshell, and the spin of the electron. Table 2.1 lists the number of electrons that can be in each shell and subshell.

The energy level of an electron depends on the shell and subshell location, as shown in Figure 2.2 (Callister 1985). Electrons always try to fill the lowest energy location first. As shown in Figure 2.2, for a given subshell, the larger the principal quantum number the higher the energy level (e.g., the energy of subshell 1s is less than subshell 2s). Within a shell the energy level increases with the subshell location: Subshell f is a higher energy state than subshell d, and so on. There is an energy overlap between the

**TABLE 2.1** Number of Available Electron States in Some of the Electron Shells and Subshells

| Principal Quantum Number | Shell Designation | Subshell | Number of States | *Number of Electrons* | |
| --- | --- | --- | --- | --- | --- |
| | | | | per Subshell | per Shell |
| 1 | K | s | 1 | 2 | 2 |
| 2 | L | s | 1 | 2 | 8 |
| | | p | 3 | 6 | |
| 3 | M | s | 1 | 2 | 18 |
| | | p | 3 | 6 | |
| | | d | 5 | 10 | |
| 4 | N | s | 1 | 2 | 32 |
| | | p | 3 | 6 | |
| | | d | 5 | 10 | |
| | | f | 7 | 14 | |

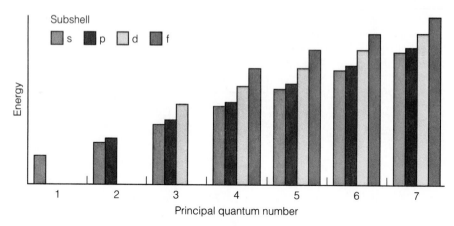

**FIGURE 2.2** Energy levels of electrons in different shells and subshells.

shells: Subshell 4s is at a lower energy state than subshell 3d. The electron configuration, or structure of an atom, describes the manner in which the electrons are located in the shells and subshells. The convention for listing the location of shells in the atom is to list the quantum number, followed by the subshell designation, followed by the number of electrons in the subshell raised to a superscript. This sequence is repeated for each subshell that contains electrons. Table 2.2 provides some examples.

The electrons in the outermost filled shell are the *valence electrons*. These are the electrons that participate in the formation of primary bonds between atoms. The eight electrons needed to fill the s and p subshells are particularly important. If these subshells are completely filled, the atoms are very stable and virtually nonreactive, as is the case for the noble gasses neon and argon. In many cases atoms will either release, attract, or share electrons to complete these subshells and reach a stable configuration. Calcium and chromium are examples of atoms with electrons that fill

**TABLE 2.2** Sample Electron Configurations

| Element | Atomic Number | Electron Configuration |
|---|---|---|
| Hydrogen | 1 | $1s^1$ |
| Helium | 2 | $1s^2$ |
| Carbon | 6 | $1s^2 2s^2 2p^2$ |
| Neon | 10 | $1s^2 2s^2 2p^6$ |
| Sodium | 11 | $1s^2 2s^2 2p^6 3s^1$ |
| Aluminum | 13 | $1s^2 2s^2 2p^6 3s^2 3p^1$ |
| Silicon | 14 | $1s^2 2s^2 2p^6 3s^2 3p^2$ |
| Sulfur | 16 | $1s^2 2s^2 2p^6 3s^2 3p^4$ |
| Argon | 18 | $1s^2 2s^2 2p^6 3s^2 3p^6$ |
| Calcium | 20 | $1s^2 2s^2 2p^6 3s^2 3p^6 4s^2$ |
| Chromium | 24 | $1s^2 2s^2 2p^6 3s^2 3p^6 3d^5 4s^1$ |
| Iron | 26 | $1s^2 2s^2 2p^6 3s^2 3p^6 3d^6 4s^2$ |
| Copper | 29 | $1s^2 2s^2 2p^6 3s^2 3p^6 3d^{10} 4s^1$ |

the 4s subshell, while the 3d subshell incomplete or empty, as would be expected from Figure 2.2. Copper demonstrates that there are exceptions to the energy rule. We would expect copper to have nine electrons in the 3d subshell and only one in the 4s subshell, but it has ten in the 3d subshell. A similar disparity exists for chromium. Note that iron has two electrons in the 4s subshell; thus it has two electrons more than it needs for a stable configuration. These two electrons are readily released to form iron molecules. Aluminum, too, is an exception, since it has an excess of three electrons.

## Bonding

As two atoms are brought together, both attractive and repulsive forces develop. The effects of these forces are additive, as shown in Figure 2.3, such that once the atoms are close enough to interact, they will reach a point where the attractive and repulsive forces are balanced and an equilibrium is reached. Energy is required either to bring the atoms closer together (*compression*) or separate them (*tension*). The distance at which the net force is zero corresponds to the minimum energy level and is called the *equilibrium spacing*. The minimum energy is represented by a negative sign. The largest negative value is defined as the bonding energy. The bonding energy can be computed from equations for the attractive and repulsive forces. Based on the strength of the bonds, the theoretical strength of a material can be estimated. However, this theoretical strength grossly overestimates the actual strength due to flaws in the molecular structure (Van Vlack 1964, 1989).

The bonding energy depends on the molecular mechanism holding the atoms together. There are two basic categories of bonds: primary and secondary. *Primary bonds* form when atoms interact to change the number of electrons in their outer

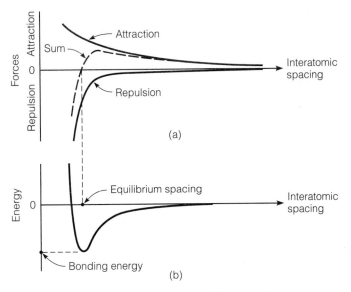

**FIGURE 2.3** Attractive and repulsive (a) forces and (b) energies between atoms.

---

shells to achieve a stable and nonreactive electron structure similar to that of a noble gas. *Secondary bonds* are formed when the physical arrangement of the atoms in the molecule results in an imbalanced electric charge; one side is positive and the other is negative. The molecules are then bonded together through electrostatic force.

**Primary Bonds** Three types of primary bonds are defined based on the manner in which the valence electrons interact with other atoms

1. *ionic bonds*—transfer of electrons from one elemental atom to another (Figure 2.4)
2. *covalent bonds*—sharing of electrons between specific atoms (Figure 2.5)
3. *metallic bonds*—mass sharing of electrons between several atoms (Figure 2.6)

Ionic bonds are the result of one atom releasing electrons to other atoms that accept the electrons. Each of the elements reaches a stable electron configuration of the outer s and p subshells. All of the atoms are ions since they have an electrical charge. When an atom releases an electron, the atom becomes positively charged; the atom receiving the electron becomes negatively charged. An ion with a positive charge is a *cation* and one with a negative charge is an *anion*. The ionic bond results from the electrostatic attraction of the negatively and positively charged atoms. Since these bonds are based on the transfer of electrons, they have no directional nature.

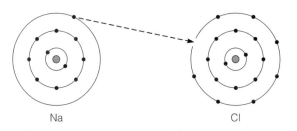

Na              Cl

**FIGURE 2.4** Ionic bonding.

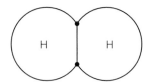

**FIGURE 2.5** Covalent bonding.

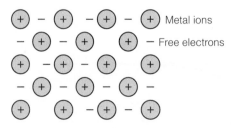

**FIGURE 2.6** Metallic bonding.

Covalent bonds occur when two similar atoms share electrons in the outer sub-shell. The atomic orbitals of the atoms overlap and an electron in each atom can exchange with an electron in its partner atom. If the shared electron is considered to be attached to either of the atoms, that atom would have the s and p subshells filled and would therefore be a stable atom. Since the orbital paths of the atoms must overlay for the covalent bond to form, these bonds are highly directional. In materials such as diamond, the covalent bonds are very strong. However, the carbon chain structure of polymers is also formed by covalent bonds and these elements display a wide range of bond energy. The number of bonds that form depends on the number of valence electrons. An atom with $N$ electrons in the valence shell can bond with only $8-N$ neighbors by sharing electrons with them. When the number of electrons $N$ equals 7, the atoms join in pairs. When $N$ equals 6, as in sulfur, long chains can form since the atom can bond with two neighbors. When $N$ equals 5, a layered structure can be developed. If there are 4 valence electrons, three-dimensional covalent bonds can result (e.g., the structure of carbon in diamonds) (Jastrzebski 1987). The calcium-silicate chain in portland cement concrete is based on covalent bonds.

Most interatomic bonds are partially ionic and partially covalent, and few compounds have pure covalent or ionic bonds. In general, the degree of either type of bonding depends on their relative position in the periodic table. The wider the separation, the more ionic bonds form. Since the electrons in ionic and covalent bonds are fixed to specific atoms, these materials have good thermal and electrical insulation properties.

Metallic bonds are the result of the metallic atoms having loosely held electrons in the outer s subshell. When similar metallic atoms interact, the outer electrons are released and are free to float between the atoms. Thus the atoms are ions that are electrically balanced by the free electrons. In other words, the free electrons disassociate themselves from the original atom and do not get attached to another atom. Metallic bonds are not directional and the spacing of the ions depends on the physical characteristics of the atoms. The atoms tend to pack together to give simple, high density structures, like marbles shaken down in a box. The easy movement of the electrons leads to the high electrical and thermal conductivity of metals.

**Secondary Bonds** Secondary bonds are much weaker than primary bonds, but they are important in the formation of links between polymer materials. These bonds result from a dipole attraction between uncharged atoms. As the electrons move about the nucleus, at any instant the charge is distributed in a nonsymmetrical manner relative to the nucleus. Thus, at any instant one side of the atom has a negative charge and the other side of the atom has a positive charge. Secondary bonds result from electrostatic attraction of the dipoles of the atoms. These dipole interactions occur between induced dipoles or polar molecules that have permanent dipoles; both types of interactions are classified as van der Waals forces. Hydrogen, however, has only one proton and one electron; thus it tends to form a polar molecule when bonded with other atoms. The electrostatic bond formed due to the hydrogen bond is generally stronger than van der Waals forces, so the hydrogen bond, shown in Figure 2.7, is a special form of the secondary bond.

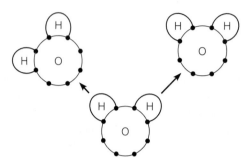

**FIGURE 2.7** Secondary bond: hydrogen bridge.

## Material Classification by Bond Type

Based on the predominant type of bond a material's atoms may form, materials are generally classified as metal, inorganic solids, and organic solids. These have predominantly metallic, covalent and ionic, and secondary bonds, respectively. Solids with these types of bonds have distinctly different characteristics. Metals and inorganic solids generally have a *crystalline* structure, a repeated pattern or arrangement of the atoms. On the other hand, organic solids usually have a random molecular structure. The predominant materials used in civil engineering in each category are

> Metallic
> >steel
> >iron
> >aluminum
> Inorganic solids
> >portland cement concrete
> >bricks and cinder blocks
> >glass
> >aggregates (rock products)
> Organic solids
> >asphalt
> >plastics
> >wood

## Metallic Materials

The chemical definition of a metal is an element with one, two, or three valence electrons. These elements bond into a mass with metallic bonds. Due to the nature of metallic bonds, metals have a very regular and well-defined structure. Since the metallic bonds are nondirectional, the atoms are free to pack into a dense configuration. The regular three-dimensional geometric pattern of the atoms in a metal is called a *unit cell*. Repeated coalescing of unit cells form a space lattice of the material as defined later in this chapter. However, in a mass of material a perfect structure can be achieved only through carefully controlled conditions. Generally, metallic solids are formed by cooling a mass of molten material. As the material cools, the atoms are vibrating. This can cause one atom to occupy the space of two atoms, generating a defect in the lattice structure. In addition, during the cooling process crystals grow simultaneously from several nuclei. As the material continues to cool, these crystals grow together with a boundary forming between the grains. This produces flaws or slip planes in the structure that have an important influence on the behavior and characteristics of the material. In addition, rarely are pure elemental metals used for engineering applications. At the least, the materials contain impurities that were not removed in the refining process. In addition, the iron and aluminum used for structural applications have alloying elements that impart special characteristics to the metal. As a result, understanding the nature of metals at the molecular level requires an examination of the primary structure of the metal, the effect of cooling rates, and the impact of impurities and alloying elements.

## Lattice Structure

As metals cool from the molten phase, the atoms are arranged into definite structures dependent on the size of the atom and the valence electrons. Certain characteristics are apparent in the three-dimensional array of points formed by the intersection of the parallel lines shown in Figure 2.8. In this configuration the arrangement of neighboring points about any specific point is identical with the arrangement around any other internal point. This property can be described mathematically by three vectors, **a**, **b**, and **c**. The location of any point, r′, relative to a reference point, can be defined in terms of an integer number of vector movements.

$$r' = r + n_1\mathbf{a} + n_2\mathbf{b} + n_3\mathbf{c} \tag{2.1}$$

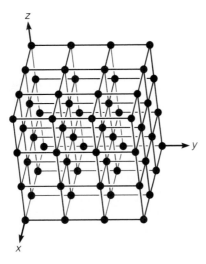

**FIGURE 2.8** Parallelogram structure for crystal lattices.

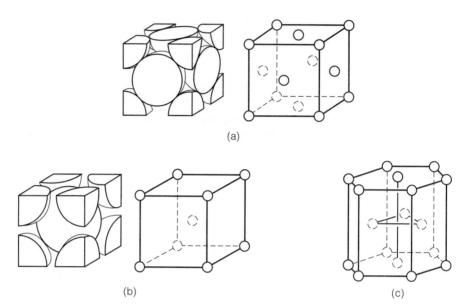

(a)

(b)                                    (c)

**FIGURE 2.9** Common lattice structures for metals: (a) face center cubic (FCC) (b) body center cubic (BCC), and (c) hexagonal close pack (HCP).

A continuous repetition of Equation 2.1 using incremental integers for $n_1$, $n_2$, and $n_3$ will result in the parallelogram shown in Figure 2.8. It should be noted that the angles between the axes need not be 90 degrees. Such a network of lines is called a *space lattice*. There are 14 possible space lattices in three dimensions that can be described by vectors **a**, **b**, and **c**. However, the space lattices of common engineering materials can be described by only a few of these space lattices; there are two cubic structures and one based on a hexagonal structure, as shown in Figure 2.9.

A simple cubic lattice structure has one atom on each corner of a cube that has axes at 90 degrees and equal vector lengths. However, this is not a common structure, although it does exist in some metals. There are two important variations on the cubic structure, the *face center cubic* and the *body center cubic*. The face center cubic (FCC) structure has an atom at each corner of the cube plus an atom on each of the faces, as shown in Figure 2.9(a). The body center cubic (BCC) structure has one atom on each of the corners plus one in the center of the cube, as shown in Figure 2.9(b).

The third common metal lattice structure is the *hexagonal close pack* (HCP). As shown in Figure 2.9(c), the HCP has top and bottom layers, with atoms at each of the corners of the hexagon and one atom in the center of the top and bottom planes; in addition, there are three atoms in a center plane. The atoms in the center plane are equidistant from all neighboring atoms. See Table 2.3 for the crystal structures and the atomic radii of some metals.

Two of the important characteristics of the crystalline structure are the coordination number and the atomic packing factor. The *coordination number* is the number of "nearest neighbors." The coordination number is 12 for FCC and HCP structures and 8 for BCC structures. This can be confirmed by examination of Figure 2.9. The *atomic*

**TABLE 2.3** Lattice Structure of Metals

| Metal | Crystal Structure | Atomic Radius, nm | Metal | Crystal Structure | Atomic Radius, nm |
|---|---|---|---|---|---|
| Aluminum | FCC | 0.1431 | Molybdenum | BCC | 0.1363 |
| Cadmium | HCP | 0.1490 | Nickel | FCC | 0.1246 |
| Chromium | BCC | 0.1249 | Platinum | FCC | 0.1387 |
| Cobalt | HCP | 0.1253 | Silver | FCC | 0.1445 |
| Copper | FCC | 0.1278 | Tantalum | BCC | 0.1430 |
| Gold | FCC | 0.1442 | Titanium | HCP | 0.1445 |
| Iron | BCC | 0.1241 | Tungsten | BCC | 0.1371 |
| Lead | FCC | 0.1750 | Zinc | HCP | 0.1332 |

*packing factor* (APF) is the fraction of the volume of the unit cell that is occupied by the atoms of the structure.

$$\text{APF} = \frac{\text{volume of atoms in the unit cell}}{\text{total unit volume of the cell}} \tag{2.2}$$

In order to compute the atomic packing factor, we must know the *equivalent number of atoms* associated with each unit cell and the atomic radius of the atoms, which are given in Table 2.3. The equivalent number of atoms associated with a cell would be the number of atoms in a large block of material divided by the number of unit cells in the block. However, by properly considering the fraction of atoms in Figure 2.9, we can count the number of "whole" atoms in each unit cell. The FCC structure has 8 corner atoms, each of these is shared with 7 other unit cells; thus all the corner atoms contribute 1 atom to the atom count for each unit cell. The face atoms are shared only with one other unit cell so that each of the face atoms contribute one-half atom; since there are six faces, the face atoms add 3 atoms to the count. Adding the face and corner atoms yields a total of 4 atoms. The BCC structure has the equivalent of only 2 atoms. The HCP structure has an equivalent of 6 atoms. Each of the 12 corner atoms is shared between six unit cells, the face atoms are shared between two unit cells, and the 3 atoms on the center planes are not shared with any other unit cells, so $12/6 + 2/2 + 3 = 6$.

The volume of the atoms in the unit cell can then be computed as the volume of 1 atom times the number of equivalent atoms. The volume of a sphere is $V = (4/3)\pi r^3$, where $r$ is the radius. The volume of the unit cell can be determined by knowing the radius of the atoms and the fact that the atoms are in contact with each other.

**SAMPLE PROBLEM 2.1**

Show that the atomic packing factor for the FCC lattice structure is 0.74.

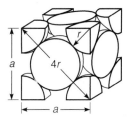

**Solution:**

Number of equivalent whole atoms in each unit cell $= 4$

Volume of the sphere $= \left(\frac{4}{3}\right)\pi r^3$

Volume of atoms in the unit cell $= 4 \times \left(\frac{4}{3}\right)\pi r^3 = \left(\frac{16}{3}\right)\pi r^3$

By inspection, the diagonal of the face of a FCC unit cell $= 4r$

Length of each side of the unit cell $= 2\sqrt{2}r$

Volume of the unit cell $= (2\sqrt{2}r)^3$

$$\text{APF} = \frac{\text{volume of atoms in the unit cell}}{\text{total unit volume of the cell}} = \frac{\left(\frac{16}{3}\right)\pi r^3}{(2\sqrt{2}r)^3} = 0.74 \qquad \blacklozenge$$

Similar geometric considerations allow for the calculation of the atomic packing factor for the BCC and HCP structures as 0.68 and 0.74, respectively.

The density of a metal is a function of the type of lattice structure and can be determined from:

$$\rho = \frac{nA}{V_C N_A} \tag{2.3}$$

where

$\rho$ = density of the material

$n$ = number of equivalent atoms in the unit cell

$A$ = atomic mass of the element (grams/mole)

$V_C$ = volume of the unit cell

$N_A$ = Avogadro's number ($6.023 \times 10^{23}$ atoms/mole)

**SAMPLE
PROBLEM 2.2** Calculate the radius of the aluminum atom given that aluminum has an FCC crystal structure, a density of 2.70 Mg/m³, and an atomic mass of 26.98 g/mole. Note that the APF for the FCC lattice structure is 0.74.

**Solution:**

$$\rho = \frac{nA}{V_C N_A}$$

For FCC lattice structure, $n = 4$.

$$V_C = \frac{4 \times 26.98}{2.70 \times 10^6 \times 6.023 \times 10^{23}} = 6.636 \times 10^{-29} \text{ m}^3$$

$$\text{APF} = 0.74 = \frac{4 \times \left(\frac{4}{3}\right)\pi r^3}{6.636 \times 10^{-29}}$$

$$r^3 = 0.293 \times 10^{-29} \text{ m}^3$$

$$r = 0.143 \times 10^{-9} \text{ m} = 0.143 \text{ nm} \qquad \blacklozenge$$

Knowing the type of lattice structure is important when determining the mechanical behavior of a metal. Under elastic behavior the bonds of the atoms are stretched, but when the load is removed, the atoms return to their original position. On the other hand, plastic deformation, by definition, is a permanent distortion of the materials;

therefore, plastic deformation must be associated with a change in the atomic arrangement of the metal. Plastic deformation is the result of planes of atoms slipping over each other due to the action of shear stresses. Naturally, the slip will occur on the planes that are the most susceptible to distortion. Since the basic bonding mechanism of the various metals is similar, differences in the theoretical strength of the material are attributed to the differences in the number and orientation of the slip planes that result from the different lattice structures.

## Lattice Defects

Even under special circumstances it is very difficult to grow perfect crystalline structures. Generally, pure crystalline structures are limited to 1 micron in diameter. These pure materials have a strength and modulus of elasticity approaching the maximum theoretical values based on the bonding characteristics. However, the strength and deformation of all practical materials are limited by defects. There are several causes for the development of defects in the crystal structure. These can be classified as

1. point defects or missing atoms
2. line defects or rows of missing atoms, commonly called an *edge dislocation*
3. area defects or grain boundaries
4. volume defects or cavities in the material

In the case of point defects, single atoms can be missing in the lattice structure because the atoms are vibrating as they transition from liquid to solid. As a result, one atom may vibrate in the area where two atoms should be in the lattice. Vacancies have little effect on the properties of the material.

When considering the differences between manufactured and theoretical material behavior understanding line defects becomes important. A typical line defect is shown in Figure 2.10, where a line of missing atoms extends back into the illustration (Van Vlack 1989; Guy and Hran 1974). The atoms above the dislocation are in compression and those below the dislocation are in tension. As a result, the atoms are not at their natural spacing, so the bonds of these atoms are not at the point of minimum energy, as shown in Figure 2.2. Thus, when a shear stress is applied to this location, there will be a tendency for the atoms to slip in a progressive manner from position (a) to (b) to (c), as shown in Figure 2.11 (Flinn and Trojan 1986; Budinski 1996).

Volume defects are flaws in the manufactured material; they will not be discussed further. Area defects are discussed next.

## Grain Structure

We have described the structure of metals in terms of the unit cell or the repeated crystalline structure. However, equally important to the behavior of the material is the size and arrangement of the grains in the material. The grain structure (microscopic structure) of the material should be distinguished from the atomic structure. Figure 2.12, for example, is an optical photomicrograph of a low-carbon steel. This photomicrograph was obtained using a scanning electron microscope with a magnification of 500 times. Note that this microscopic scale is much different than the atomic scale of Figure 2.9, which has a magnification in the order of 10,000,000 times.

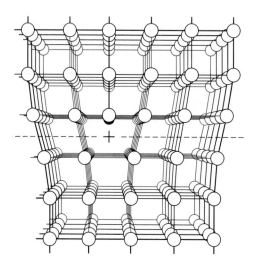

**FIGURE 2.10** Atomic packing at a line defect.

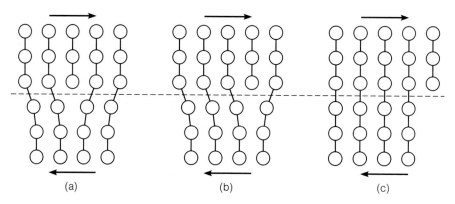

(a)                              (b)                              (c)

**FIGURE 2.11** Plastic deformation involving movement of atoms along a slip plane.

In metals manufacturing, the material is heated to a liquid, impurities are removed from the stock material, and alloying agents are added. An *alloy* is simply the addition of a second element to a metal. As the material cools from the liquid state, crystals form. Under normal cooling conditions, multiple nuclei will form, producing multiple crystals. As these crystals grow they will eventually contact each other, forming boundaries. For a given material, the size of the grains depends primarily on the rate of cooling. Under rapid cooling, multiple nuclei are formed, resulting in small grains with extensive boundaries.

There are four types of grain boundaries: coherent, coherent strain, semicoherent, and incoherent, as shown in Figure 2.13. At the coherent boundary, the lattices of the two grains align perfectly, and, in essence, there is no physical boundary. Coherent strain boundaries have a different spacing of atoms on each side of the boundary,

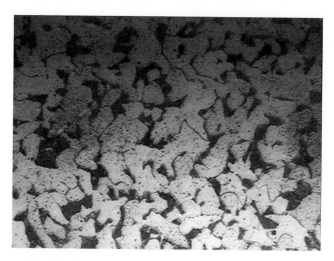

**FIGURE 2.12** Optical photomicrograph of low-carbon steel (magnification: 500×).

which may be the effect of an alloying agent. As with the line dislocation, there will be a strain in the atoms on each side of the boundary due to the difference in the spacing of the atoms. The semicoherent boundary has a different number of atoms on each side of the boundary; therefore, not all of the boundaries can match up. Again, this is similar to line dislocation with strain occurring on each side of the boundary. The incoherent boundary has a different orientation of the crystals on each side of the boundary; thus, the atoms do not match up in a natural manner.

Grain boundaries have an important effect on the behavior of a material. Although the bonds across the grain boundary do not have the strength of the pure crystal structure, the grain boundaries are at a higher energy state than the atoms away from the grain boundary. As a result, when slip occurs along one of the slip planes in the crystal, it is blocked from crossing the grain boundary, and can be diverted to run along the grain boundary. This increases the length of the slip path, thus requiring more energy to deform or fracture the material. Therefore, reducing grain size increases the strength of a material.

The grain structure of metals is affected by plastic strains. Fabrication methods frequently involve plastic straining to produce a desired shape (e.g., wire is produced

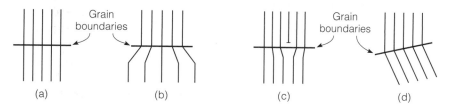

**FIGURE 2.13** Types of grain boundaries: (a) coherent, (b) coherent strain, (c) semicoherent, and (d) incoherent.

by forcing a metal through successively smaller dies to reduce the dimension of the metal from the shape produced in the mill to the desired wire diameter.)

Heat treatments are used to refine grain structure. There are two basic heat treatment methods, *annealing* and *hardening*. Both processes involve heating the material to a point where the existing grain boundaries will break down and reform upon cooling. The difference between the processes include the temperature to which the material is heated, the amount of time it is held at the elevated temperature, and the rate of cooling. In hardening, rapid cooling is achieved by immersing the material in a liquid. In annealing, the material is slowly cooled. The slowest rate of cooling is achieved by leaving the material in the furnace and gradually reducing the temperature. This is an expensive process since it ties up the furnace. More commonly the material is cooled in air. Heat treatments of steel and aluminum are discussed in Chapters 3 and 4.

The grain size is affected by the rate of cooling. Rapid cooling limits the time available for the growth of grains, thus the resulting grains are small, whereas slow cooling results in large grains.

## Alloys

The engineering characteristics of most elements make them unsuitable for use in a pure form. In most cases the properties of the materials can be significantly improved with the addition of alloying agents. An alloying agent is simply a chemical that is compounded or in solution in the crystalline structure of a metal. Steel, composed primarily of iron and carbon, is perhaps the most common alloy. The way the alloying agent fits into the crystalline structure is extremely important. The alloying atoms can either fit into the voids between the atoms, called *interstitial* atoms, or can replace the atoms in the lattice structure, *substitutional* atoms.

Because metals have a close pack structure, the radius of the interstitial atom must be less than 0.6 of the radius of the host element (Derucher et. al. 1994). Also, the solubility limit of interstitial atoms is relatively low, less than 6%. Interstitial atoms can be larger than the size of the void in the lattice structure, but this will result in a strain of the structure and, therefore, limit solubility.

If the characteristics of two metal elements are sufficiently similar, the metals can have complete miscibility; that is, there is no solubility limit; the atoms of the elements are completely interchangeable. The similarity criteria are defined by the Hume-Rothery rules (Shackelford 1996). Under these rules the elements must have

1. less than 15% difference in the atomic radius,
2. the same crystal structure,
3. similar electronegatives (the ability of the atom to attract an electron),
4. the same valence.

Violation of any of the Hume-Rothery rules reduces the solubility of the atoms. Atoms that are either too large or too small will result in a strain of the lattice structure.

The arrangement of the alloy atoms in the structure can be either random or ordered. In a random arrangement there is no pattern to the placement of the alloy atoms. An ordered arrangement can develop if the alloy element has a preference for a certain location in the lattice structure. For example, in a gold-copper alloy, a FCC structure, the copper preferably occupies the face positions and the gold preferably occupies the corner positions.

Frequently, more than one alloying agent is used to modify the characteristics of a metal. Steel is a good example of a multiple element alloy. By definition, steel contains iron with carbon; however, steel frequently includes chromium, copper, nickel, phosphorous, etc.

## Phase Diagrams

To produce alloys metals, the components are heated to a molten state, mixed, and then cooled. The temperature at which the material transitions between a liquid and a solid is a function of the percentages of the components. The liquid and solid states of a material are called *phases*, and a phase diagram displays the relationship between the percentages of the elements and the transition temperatures.

**Soluble Materials** The simplest type of phase diagram is for two elements that are completely soluble in both the liquid and solid phases. Solid solutions occur when the elements in the alloys remain dispersed throughout the matrix of the material in the solid state. A two-element or *binary* phase diagram is shown in Figure 2.14(a) for two completely soluble elements. In this diagram, temperature is plotted on the vertical axis and the percent weight of each element is plotted on the horizontal axis. In this case, the top axis is used for element A and the bottom axis is used for element B. The percentage of element B increases linearly across the axis, while the percentage of element A starts at 100% on the left and decreases to 0% on the right. Since this is a binary phase diagram, the sum of the percentage of elements A and B must equal 100%. In Figure 2.14(a) there are three areas. The areas at the top and bottom of the diagram have a single phase of liquid and solid material, respectively. Between the two single-phase areas there is a two-phase area where the material is both liquid and solid. The line between the liquid and two-phase area is the *liquidus* and the line between the two-phase area and the solid area is the *solidus*. For a given composition of elements A and B, the liquidus defines the temperature where, upon cooling, the first solid crystals form. The solidus defines the temperature where all material has crystallized. It should be noted that for a pure element, the transition between liquid and solid occurs at a single temperature. This is indicated on the phase diagram by the convergence of the liquidus and solidus on the left and right sides of Figure 2.14(a), where there is pure element A and B, respectively.

A specific composition of elements at a specific temperature is defined as the *state point*, as shown in Figure 2.14(b). If the state point is above the liquidus, all the material is liquid and composition of the liquid is the same as the total composition of the material. Similarly, if the state point is below the solidus, all the material is solid and the composition of the solid is the same as for the material. In the two-phase region between the liquidus and the solidus, the percent of material that is in either the liquid or solid phase varies with the temperature. In addition, the composition of the liquid and solid phases in this region changes with temperature. The compositions of the liquid and solid can be determined directly from the phase diagram. First a *tie line* is established by connecting the liquidus and solidus with a horizontal line that passes

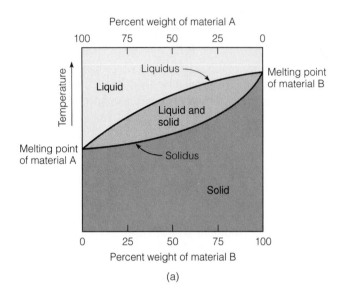

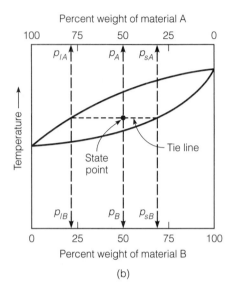

**FIGURE 2.14** Binary phase diagram, two soluble elements.

through the state point, as shown on Figure 2.14(b). A vertical projection from the intersection of the tie line and the liquidus defines the composition of the liquid phase. A vertical projection intersection of the tie line and the solidus defines the composition of the solid phase. For the example, in Figure 2.14(b) the alloy is composed of 50% material A and 50% material B. For the defined state point, 79% of the liquid material is element A and 21% is element B, and 31% of the solid phase material is element A and 69% of the solid material is element B. In addition, the percent of the material in the liquid and solid phases can be determined from the phase diagram. From mass balance, the total material must equal the sum of the masses of the components

$$m_t = m_l + m_s \tag{2.4}$$

where

$m_t$ = mass of total material
$m_l$ = mass of the total material that is in the liquid phase
$m_s$ = mass of the total material that is in the solid phase

This mass balance also applies to each of the component materials, that is,

$$p_B m_t = p_{lB} m_l + p_{sB} m_s \tag{2.5}$$

where

$p_{lB}$ = percent of the liquid phase that is composed of material B
$p_{sB}$ = percent of the solid phase that is composed of material B
$p_B$  = percent of the material that is component B

Using these two equations, the amount of material in the liquid and solid phase can be derived as

$$p_B m_t = p_{lB} m_l + p_{sB}(m_t - m_l)$$
$$p_B m_t = p_{lB} m_l + p_s m_t - p_{sB} m_l$$
$$p_B m_t - p_{sB} m_t = p_{lB} m_l - p_{sB} m_l$$
$$m_l = \frac{(p_B - p_{sB})}{(p_{lB} - p_{sB})} m_t$$
$$m_s = m_t - m_l \tag{2.6}$$

**SAMPLE PROBLEM 2.3**

Considering an alloy of the two soluble components A and B described by a phase diagram similar to that shown in Figure 2.14, determine the masses of the alloy that are in the liquid and solid phases at a given temperature if the total mass of the alloy is 100 grams, component B represents 40% of the alloy, 20% of the liquid is component B, and 70% of solid is component B.

*Solution:*

$$m_t = 100 \text{ g}$$
$$P_B = 40\%$$
$$P_{lB} = 20\%$$
$$P_{sB} = 70\%$$

From Equations 2.4 and 2.5

$$m_l + m_s = 100$$
$$20 m_l + 70 m_s = 40 \times 100$$

Solving the two equations simultaneously, we get

$$m_l = \text{mass of the alloy that is in the liquid phase} = 60 \text{ g}$$
$$m_s = \text{mass of the alloy that is in the solid phase} = 40 \text{ g}$$

The same answer can also be obtained using Equation 2.6. ◆

**Insoluble Materials** Our discussion thus far has dealt with two completely soluble materials. It is equally important to understand the phase diagram for immiscible materials, that is, for components that are so dissimilar that their solubility in each other is nearly negligible in the solid phase. Figure 2.15 shows the phase diagram for this situation. The intersections of the liquidus with the right and left axes are the melting points for each of the components. As the materials are blended together, the liquidus forms a V shape. The point of the V defines the combination of the components that will change from liquid to solid without the formation of two phases. This point defines the *eutectic temperature* and the *eutectic composition* for the components. The solidus is horizontal and passes through the eutectic temperature. There are two areas on the graph with two phases. The area to the left of the liquidus will have solid component A, and the liquid phase will be a mixture of the A and B components (vice versa for the area to the right of the liquidus).

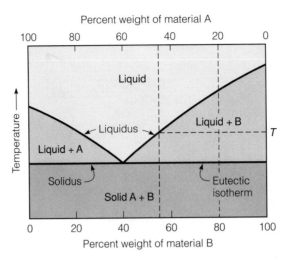

**FIGURE 2.15**  Binary phase diagram, insoluble solids.

The sudden phase transformation at the eutectic temperature means that the grains do not have time to grow as the material cools. Thus, the eutectic material will have a fine grain structure.

As the material cools at a temperature other than the eutectic temperature, one of the components becomes solid as the temperature drops below the liquidus. As a result, the amount of this component in the liquid continuously decreases as the material cools in the two-phase area. In fact, the composition of the liquid will follow the liquidus as the temperature decreases. When the eutectic temperature is reached, the remainder of the liquid becomes solid and has a fine grain structure. Starting with a composition of 80% B and 20% A, as in Figure 2.15, when the temperature is lowered to $T$ the liquid will contain 55% B and 45% A. The amount of each component in the liquid and solid phases can be determined by mass balance (Equations 2.4 and 2.5). At $T$ this will yield 44.4% liquid and 55.6% solid; all of the solid will be component B. Similarly, the composition of the solid at the eutectic temperature can be computed to be 66.7% solid B and 33.3% eutectic mixture. Since these materials are insoluble in the solid state, the eutectic mixture will actually be composed of an intimate mixture of fine crystals of A and B, usually in a platelike structure.

**Partially Soluble Materials**  In between purely soluble and insoluble materials are the materials that are partially soluble. In other words, there is a solubility limit between the components of A and B. If the percent of component B is less than or equal to the solubility limit, on cooling all of the B atoms will be in solution with the A component. If the percent of the B component is above the solubility limit, the atoms in excess of the amount that will go into solution will form separate grains of the component B. The result is shown on the phase diagram in Figure 2.16. Note that the only difference between this phase diagram and the one shown in Figure 2.15 is the presence of the solid solution regions on each side of the graph. The composition analysis of the two-phase region is the same as was described for Figure 2.15.

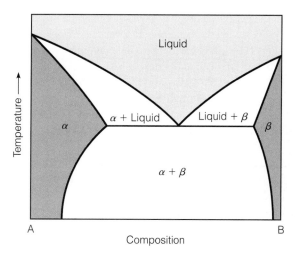

**FIGURE 2.16** Binary phase diagram, partially soluble material.

**Eutectoid Reaction** Up to now, the phase diagram has been used to describe the transition between liquid and solid phases of materials. However, the lattice structure of some elements (e.g., iron) is a function of the temperature of the solid. As a result of the lattice transformation, the microstructure of the solid material changes as a function of temperature, as shown by the phase diagram in Figure 2.17. When this occurs, a eutectoid reaction occurs on the phase diagram; it has characteristics similar to the eutectic reaction, but is for a lattice structure transformation of the material rather than for a liquid-solid transformation. The rules for the analysis of the components of the material are the same as those discussed for the eutectic material. As with the phase transformation at the eutectic temperature, the transformation of the lattice structure at the eutectoid temperature will result in fine-grained materials.

## Combined Effects

The topics of lattice structure, grain size, heat treatments, and alloying are closely interrelated. The behavior of a metal is dictated by the combination of each of these factors. Clearly, the properties of a metal depend on the elemental make up, the refining and production process, and the types and extent of alloys used in the metal. These topics are too complex for detailed treatment in this text. A practicing engineer in this area must devote considerable study to material characteristics and the impact of alloying and heat treatments.

# Inorganic Solids

Inorganic solids include all materials composed of nonmetallic elements or a combination of metallic and nonmetallic elements. This class of materials is sometimes referred to as ceramic materials. By definition, ceramic elements have five, six, and seven valence electrons. Ceramic materials are formed by a combination of ionic and covalent bonds. In the generic sense, ceramics encompass a broad range of materials, including glass, pottery, inorganic cements, and various oxides. Fired clay products, including bricks and pottery, are some of the oldest ceramic products made by humans. In terms of tonnage, portland cement concrete is the most widely used manufactured material. In the 1980s the search for highly durable products with unique strength and thermal properties initiated the rapid development of advanced ceramics, such as zirconias, aluminum oxides, silicon carbides, and silicon nitrates. These materials have high strength, stiffness, and wear resistance, and are not sensitive to corrosion. The high-performance or engineered ceramics have found many applications in machine and tool design. Although the availability of high-performance ceramics has grown rapidly, the civil and construction engineering applications of these materials have been generally limited due to the cost of the sophisticated ceramics and their lack of fracture toughness that is needed for structural design.

Five classes of ceramic materials have been defined (Ashby and Jones 1986).

1. glasses—based on silica
2. vitreous ceramics—clay products used for pottery, bricks, etc.
3. high-performance ceramics—highly refined inorganic solids used for specialty applications where special properties, not available from other materials, compensate for the high cost
4. cement and concrete—a multiphase material widely used in civil engineering applications
5. rocks and minerals

Ceramics can also be classified by the predominant type of their atomic bonding. Materials composed of a combination of nonmetallic and metallic elements have predominantly ionic bonds. Materials composed of two nonmetals have predominantly

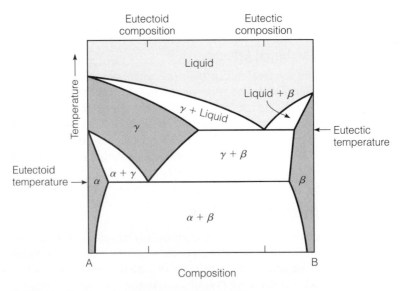

**FIGURE 2.17** Phase diagram for a eutectoid reaction.

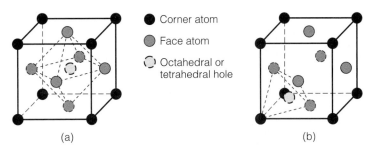

Corner atom
Face atom
Octahedral or
tetrahedral hole

(a)                    (b)

**FIGURE 2.18** Lattice structure for simple ionic-bonded ceramic materials:
(a) octahedral hole, and (b) tetrahedral hole.

covalent bonds. The type of bond dictates the crystal structure of the compound. As with metals, inorganic solids have a well-defined, although more complicated, unit cell structure. This structure is repeated for the formation of the crystals.

In a simple ionic compound the nonmetallic element will form either a face centered cubic or hexagonal close pack structure, as shown in Figure 2.18. Octahedral holes fit in the middle of the octahedron that is formed by connecting all the face atoms of the FCC unit cell. Tetrahedral holes are in the middle of the tetrahedron formed by connecting a corner atom with the adjacent face atoms. In the FCC structure there is one octahedral hole and six tetrahedral holes. The number of atoms and the way they fit into the holes in the close pack structure is determined by the number of valence electrons in the metallic and nonmetallic elements. The atoms pack to maximize density, with the constraint that like ions are not nearest neighbors (Ashby and Jones 1986).

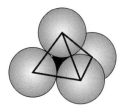

**FIGURE 2.19** Silicate tetrahedron.

Simple covalent bonds form materials that are highly durable and strong. Diamond is the preeminent example of an elemental high-strength material. Diamond is widely used in industrial applications needing wear resistance, such as cutting tools. However, the most important of the predominately covalent bond materials used by civil engineers are the silicate compounds of portland cement concrete. Silicon atoms link with four oxygen atoms to form a stable tetrahedron (Figure 2.19) that is the basic building block for all silicates. When combined with metal oxides, MO, with a ratio of $MO/SiO_2$ of 2 or greater, the resulting silicate is made up of separate $SiO_4$ monomers linked by the MO molecules. Figure 2.20 shows the calcium silicate crystal structure. The primary reaction compounds of portland cement are tricalcium silicates and dicalcium silicates. Each of these have two or more metal oxides, CaO, per silicate molecule. An ionic bond forms between the metal oxide and the silicate.

Ceramics have only about one-fiftieth the fracture toughness of metals. Due to the nature of the ionic and covalent bonds, ceramic compounds tend to fracture in a brittle manner rather than have plastic deformation, as was the case for metals. Like metals, ceramic compounds form distinct grains as a result of multiple nuclei forming multiple crystals during the production of the compound. In addition, due to the production process, ceramic materials tend to have internal cracks and flaws. Stress concentrations occur at the cracks, flaws, and grain boundaries, all of which lower the strength and toughness of the material. As a result, ceramic materials must be reinforced when they are used for structural applications.

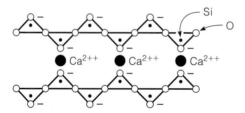

**FIGURE 2.20**  Simple inorganic solid structure (calcium silicate).

Glasses are a special type of inorganic solids; they do not develop a crystalline structure. Commercial glasses are based on silica. In glass, the silica tetrahedrons link at the corners, resulting in a random or amorphous structure. Glass can be made of pure silica; it has a high softening point and low thermal expansion. However, due to the high softening point, pure silica glass is hard to form. Metal oxides are added to reduce the cross-linking of the silicates, improving the ability to form the glass into the desired shape. Although glass has an amorphous structure, it is a very stable compound at atmospheric temperatures. It does not flow despite the often cited example of glass windows in European churches.

## Organic Solids

All organic solids are composed of long molecules of covalent bonded carbon atoms. These molecules are chains of carbon and hydrogen combined with various radical components. The radical component can be a hydrogen atom, another hydrocarbon, or another element. These long molecules are bound together by secondary bonds; in many cases the molecules are also cross-linked with covalent bonds. There are a wide range of organic solids used in engineering. These can be classified as follows (Ashby and Jones 1986):

1. *Thermoplastics* are characterized by linear carbon chains that are not cross-linked; at low temperatures secondary bonds adhere the chains. Upon heating, the secondary bonds melt and the thermoplastics become a viscous material. Asphalt is a natural thermoplastic. It is obtained primarily by refining petroleum. In addition, there are many manufactured thermoplastics that have broad engineering applications. These include polyethylene, polypropylene, polytetrafluroethylene, polystyrene, and polyvinyl chloride. Polyethylene and polypropylene are used in tubing, bottles, and electrical insulation. Polytetrafluroethylene is commonly known as Teflon. In addition to cookware applications, Teflon is widely used for bearings and seals due to its very low friction and good adhesion characteristics. (Polytetrafluroethylene is a carbon-fluorine chain.) Polystyrene is used for molded objects. It is foamed with carbon dioxide to make packing materials and thermal insulation. Polyvinyl chloride is used for low-pressure waterlines.

2. *Thermosets* are characteristically made of a resin and a hardener that chemically react to harden. In the formation of the solid, the carbon chains are cross-linked to form stable compounds that do not soften upon heating. The three generic types of thermosets are epoxy, polyester, and phenol-formaldehyde. Epoxies are used as glues and as the matrix material in plastic composites. Polyester is a fibrous material used in the reinforcing phase of fiberglass. Phenol-formaldehydes are brittle plastics such as bakelite and formica.

3. *Elastomers or rubbers* are characterized as linear polymers with limited cross-linking. At atmospheric temperatures the secondary bonds have melted. The cross-linking enables the material to return to its original shape when unloaded. Three forms of elastomers are polyisoprene (natural rubber), polybutadiene (synthetic rubber), and polychloroprene (Neoprene).

4. *Natural materials* are characterized as being grown in all plant matter. The primary material of interest is wood, which is composed of cellulose, lignin, and protein.

Other than the natural polymers, the balance of the organic solids are produced from refining and processing crude oil. In general, these products are classified as plastics. The properties of these materials are highly variable. Mechanical properties depend on the length of the polymer chains, the extent of cross-linking, and the type of radical compound. All of these factors can be controlled and altered in the production process to alter the material properties.

## Polymer Development, Structure, and Cross-Linking

The physical structure of the polymer chain grossly affects the mechanical response of plastics. The word polymer literally means multiple "mer" units. The mer is a base molecule that can be linked together to form the polymers. Figure 2.21 shows the structure of a simple ethylene molecule and the development of a polymer. The square boxes are carbon atoms and the open circles are hydrogen. Forming the polymer requires breaking the double bond, activating the monomer, and allowing it to link to others to form a long chain. The end of the chain either links to other chains or a terminator molecule, such as OH. Chains with useful mechanical properties require at least 500 monomers. The number of monomers in the chain defines the degree of polymerization; commercial polymers have a degree of polymerization of $10^3$ to $10^5$.

The complexity of the linear chain polymer is increased by replacing hydrogen atoms with side groups or radicals, as shown in Figure 2.21. The radicals or side groups can be aligned on one side of the chain (*isotactic*), symmetrically on alternate sides of the chain (*sindiotactic*), or in a random fashion (*atactic)*. The radicals can range from simple to complex molecules. For example, polyvinal chloride has a Cl radical, polypropylene has a $CH_3$ radical, and polystyrene has $C_6H_5$. The ability of the polymer chains to stack together is determined by the arrangement of the side chains. The simple chains can fold together into an orderly arrangement, whereas the complex side groups prevent stacking, leading to the amorphous nature of these materials.

More complex linear polymers are formed when two of the hydrogen atoms are replaced by different radicals. Polymethylmethacrylate, (Plexiglas), has the radicals

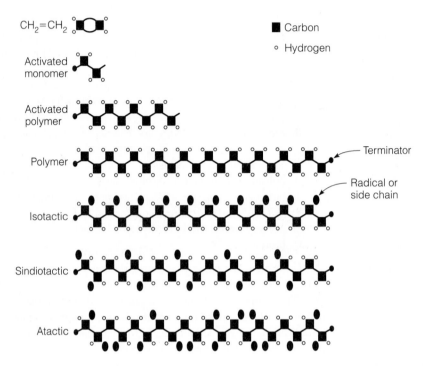

**FIGURE 2.21** Polymer structures.

$CH_3$ and $COOCH_3$. As the complexity of the radicals and the substitution for hydrogen increase, it becomes more difficult to form regular patterns.

In thermoplastics, formed from linear polymers, the structure of the molecules is a blend of amorphous and crystal structure. When there are few side groups an ordered structure is produced, Figure 2.22(a). As the number of side groups increases, the structure becomes increasingly random and cross-links develop. These structures are shown in Figure 2.22(b). Thermosets are formed from polyfunctional monomers. They are formed in a condensation reaction; in essence, the reaction bonds two chains together. Since the chains are formed from polyfunctional crystals, they have an amorphous structure with extensive cross-linking. Elastomers are formed with linear chains that have a limited number of cross-links.

## Melting and Glass Transition Temperature

The reaction of polymers to temperature depends on the degree to which the material has crystallized. Highly ordered polymers have a fairly well-defined transition between elastic and viscous behavior. As the percent of crystallization decreases, the melting point is not well defined. However, the point where these polymers transition to a glass phase is well defined. At elevated temperatures, the motion of the molecules forces a separation between them, resulting in a volume that is greater than required for tightly packed, motionless molecules. This excess volume is termed the *free volume*. As the material cools, the motion of the molecules is reduced and the viscosity increases. At a low enough temperature, the molecules are no longer free to rearrange; thus their

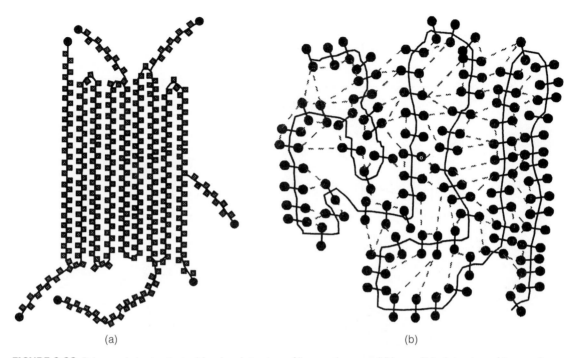

(a)                                                                                  (b)

**FIGURE 2.22** Polymer chain structures: (a) ordered structure of linear polymer and (b) cross-linked structure of linear polygon.

position is fixed and the free volume becomes zero. This is the glass transition temperature. Below this temperature the secondary bonds bind the material into an amorphous solid; above this temperature the material behaves in an elastic manner. Figure 2.23 illustrates the concept of melting and glass transition temperatures, $T_m$ and $T_g$. The glass transition temperature of Plexiglass is 100°C; at room temperature it is a brittle solid. Above $T_g$ it becomes leathery and then rubbery. $T_g$ for natural rubber is –70°C; it is flexible at all atmospheric temperatures. However, when frozen, say in liquid nitrogen, it becomes a brittle solid.

## Mechanical Properties

The mechanical behavior of polymers is directly related to the degree of orientation of the molecules and the amount of cross-linking by covalent bonds. The modulus of a polymer is the average of the stiffness of the bonds. The modulus can be estimated based on volume fractions of the covalent bonding as (Ashby and Jones 1986)

$$\epsilon = f\frac{\sigma}{E_1} + (1-f)\frac{\sigma}{E_2} \qquad (2.7)$$

where

$\epsilon$ = strain of the material

$\sigma$ = stress of the material

$f$ = fraction of covalent bonds

$E_1$ = stiffness of covalent bonds, about 1000 GPa for diamond

$E_2$ = stiffness of secondary bonds, about 1 GPa for paraffin wax

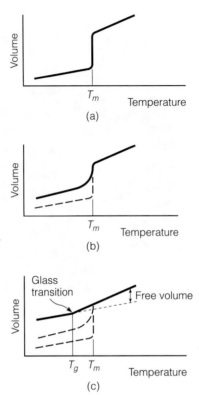

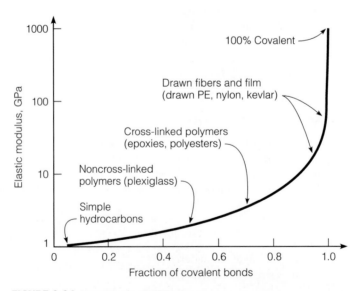

**FIGURE 2.23**  Melting point and glass transition temperatures: (a) perfect crystallization, (b) imperfect crystallization, and (c) glass formation.

**FIGURE 2.24**  Modulus of polymers.

Based on Equation 2.7, the expected modulus of various polymers can be computed for temperatures less than the glass transition temperature, as shown in Figure 2.24.

**SUMMARY**

The behavior of materials important to engineering is directly related to their microscopic and macroscopic structure. Although our understanding of these materials is imperfect at this time, much of the their behavior can be attributed to the bonding and arrangement of the materials at the atomic level. This chapter provides only a broad overview of the subject. For more information consult references with more in-depth treatments of these subjects.

**QUESTIONS AND PROBLEMS**

**2.1.** Define elastic and plastic behaviors at the micro and macro levels.

**2.2.** Describe the parts of an atom. Define proton, electron, atomic number, and atomic mass.

**2.3.** What are the valence electrons and why are they important?

**2.4.** Describe the order in which electrons fill the shells and subshells.

**2.5.** Describe the different types of bonds.

**2.6.** Why do atoms maintain specific separations?

**2.7.** Materials are generally classified into three categories based on the predominant types of bond. What are these three categories and what are the predominant types of bond in each category? In each category provide two examples of common materials used by civil engineers.

**2.8.** What is the atomic packing factor? What information do you need to compute it?

**2.9.** Describe FCC, BCC, and HCP lattice structures.

**2.10.** If the volume of the sphere is $4/3\pi r^3$, show that the atomic packing factor for the BCC lattice structure is 0.68.

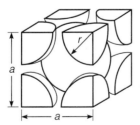

**2.11.** Calculate the volume of the unit cell of iron in cubic meters given that iron has a BCC crystal structure and an atomic radius of 0.124 nm.

**2.12.** Calculate the density of iron given that it has a BCC crystal structure, an atomic radius of 0.124 nm, and an atomic mass of 55.9 g/mole.

**2.13.** Calculate the radius of the copper atom given that copper has an FCC crystal structure, a density of 8.89 g/cm$^3$, and an atomic mass of 63.55 g/mole.

**2.14.** What are the classes of defects in crystal structures?

**2.15.** Why do grains form in crystal structures?

**2.16.** Explain the slipping of atoms and the effect on material deformation.

**2.17.** Sketch a phase diagram for two soluble components.

**2.18.** What is the eutectic composition and why is it important?

**2.19.** Considering an alloy of the two soluble components A and B described by a phase diagram similar to that shown in Figure 2.14, determine the masses of the alloy that are in the liquid and solid phases at a given temperature if the total mass of the alloy is 100 grams, component B represents 65% of the alloy, 30% of the liquid is component B, and 80% of solid is component B.

**2.20.** The figure shows a portion of the $H_2O$-NaCl phase diagram.

**a.** Using the diagram, briefly explain how spreading salt on ice causes the ice to melt. Show numerical examples in your discussion.

**b.** At a salt composition of 10%, what is the temperature at which ice will start melting?

**c.** What is the eutectic temperature of the ice and salt combination?

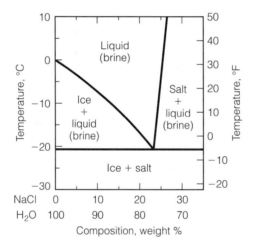

**2.21.** What are the five classes of ceramic materials?

**2.22.** What are the four types of organic solids used in engineering applications?

**REFERENCES**

Ashby M. F., and D. R. H. Jones. 1986. *Engineering materials 2, an introduction to micro-structures, processing and design.* International Series on Materials Science and Technology, Vol. 39. Oxford, England: Pergamon Press.

Budinski, K. G. 1996. *Engineering materials, properties and selection.* 5th ed. Upper Saddle River, NJ: Prentice Hall.

Callister, W. D., Jr. 1985. *Material science and engineering, an introduction.* New York: John Wiley & Sons.

Derucher, K. N., G. P. Korfiatis, and A. S. Ezeldin. 1994. *Materials for civil & highway engineers.* 3rd ed. Englewood Cliffs, NJ: Prentice Hall.

Flinn, R. A., and P. K. Trojan. 1986. *Engineering materials and their applications.* 3rd ed. Boston, MA: Houghton Mifflin.

Guy, A. G. and J. J. Hren. 1974. *Elements of physical metallurgy.* 3rd ed. Reading, MA: Addison-Wesley.

Jackson, N. and R. K. Dhir, eds. 1988. *Structural engineering materials.* 4th ed. New York: Hemisphere.

Jastrzebski, Z. D. 1987. *The nature and properties of engineering materials.* 3rd ed. New York: John Wiley & Sons.

Shackelford, J. F. 1996. *Introduction to materials science for engineers.* 4th ed. New York: Macmillan.

Van Vlack, L. H. 1964. *Elements of materials science.* 2nd ed. Reading, MA: Addison-Wesley.

Van Vlack, L. H. 1989. *Elements of materials science and engineering.* 6th ed. Reading, MA: Addison-Wesley.

# 3 Steel

The use of iron dates back to about 1500 B.C. when primitive furnaces were used to heat the ore in a charcoal fire. Ferrous metals were produced on a relatively small scale until the blast furnace was developed in the eighteenth century. Iron products were widely used in the latter half of the eighteenth century and the early part of the nineteenth century. Steel production started in mid-1800s when the Bessemer converter was invented. In the second half of the nineteenth century steel technology advanced rapidly due to the development of the basic oxygen furnace and continuous casting methods. More recently, computer-controlled manufacturing has increased the efficiency and reduced the cost of steel production.

Currently, steel and steel alloys are used widely in civil engineering applications. In addition, wrought iron is still used on a smaller scale for pipes, as well as general blacksmith work. Cast iron is used for pipes, hardware, and machine parts not subjected to tensile or dynamic loading.

Steel products used in construction can be classified (Cordon 1979) as

1. *structural steel* for use in plates, bars, pipes, structural shapes, etc.,
2. *reinforcing steel* (rebars) for use in concrete reinforcement,
3. miscellaneous for use in such applications as forms and pans.

Civil and construction engineers rarely have the opportunity to formulate steel with specific properties. Rather, they must select existing products from suppliers. Even the shapes for structural elements are generally restricted to those readily available from manufacturers. While specific shapes can be made to order, the cost to fabricate low-volume members is generally prohibitive. Therefore, the majority of civil engineering projects are designed using standard steel types and structural shapes.

Even though civil and construction engineers are not responsibile for formulating steel products, they still must understand how steel is manufactured and treated and how it responds to loads and environmental conditions. This chapter reviews steel production, the iron-carbon phase diagram, heat treatment, steel alloys, structural steel, and reinforcing steel. This chapter also presents common tests used to characterize the mechanical properties of steel. The topics of welding and corrosion of steel are also introduced.

## Steel Production

The overall process of steel production is shown in Figure 3.1 This process may be described in three phases

1. reduction of iron ore to pig iron
2. refining pig iron to steel
3. forming the steel into products

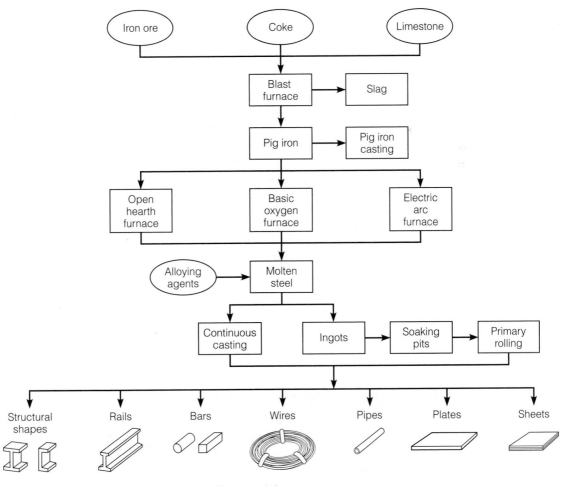

**FIGURE 3.1** Conversion of raw materials into different steel shapes.

The materials used to produce pig iron are coal, limestone, and iron ore. The coal, after transformation to coke, supplies carbon used to reduce iron oxides in the ore. Limestone is used to help remove impurities. Prior to reduction, the concentration of iron in the ore is increased by crushing and soaking the ore. The iron is magnetically extruded from the waste, and the extracted material is formed into pellets and fired. The processed ore contains about 65% iron.

Reduction of the ore to pig iron is accomplished in a blast furnace. The ore is heated in the presence of carbon. Oxygen in the ore reacts with carbon to form gases. A flux is used to help remove impurities. The molten iron, with an excess of carbon in solution, collects at the bottom of the furnace. The impurities, slag, float on top of the molten pig iron.

The excess carbon, along with other impurities, must be removed to produce high-quality steel. Using the same refining process, scrap steel can be recycled. Three types of furnaces are used for refining pig iron to steel:

1. open hearth
2. basic oxygen
3. electric arc

The open hearth and basic oxygen furnaces remove excess carbon by reacting the carbon with oxygen to form gases. Lances circulate oxygen through the molten material. The process is continued until all impurities are removed and the desired carbon content is achieved. Open hearth furnaces have been used since the early 1900s. Now, due to greater efficiency and productivity, basic oxygen furnaces are the industry standard for high-production mills. A basic oxygen furnace can refine 280,000 kg (300 tons) of steel in 25 minutes, compared to the 8 hours it takes to refine the same quantity of steel in an open-hearth furnace.

Electric furnaces use an electric arc between carbon electrodes to melt and refine the steel. These plants require a tremendous amount of energy and are primarily used to recycle scrap steel. Electric furnaces are frequently used in minimills, which produce a limited range of products. In this process, molten steel is transferred to the ladle. Alloying elements and additional agents can be added either in the furnace or the ladle.

Regardless of the refining process, the molten steel, with the desired chemical composition, is then either cast into ingots (large blocks of steel) or cast continuously into a desired shape. Continuous casting is becoming the standard production method since it is more energy efficient than casting ingots as the ingots must be reheated prior to shaping the steel into the final product.

## Iron-Carbon Phase Diagram

In refining steel from iron ore, the quantity of carbon used must be carefully controlled in order for the steel to have the desired properties. The reason for the strong relationship between steel properties and carbon content can be understood by examining the iron-carbon phase diagram.

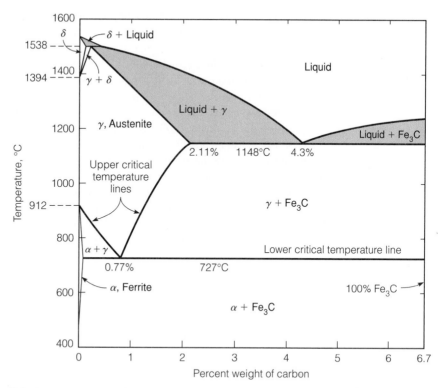

**FIGURE 3.2** The iron-carbon phase diagram.

Figure 3.2 presents a commonly accepted iron-carbon phase diagram. One of the unique features of this diagram is that the abscissa extends only to 6.7% rather than 100%. This is a matter of convention. In an iron rich material, each carbon atom bonds with three iron atoms to form iron carbide, $Fe_3C$, also called cementite. Iron carbide is 6.7% carbon by weight. Thus, on the phase diagram, a carbon weight of 6.7% corresponds to 100% iron carbide. Therefore, Figure 3.2 is technically an iron-iron carbide phase diagram. A complete iron-carbon phase diagram should extend to 100% carbon. However, only the iron rich portion, as shown on Figure 3.2, is of practical significance (Callister 1985).

The left side of Figure 3.2 demonstrates that pure iron goes through two transformations as temperature increases. Pure iron below 912°C has a BCC crystalline structure called ferrite. At 912°C the ferrite undergoes a polymorphic change to a FCC structure called austenite. At 1394°C another polymorphic change occurs, returning the iron to a BCC structure. The high and low temperature ferrites are identified as $\delta$ and $\alpha$ ferrite, respectively. The $\delta$ ferrite does not have practical significance for this book.

Carbon goes into solution with $\alpha$ ferrite at temperatures between 400°C and 912°C. However, the solubility limit is very low, with a maximum of 0.022% at 727°C. At low temperatures and to the right of the solubility limit line, $\alpha$ ferrite and iron carbide coexist. From 727°C to 1148°C the solubility of carbon in the austenite increases from 0.77% to 2.11%. The solubility of carbon in austenite is greater than in $\alpha$ ferrite because of the crystalline structure of the austenite.

At 0.77% carbon and 727°C a eutectoid reaction occurs; that is, a solid phase change occurs when either the temperature or carbon content changes. At 0.77% carbon, and above 727°C, the carbon is in solution with the FCC structure austenite. A temperature drop to below 727°C, which happens slowly enough to allow the atoms to reach an equilibrium condition, results in a two-phase material, $\alpha$ ferrite and iron carbide. The $\alpha$ ferrite will have 0.022% carbon in solution, and the iron carbide will have a carbon content of 6.7%. The ferrite and iron carbide will form as thin plates, a lamellae structure. This eutectoid material is called pearlite.

At carbon contents less than the eutectoid composition, hypoeutectoid alloys are formed. Consider a carbon content of 0.25%. Above approximately 810°C, solid austenite exists and the carbon is in solution; the austenite consists of grains of uniform material. (The grains form when the steel cools from liquid to a solid.) Under equilibrium temperature drop conditions, $\alpha$ ferrite is formed as the temperature drops from 810°C to 727°C and accumulates at the grain boundaries of the austenite. At temperatures slightly above 727°C, the ferrite will have 0.022% carbon in solution and austenite will have 0.77% carbon. When the temperature drops below 727°C the austenite will transform to pearlite. When ferrite is formed above 727°C, proeuctoid ferrite results.

When the carbon content is above the eutectoid composition, iron carbide forms at the grain boundaries at temperatures above 727°C. The resulting microstructure consists of grains of pearlite surrounded by iron carbide.

Figure 3.3 shows an optical photomicrograph of a hot-rolled mild steel plate with a carbon content of 0.18% by weight that was etched with 3% nitol. The photomicrograph is magnified at 50×. The light etching phase is proeutectiod ferrite and the dark constituent is pearlite. Note the banded structure resulting from the rolling processes. Figure 3.4 is the same as Figure 3.3, except that the magnification is 400×. At this magnification the alternating layers of ferrite and cementite in the pearlite can be seen.

The significance of ferrite, pearlite, and iron carbide formation is that the properties of the steel are highly dependent on the relative proportions of ferrite and iron carbide. Ferrite has relatively low strength but is very ductile. Iron carbide has high strength but has virtually no ductility. Combining these materials in different proportions alters the mechanical properties of the steel. Increasing the carbon content increases strength but reduces ductility. However, the modulus of elasticity of steel does not change by altering the carbon content.

All of the reactions described here were for equilibrium temperature drops. Cooling at more rapid rates greatly alters the microstructure. Moderate cooling rates produce bainite, a fine-structure pearlite without a proeuctoid phase. Rapid quenching produces martensite; the carbon is supersaturated in the iron causing a body center tetragonal lattice structure. The specific changes of steel properties that can be accomplished by altering the rate of cooling are presented in the next section.

**FIGURE 3.3** Optical photomicrograph of hot-rolled mild steel plate (magnification: 50×).

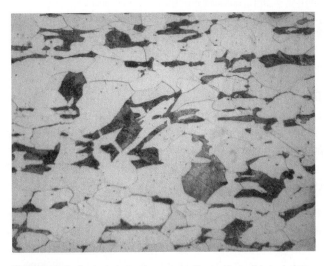

**FIGURE 3.4** Optical photomicrograph of hot-rolled mild steel plate (magnification: 400×).

## Heat Treatment of Steel

Properties of steel can be altered by applying a variety of heat treatments. For example, steel can be hardened or softened by using heat treatment; the response of steel to heat treatment depends upon its alloy composition. Common heat treatments employed for steel include annealing, normalizing, hardening, and tempering.

## Annealing

Annealing is performed by heating the metal to the austenite stable range, usually 10°C (18°F) above the austenite line, and holding it at that temperature for the proper period. The material is then slowly cooled to room temperature. During the cooling process, pearlite and ferrite or cementite form. The objectives of annealing are to refine the grain, soften the steel, remove internal stresses, remove gases, increase ductility and toughness, and change electrical and magnetic properties.

## Normalizing

Normalizing is similar to annealing, with a slight difference in heating temperature. Steel is normalized by heating it into the austenizing range, usually 40°C (72°F) above the austenite line. The material is then air cooled. Normalizing produces a uniform, fine-grained microstructure. Therefore, normalizing is regarded as a corrective treatment, and not a strengthening or hardening treatment. Normalizing is used in structural plate production to produce high-fracture toughness.

## Hardening

Steel is hardened by heating it to a temperature above the transformation range and holding it until austenite is formed. The steel is then quenched (cooled rapidly) by plunging it into water, brine, or oil. Quenching hardens the steel, and hardening puts the steel in a state of strain. This strain sometimes causes steel pieces with sharp angles or grooves to crack immediately after hardening. Thus, hardening must be followed by tempering.

## Tempering

Tempering involves reheating a hardened steel to a definite temperature below the critical temperature, holding it for a time, and cooling it, usually by quenching. Tempering increases ductility and toughness.

# Steel Alloys

Alloy metals can be used to alter the characteristics of steel. By some counts there are as many as 250,000 different alloys of steel produced. Of these as many as 200 may be used for civil engineering applications. Rather than go into the specific characteristics of selected alloys, the general effect of different alloying agents will be presented. Alloy agents are added to improve one or more of the following properties:

1. hardenability
2. corrosion resistance
3. machineability
4. ductility
5. strength

Common alloy agents, their typical percentage range, and their effects are summarized in Table 3.1.

By altering the carbon and alloy content and by using different heat treatments, steel can be produced with a wide variety of characteristics. These are classified as

**1.** Low alloy
  - Low carbon
    Plain
    High strength–low alloy
  - Medium carbon
    Plain
    Heat treatable
  - High carbon
    Plain
    Tool

**2.** High alloy
  - Tool
  - Stainless

**TABLE 3.1**  Common Steel Alloying Agents[*]

| | Typical Ranges in Alloy Steels, % | Principal Effects |
|---|---|---|
| Aluminum | < 2 | Aids nitriding<br>Restricts grain froth<br>Removes oxygen in steel melting |
| Sulfur | < 0.5 | Adds machinability<br>Reduces weldability and ductility |
| Chromium | 0.3 to 0.4 | Increases resistance to corrosion and oxidation<br>Increases hardenability<br>Increases high-temperature strength<br>Can combine with carbon to form hard wear-resistant microconstituents |
| Nickel | 0.3 to 5.0 | Promotes an austenitic structure<br>Increases hardenability<br>Increases toughness |
| Copper | 0.2 to 0.5 | Promotes tenacious oxide film to aid atmospheric corrosion resistance |
| Manganese | 0.3 to 2.0 | Increases hardenability<br>Promotes an austenitic structure<br>Combines with sulfur to reduce its adverse effects |
| Silicon | 0.2 to 2.5 | Removes oxygen in steel-making<br>Improves toughness<br>Increases hardenability |
| Molybdenum | 0.1 to 0.5 | Promotes grain rennement<br>Increases hardenability<br>Improves high-temperature strength |
| Vanadium | 0.1 to 0.3 | Promotes grain rennement<br>Increases hardenability<br>Combines with carbon to form wear-resistant microconstituents |

*Budinski, K. *Engineering Materials, Properties, and Selection*. 5th ed. 1996. Reprinted by permission of Prentice-Hall Inc., Upper Saddle River, NJ.

Steels used for construction projects are predominantly low and medium carbon plain steels. Stainless steel has been used in some highly corrosive applications, such as dowel bars in concrete pavements and steel components in swimming pools and drainage lines. As mentioned earlier in this chapter, steel used in construction can be classified mainly as structural steel and reinforcing steel.

# Structural Steel

Structural steel is used in hot-rolled structural shapes, plates, and bars. Structural steel is used for various types of structural members such as columns, beams, bracings, frames, trusses, bridge girders, and other structural applications.

## Structural Steel Grades

Structural steel is produced in the United States in six grades: A36, A529, A572, A242, A588, and A514. Table 3.2 shows the chemical and tensile requirements of these grades as specified by ASTM standards. Structural steels have a carbon content in the range of 0.15% to 0.27%. All structural steels are alloyed with a small amount of copper. All but A36 contain manganese.

Grade A36 is a structural carbon steel with a tensile yield stress of 250 MPa (36 ksi). Grade A529 is a structural steel with a minimum yield stress of 290 MPa (42 ksi), while other grades are high-strength steels with higher yield stresses, as shown in Table 3.2. Grades A242 and A588 are corrosion-resistant, high-strength, low-alloy structural steels.

Grade A36 is the most commonly used steel in buildings, bridges, transmission towers, and other structures. It is available in plates, structural shapes, and bars. It offers satisfactory performance in various temperature conditions since it does not experience brittle fracture at low temperatures. High-strength steel results in lighter sections that can prove to be economical, especially for tension members and beams in continuous and composite construction. Because of their corrosion resistance, Grades A242 and A588 can be used in the uncoated condition in most atmospheres. When these two grades are coated, they produce longer coating lives than other grades. Although corrosion-resistant steel grades have high initial cost, they are used to lower maintenance costs over the life of the structure.

## Sectional Shapes

Figure 3.5 illustrates structural cross-sectional shapes commonly used in structural applications. These shapes are produced in different sizes and designated with the letters W, HP, M, S, C, MC, and L. W shapes are doubly symmetric wide-flange shapes whose flanges are substantially parallel. HP shapes are also wide-flange shapes whose flanges and webs are of the same nominal thickness and whose depth and width are essentially the same. The S shapes are doubly symmetric shapes whose inside flange surfaces have approximately 16.67% slope. The M shapes are doubly symmetric shapes that cannot be classified as W, S, or HP shapes. C shapes are channels with inside flange surfaces having a slope of an approximately 16.67%. MC shapes are channels

**TABLE 3.2** Grades and Properties of Structural Steel According to ASTM*

| ASTM Designation | Type | Grade† | Chemical Requirements, % | | | Tensile Requirements | | Availability |
|---|---|---|---|---|---|---|---|---|
| | | | Carbon (max.) | Manganese (max.) | Copper (min.) | Tensile strength, MPa (ksi) | Yield Point (min.), MPa (ksi) | |
| A36 | Structural carbon steel | 36 | 0.26 | — | 0.20 | 400–550 (58–80) | 250 (36) | All shapes, plates, and bars |
| A529 | Structural steel with 42 ksi min. yield point | 42 | 0.27 | 1.2 | 0.20 | 415–485 (60–85) | 290 (42) | Selected shapes, plates, and bars ≤ 1/2 in. thick |
| A572 | High-strength low-alloy steel of structural quality | 42 | 0.21 | 1.35 | 0.20 | 415 (60) | 290 (42) | All shapes, sheet piling, and tees |
| | | 50 | 0.23 | 1.35 | 0.20 | 450 (65) | 345 (50) | All shapes, sheet piling, and tees |
| | | 60 | 0.26 | 1.35 | 0.20 | 520 (75) | 415 (60) | Limited shapes, all sheet piling, and tees |
| | | 65 | 0.26 | 1.35 | 0.20 | 550 (80) | 450 (65) | Limited shapes and all tees |
| A242 | High-strength low-alloy structural steel (corrosion resistant) | 42–50 | 0.15 | 1.0 | 0.20 | 435–480 (63–70) | 290–345 (42–50) | Limited shapes, plates, and bars |
| A588 | High-strength low-alloy structural steel with 50 ksi min. yield point (corrosion resistant) | 50 | 0.17–0.19 | 0.5–1.25 | 0.2–0.5 | 485 (70) 435–485 (63–70) | 345 (50) 290–345 (42–50) | All shapes, plates, and bars |
| A514 | High-yield strength quenched and tempered alloy steel | 90–100 | 0.12–0.21 | 0.4–1.10 | 0.15–0.50 | 690–895 (100–130) | 290–690 (90–110) | Plates |

*Somayaji, S. *Civil Engineering Materials.* 1995. Reprinted by permission of Prentice-Hall Inc., Upper Saddle River, NJ.
†Equals yield stress in ksi.

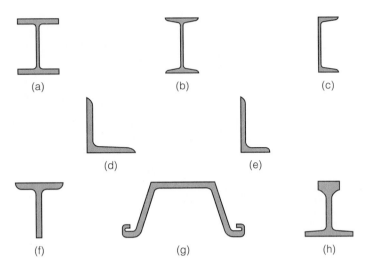

(a)                        (b)                        (c)

(d)                        (e)

(f)                        (g)                        (h)

**FIGURE 3.5** Shapes commonly used in structural applications: (a) wide-flange
(W, HP, and M shapes), (b) I-beam (S shape), (c) channel (C and MC shapes),
(d) equal-legs angle (L shape), (e) unequal-legs angle (L shape), (f) tee, (g) sheet piling,
and (h) rail.

that cannot be classified as C shapes. L shapes are angle shapes with either equal or
unequal legs. In addition to these shapes, other structural sections are available, such
as tee, sheet piling, and rail, as shown in Figure 3.5.

The W, M, S, HP, C, and MC shapes are designated by a letter, followed by two
numbers separated by an ×. The letter indicates the shape, while the two numbers
indicate the nominal depth and the weight per linear unit length. For example, W 44×
335 means W shape with a nominal depth of 44 in. and a weight of 335 lb/linear foot.
An angle is designated with the letter L, followed by three numbers that indicate the
leg dimensions and thickness in inches, such as L 4 × 4 × 1/2. Dimensions of these
structural shapes are controlled by ASTM A6/A6M.

W shapes are commonly used as beams and columns, HP shapes are used as
bearing piles, and S shapes are used as beams or girders. Composite sections can also
be formed by welding different shapes to use in various structural applications. Sheet
piling sections are connected to each other and used as retaining walls.

## Reinforcing Steel

Since concrete has negligible tensile strength, structural concrete members subjected
to tensile and flexural stresses must be reinforced. Either conventional or prestressed
reinforcing can be used, depending on the design situation. In conventional reinforcing, the stresses fluctuate with loads on the structure. This does not place any special
requirements on the steel. On the other hand, in prestressed reinforcement the steel is
under continuous tension. Any stress relaxation will reduce the effectiveness of the
reinforcement. Hence special steels are required.

**FIGURE 3.6** Deformed reinforcing bars.

Reinforcing steel (rebars) is manufactured in three forms: *plain bars, deformed bars,* and *plain and deformed wire fabrics.* Plain bars are round without surface deformations. Plain bars provide only limited bond with the concrete and, therefore, are not typically used in sections subjected to tension or bending. Deformed bars have protrusions (deformations) at the surface, as shown Figure 3.6; thus they ensure a good bond between the bar and the concrete. The deformed surface of the bar prevents slipping, allowing the concrete and steel to work as one unit. Wire fabrics are flat sheets in which wires pass each other at right angles, and one set of elements is parallel to the fabric axis. Plain wire fabrics develop the anchorage in concrete at the welded intersections, while deformed wire fabrics develop anchorage through deformations and at the welded intersections.

Deformed bars are used in concrete beams, slabs, columns, walls, footings, pavements, and other concrete structures, as well as in masonry construction. Welded wire fabrics are used in some concrete slabs and pavements, mostly to resist temperature and shrinkage stresses. Welded wire fabrics can be more economical to place, and thus allow for closer spacing of bars than is practical with individual bars.

Reinforcing steel is produced in the standard sizes shown in Table 3.3. Bars are made of four types of steel: A615 (billet), A616 (rail), A617 (axle), and A706 (low-alloy), as shown in Table 3.4. Billet steel is the most widely used. Reinforcing steel is produced in four grades: 40, 50, 60, and 75, with yield stresses of 276 MPa, 345 MPa, 414 MPa, and 517 MPa (40 ksi, 50 ksi, 60 ksi, and 75 ksi), respectively.

Prestressed concrete requires special prestressing wires, strands, cables, and bars. Steel for prestressed concrete reinforcement must be of high strength and low relaxation properties. High-carbon steels and high-strength alloy steels are used for this purpose. Properties of prestressed concrete reinforcement are presented in ASTM specification A416 and AASHTO specification M203. These specifications define the requirements for a seven-wire uncoated steel strand. The specifications allow two types of steel, stress-relieved (normal-relaxation) and low-relaxation. Relaxation refers to the percent of stress reduction that occurs when a constant amount of strain is

**TABLE 3.3**   Standard-Size Reinforcing Bars According to ASTM A615[*]

| Bar Designation Number[†] | Nominal Mass, kg/m | Nominal Dimensions[‡] | | | Deformation Requirements (mm) | | |
|---|---|---|---|---|---|---|---|
| | | Diameter, mm | Cross-sectional Area, mm² | Perimeter, mm | Maximum Average Spacing | Minimum Average Height | Maximum Gap[**] |
| 10 (3) | 0.560 | 9.5 | 71 | 29.9 | 6.7 | 0.38 | 3.6 |
| 13 (4) | 0.994 | 12.7 | 129 | 39.9 | 8.9 | 0.51 | 4.9 |
| 16 (5) | 1.552 | 15.9 | 199 | 49.9 | 11.1 | 0.71 | 6.1 |
| 19 (6) | 2.235 | 19.1 | 284 | 59.8 | 13.3 | 0.97 | 7.3 |
| 22 (7) | 3.042 | 22.2 | 387 | 69.8 | 15.5 | 1.12 | 8.5 |
| 25 (8) | 3.973 | 25.4 | 510 | 79.8 | 17.8 | 1.27 | 9.7 |
| 29 (9) | 5.059 | 28.7 | 645 | 90.0 | 20.1 | 1.42 | 10.9 |
| 32 (10) | 6.404 | 32.3 | 819 | 101.3 | 22.6 | 1.63 | 12.4 |
| 36 (11) | 7.907 | 35.8 | 1006 | 112.5 | 25.1 | 1.80 | 13.7 |
| 43 (14) | 11.38 | 43.0 | 1452 | 135.1 | 30.1 | 2.16 | 16.5 |
| 57 (18) | 20.24 | 57.3 | 2581 | 180.1 | 40.1 | 2.59 | 21.9 |

[*]Copyright ASTM. Reprinted with permission.
[†]Bar numbers approximating the number of millimeters of the nominal diameter of the bars (bar numbers based on the number of eighths of an inch of the nominal diameter of the bars).
[‡]The nominal dimensions of a deformed bar are equivalent to those of plain round bar having the same weight per meter as the deformed bar.
[**]Chord 12.5% of nominal perimeter.

applied over an extended time period. Both stress-relieved and low-relaxation steels can be specified as Grade 250 or Grade 270, with ultimate strengths of 1725 MPa (250 ksi) and 1860 MPa (270 ksi), respectively. The specifications for this application are based on mechanical properties only; the chemistry of wires is not pertinent to

**TABLE 3.4**   Types and Properties of Reinforcing Bars According to ASTM[*]

| ASTM Designation | Type | Grade | Tensile Strength (min.), MPa (ksi) | Yield Strength[†] (min.), MPa (ksi) | Size Availability No. |
|---|---|---|---|---|---|
| A615 | Billet steel bars (plain and deformed) | 40 | 483 (70) | 276 (40) | 3–6 |
| | | 60 | 620 (90) | 414 (60) | 3–18 |
| | | 75 | 689 (100) | 517 (75) | 11–18 |
| A616 | Rail steel (plain and deformed) | 50 | 552 (80) | 345 (50) | 3–11 |
| | | 60 | 620 (90) | 414 (60) | 3–11 |
| A617 | Axle steel (plain and deformed) | 40 | 483 (70) | 276 (40) | 3–11 |
| | | 60 | 620 (90) | 414 (60) | 3–11 |
| A706 | Low-alloy steel bars, (deformed) | 60 | 552 (80) | 414–538 (60–78) | 3–18 |

[*]Somayaji, S. *Civil Engineering Materials*. 1995. Reprinted by permission of Prentice-Hall Inc., Upper Saddle River, NJ.
[†]When the steel does not have a well-defined yield point, yield strength is the stress corresponding to a strain of 0.005 m/m (0.5% extension) for grades 40, 50, and 60; and a strain of 0.0035 m/m (0.35% extension) for grade 75 of A615, A616, and A617 steels. For A706 steel, grade point is determined at a strain of 0.0035 m/m.

**TABLE 3.5**   Required Properties for Seven-Wire Strand

| Property | Stress-Relieved | | Low-Relaxation | |
|---|---|---|---|---|
| | Grade 250 | Grade 270 | Grade 250 | Grade 270 |
| Breaking strength, MPa (ksi)[*] | 1725 (250) | 1860 (270) | 1725 (250) | 1860 (270) |
| Yield strength, 1% extension | 85% of breaking strength | | 90% of breaking strength | |
| Elongation, min. percent | 3.5 | | 3.5 | |
| Relaxation, max. percent[†] | | | | |
|   Load = 70% min. breaking strength | — | | 2.5 | |
|   Load = 80% min. breaking strength | — | | 3.5 | |

[*]Breaking strength is the stress required to break one or more wires.
[†]Relaxation is the reduction in stress that occurs when constant strain is applied over an extended time period. The specification is for a load duration of 1000 hours at a test temperature of $20 \pm 2\,^\circ C$ ($68 \pm 3\,^\circ F$).

this application. After stranding, low-relaxation strands are subjected to a continuous thermal-mechanical treatment to produce the required mechanical properties. Table 3.5 shows the required properties for seven-wire strand.

# Mechanical Testing of Steel

Many tests are available to evaluate the mechanical properties of steel. This section summarizes some laboratory tests commonly used to determine properties required in product specifications. Test specimens can take several shapes, such as bar, tube, wire, flat section, and notched bar, depending on the test purpose and the application.

Certain methods of fabrication such as bending, forming, and welding, or operations involving heating, may affect the properties of material being tested. Therefore, the product specifications cover the stage of manufacture at which mechanical testing is performed. The properties shown by testing before the material is fabricated may not necessarily be representative of the product after it has been completely fabricated. In addition, flaws in the specimen or improper machining or preparation of test specimen will give erroneous results (ASTM A370).

## Tension Test

The tension test (ASTM E8) on steel is performed to determine the yield strength, yield point, ultimate (tensile) strength, elongation, and reduction of area. Typically, the test is performed at temperatures between 10°C and 35°C (50°F to 95°F).

The test specimen can be either full sized or machined into a shape, as prescribed in the product specifications for the material being tested. It is desirable to use a small cross-sectional area at the center portion of the specimen to ensure fracture within the gauge length. Several cross-sectional shapes are permitted, such as round and rectangular, as shown in Figure 3.7. Plate, sheet, round rod, wire, and tube specimens may be used. A 12.5 (1/2 in.) diameter round specimen is used in many cases. The gauge length over which the elongation is measured typically is four times the diameter for most round-rod specimens.

**FIGURE 3.7** Tension test specimens with round and rectangular cross sections.

**FIGURE 3.8** Tension test on a round steel specimen showing grips and an extensometer with an LVDT.

Various types of gripping devices may be used to hold the specimen, depending on its shape. In all cases, the axis of the test specimen should be placed at the center of the testing machine head to ensure axial tensile stresses within the gauge length without bending. An extensometer with a dial gauge (Figure 1.24) or an LVDT (Figure 1.28) is used to measure the deformation of the entire gauge length. The test is performed by applying an axial load to the specimen at a specified rate. Figure 3.8 shows a tensile test being performed on a round steel specimen using an LVDT extensometer to measure the deformation.

As discussed in Chapter 1, mild steel has a unique stress-strain relation (Figure 3.9). Here, a linear elastic response is displayed up to the proportion limit. As the stress is increased beyond the proportion limit, the steel will yield, at which time the strain will increase without an increase in stress (actually the stress will slightly decrease). As tension increases past the yield point, strain increases following a nonlinear relation up to the point of failure.

Note that the decrease in stress after the peak does not mean a decrease in strength. In fact, the actual stress continues to increase until failure. The reason for the apparent decrease is that a neck is formed in the steel specimen, causing an appreciable decrease in the cross-sectional area. The traditional, or engineering, way of calculating the stress and strain uses the original cross-sectional area and gauge length. If the stress and stains are calculated based on the instantaneous cross-sectional area and gauge length, a *true stress-strain curve* is obtained, which is different than the *engineering stress-strain curve* (Figure 3.9).

Different carbon-content steels have different stress-strain relations. Increasing the carbon content in the steel increases the yield stress and reduces the ductility. Figure 3.10 shows the tension stress-strain diagram for hot-rolled steel bars containing carbons from 0.19% to 0.90%. Increasing the carbon content from 0.19% to 0.90% increases the yield stress from 280 MPa to 620 MPa (40 ksi to 90 ksi). Also, this increase in carbon content decreases the fracture strain from about 0.27 m/m to 0.09 m/m. Note that the increase in carbon content does not change the modulus of elasticity.

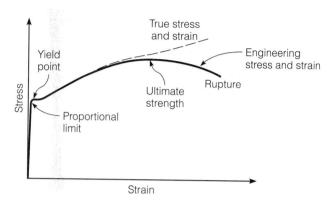

**FIGURE 3.9** Typical stress-strain behavior of mild steel.

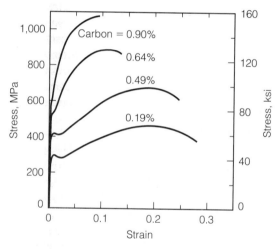

**FIGURE 3.10** Tensile stress-strain diagrams of hot-rolled steel bars with different carbon contents.

**SAMPLE PROBLEM 3.1**

A steel alloy bar 100 mm long with a rectangular cross section (10 mm × 40 mm) is subjected to tension with a load of 89 kN and experiences an increase in length of 0.1 mm. If the increase in length is entirely elastic, calculate the modulus of elasticity of the steel alloy.

*Solution:*

$$\sigma = \frac{89000}{0.01 \times 0.04} = 0.2225 \times 10^9 \text{ Pa} = 0.2225 \text{ GPa}$$

$$\epsilon = \frac{0.1}{100} = 0.001 \text{ mm/mm}$$

$$E = \frac{\sigma}{\epsilon} = \frac{0.2225}{0.001} = 222.5 \text{ GPa}$$

**SAMPLE
PROBLEM 3.2**

A steel specimen is tested in tension. The specimen is 1 in. wide by 0.5 in. thick in the test region. By monitoring the load dial of the testing machine, it was found that the specimen yielded at a load of 36 kips and fractured at 48 kips.

    **a.** Determine the tensile stresses at yield and at fracture.

    **b.** If the original gauge length was 4 in., estimate the gauge length when the specimen is stressed to 1/2 the yield stress.

*Solution:*

    **a.** Yield stress $\sigma_y = \dfrac{36}{1 \times 0.5} = 72$ ksi

         Fracture stress $\sigma_f = \dfrac{48}{1 \times 0.5} = 96$ ksi

    **b.** Assume $E = 30 \times 10^6$ psi

$$\epsilon = \frac{\left(\frac{1}{2}\right)\sigma_y}{E} = \frac{\left(\frac{1}{2}\right) \times 72 \times 10^3}{30 \times 10^6} = 0.0012 \text{ in./in.}$$

$$\Delta L = L\epsilon = 4 \times 0.0012 = 0.0048 \text{ in.}$$

         Final gauge length $= 4 + 0.0048 = 4.0048$ in.      ◆

## Torsion Test

The torsion test (ASTM E143) is used to determine the shear modulus of structural materials. The shear modulus is used in the design of members subjected to torsion, such as rotating shafts and helical compression springs. In this test a cylindrical, or tubular, specimen is loaded either incrementally or continually by applying an external torque to cause a uniform twist within the gauge length (Figure 3.11). The amount of applied torque and the corresponding angle of twist are measured throughout the test. Figure 3.12 shows the shear stress-strain curve. The shear modulus is the ratio of maximum shear stress to the corresponding shear strain below the proportional limit of the material, which is the slope of the straight line between $R$ (a pretorque stress) and $P$ (the proportional limit). For a circular cross section, the maximum shear stress ($\tau_{max}$), shear strain ($\gamma$), and the shear modulus ($G$) are determined by the equations

$$\tau_{max} = \frac{Tr}{J} \tag{3.1}$$

$$\gamma = \frac{\theta r}{L} \tag{3.2}$$

$$G = \frac{\tau_{max}}{\gamma} = \frac{TL}{J\theta} \tag{3.3}$$

where

     $T =$ torque

     $r =$ radius

     $J = \dfrac{\pi r^4}{2} =$ polar moment of inertia of the specimen about its center

     $\theta =$ angle of twist in radians

     $L =$ gauge length

**FIGURE 3.11** Torsion test apparatus.

The test method is limited to materials and stresses at which creep is negligible compared to the strain produced immediately upon loading. The test specimen should be sound, without imperfections near the surface. Also, the specimen should be straight and of uniform diameter for a length equal to the gauge length plus two to four diameters. The gauge length should be at least four diameters. During the test, torque is read from a dial gauge or a readout device attached to the testing machine, while the angle of twist may be measured using a torsiometer fastened to the specimen at the two ends of the gauge length. A curve fitting procedure can be used to estimate the straight line portion of shear stress-strain relation of Figure 3.12 (ASTM E143).

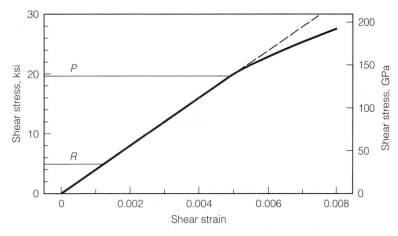

**FIGURE 3.12** Typical shear stress-strain diagram of steel (ASTM E143). Copyright ASTM. Reprinted with permission.

**SAMPLE PROBLEM 3.3**

A rod with a length of 1 m and a radius of 20 mm is made of high-strength steel. The rod is subjected to a torque $T$, which produced a shear stress below the proportional limit. If the cross section at one end is rotated 45 degrees in relation to the other end, and the shear modulus $G$ of the material is 90 GPa, what is the amount of applied torque?

**Solution:**

$$J = \frac{\pi r^4}{2} = \frac{\pi (0.02)^4}{2} = 0.2513 \times 10^{-6} \text{ m}^4$$

$$\theta = 45\left(\frac{\pi}{180}\right) = \frac{\pi}{4}$$

$$T = \frac{GJ\theta}{L} = \frac{(90 \times 10^9) \times (0.2513 \times 10^{-6}) \times \left(\frac{\pi}{4}\right)}{1} = 17.8 \times 10^3 \text{ N} \cdot \text{m} = 17.8 \text{ kN} \cdot \text{m}$$

◆

### Charpy V Notch Impact Test

The Charpy V Notch impact test (ASTM E23) is used to measure the toughness of the material or the energy required to fracture a V-notched simply supported specimen. The test is used for structural steels in tension members.

The standard specimen is $55 \times 10 \times 10$ mm ($2.165 \times 0.394 \times 0.394$ in.) with a V notch at the center, as shown in Figure 3.13. Before testing, the specimen is brought to the specified temperature for a minimum of 5 min in a liquid bath or 30 min in a gas medium. The specimen is inserted into the Charpy V notch impact-testing machine (Figure 3.14) using centering tongs. The swinging arm of the machine has a striking tip that impacts the specimen on the side opposite the V notch. The striking head is released from the pretest position, striking and fracturing the specimen. By fracturing the test specimen, some of the kinetic energy of the striking head is absorbed, thereby reducing the ultimate height the strike head attains. By measuring the height the strike head attains after striking the specimen, the energy required to fracture the specimen is computed. This energy is measured in $m \cdot N$ (ft $\cdot$ lb) as indicated on a gauge attached to the machine.

**FIGURE 3.13** Charpy V notch specimens.

**FIGURE 3.14** Charpy V notch impact testing machine.

The lateral expansion of the specimen is typically measured after the test using a dial gauge device. The lateral expansion is a measure of the plastic deformation during the test. The higher the toughness of the steel, the larger the lateral expansion.

Figure 1.13, in Chapter 1, shows the typical energy required to fracture structural steel specimens at different temperatures. The figure shows that the required energy is high at high temperatures and low at low temperatures. This indicates that the material changes from ductile to brittle as the temperature decreases.

The fracture surface typically consists of dull shear area at the edges and a shiny cleavage area at the center, as depicted in Figure 3.15. As the toughness of the steel decreases, by lowering the temperature, for example, the shear area decreases while the cleavage area increases.

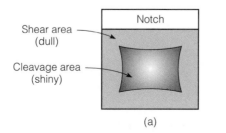
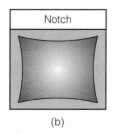

(a)  (b)

**FIGURE 3.15** Fracture surface of Charpy V notch specimen: (a) at high temperature, (b) at low temperature.

## Bend Test

In many engineering applications, steel is bent to a desired shape, especially in the case of reinforcing steel. The ductility to accommodate bending is checked by performing the semi-guided bend test (ASTM E290). The test evaluates the ability of steel, or a weld, to resist cracking during bending. The test is conducted by bending the specimen through a specified angle and to a specified inside radius of curvature. When complete fracture does not occur, the criterion for failure is the number and size of cracks found on the tension surface of the specimen after bending.

The bend test is made by applying a transverse force to the specimen in the portion that is being bent, usually at midlength. Three arrangements can be used, as illustrated in Figure 3.16. In the first arrangement, the specimen is fixed at one end and bent around a reaction pin or mandrel by applying a force near the free end, as shown in Figure 3.16(a). In the second arrangement, the specimen is held at one end and a rotating device is used to bend the specimen around the pin or mandrel, as shown in Figure 3.16(b). In the third arrangement, a force is applied in the middle of a specimen simply supported at both ends, Figure 3.16(c).

## Hardness Test

Hardness is a measure of a material's resistance to localized plastic deformation, such as a small dent or scratch on the surface of the material. A certain hardness is required for many machine parts and tools. Several tests are available to evaluate the hardness of materials. In these tests an indenter (penetrator) is forced into the surface of the material with a specified load magnitude and rate of application. The depth, or the

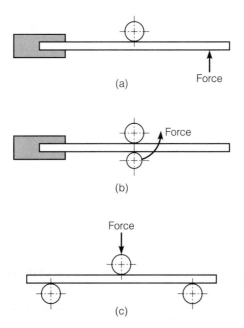

**FIGURE 3.16** Schematic fixtures for semi-guided bend test (ASTM E290). Copyright ASTM. Reprinted with permission.

size, of the indentation is measured and related to a hardness index number. Hard materials result in small impressions, corresponding to high hardness numbers. Hardness measurements depend on test conditions and are, therefore, relative. Correlations and tables are available to convert the hardness measurements from one test to another and to approximate the tensile strength of the material (ASTM A370).

One of the methods commonly used to measure hardness of steel and other metals is the *Rockwell hardness test* (ASTM E18). In this test the depth of penetration of a diamond cone, or a steel ball, into the specimen is determined under fixed conditions (Figure 3.17). A preliminary load of 10 kg is applied first, followed by an additional load. The Rockwell number, which is proportional to the difference in penetration between the preliminary and total loads, is read from the machine by means of a dial, digital display, pointer, or other device. Two scales are frequently used, namely, B and C. Scale B uses a 1.588 mm (1/16 in.) steel ball indenter and a total load of 100 kg, while scale C uses a diamond spheroconical indenter with a 120° angle and a total load of 150 kg.

To test very thin steel or thin surface layers the *Rockwell superficial hardness test* is used. The procedure is the same as the Rockwell hardness test except that smaller preliminary and total loads are used. The Rockwell hardness number is reported as a number, followed by the symbol HR, and another symbol representing the indenter and forces used. For example, 68 HRC indicates a Rockwell hardness number of 68 on Rockwell C scale.

Hardness tests are simple, inexpensive, nondestructive, and do not require special specimens. In addition, other mechanical properties such as the tensile strength can be estimated from the hardness numbers. Therefore, hardness tests are very common and are typically performed more frequently than other mechanical tests.

**FIGURE 3.17** Rockwell hardness test machine.

# Welding

Many civil engineering structures, such as steel bridges, frames, and trusses require welding during construction and repair. Welding is a technique for joining two metal pieces by applying heat to fuse the pieces together. A filler metal may be used to facilitate the process. A variety of welding methods are available, but the common types are arc welding and gas welding. Other types of welding include flux-cored arc welding, self-shielded flux arc welding, and electroslag welding (Frank and Smith 1990).

*Arc welding* uses an arc between the electrode and the grounded base metal to bring both the base metal and the electrode to their melting points. The resulting deposited weld metal is a cast structure with a composition dependent upon the base metal, electrode, and flux chemistry. *Shielded metal arc welding (stick welding)* is the most common form of arc welding. It is limited to short welds in bridge construction. A consumable electrode, which is covered with flux, is used. The flux produces a shielding atmosphere at the arc to prevent oxidation of the molten metal. The flux is also used to trap impurities in the molten weld pool. The solidified flux forms a slag that covers the solidified weld, as shown in Figure 3.18. *Submerged arc welding* is a semiautomatic or automatic arc welding process. In this process, a bare wire electrode is automatically fed by the welding machine while a granular flux is fed into the joint ahead of the electrode. The arc takes place in the molten flux, which completely

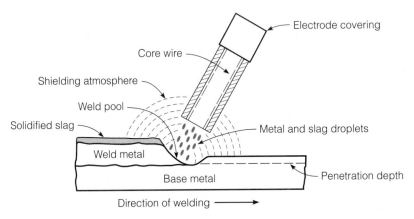

**FIGURE 3.18** Schematic drawing of arc welding.

shields the weld pool from the atmosphere. The molten flux concentrates the arc heat, resulting in deep penetration into the base metal.

*Gas welding (mig welding)* is another type of welding in which no flux is used. An external shielding gas is used, which shields the molten weld pool and provides the desired arc characteristics (Figure 3.19). This welding process is normally used for small welds due to the lack of slag formation.

Care must be taken during welding to consider the distortion that is the result of the nonuniform heating of the welding process. When the molten weld metal cools, it shrinks, causing deformation of the material and introducing residual stresses into the structure. Frequently, these residual stresses cause cracks outside the weld area. The distortion produced by welding can be controlled by proper sequencing of the welds and predeforming the components prior to welding. Finally, care must be taken by the welder and the inspector to protect their eyes and skin from the intensive ultraviolet radiation produced during welding (Frank and Smith 1990).

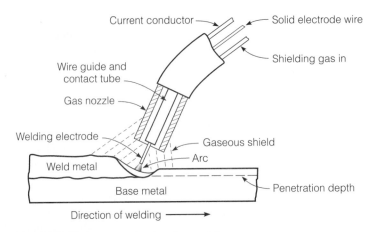

**FIGURE 3.19** Schematic drawing of gas welding.

# Steel Corrosion

Corrosion is defined as the destruction of a material by electrochemical reaction to the environment. For simplicity, corrosion of steel can be defined as the destruction that can be detected by rust formation. Corrosion of steel structures can cause serious problems and embarrassing and/or dangerous failures. For example, corrosion of steel bridges, if left unchecked, may result in lowering weight limits, costly steel replacement, or collapse of the structure. Other examples include corrosion of steel pipes, trusses, frames, and other structures. It is estimated that the cost of corrosion in the United States alone is $8 billion each year (Frank and Smith 1990).

Corrosion is an electrochemical process; that is, it is a chemical reaction in which there is transfer of electrons from one chemical species to another. In the case of steel, the transfer is between iron and oxygen, a process called *oxidation-reduction*. Corrosion requires the following four elements; without all of them corrosion will not occur.

1. an *anode*—the electrode where corrosion occurs
2. a *cathode*—the other electrode needed to form a corrosion cell
3. a *conductor*—a metallic pathway for electrons to flow
4. an *electrolyte*—a liquid that can support the flow of electrons

Steel, being a heterogeneous material, contains anodes and cathodes. Steel is also an electrical conductor. Therefore, steel contains three of the four elements needed for corrosion, while moisture is usually the fourth element (electrolyte).

The actual electrochemical reactions that occur when steel corrodes are very complex. However, the basic reactions for atmospherically exposed steel in a chemically neutral environment are dissolution of the metal at the anode and reduction of oxygen at the cathode.

Contaminants deposited on the steel surface affect the corrosion reactions and the rate of corrosion. Salt, from de-icing or a marine environment, is a common contaminant that accelerates corrosion of steel bridges and reinforcing steel in concrete.

The environment plays an important role in determining corrosion rates. Since an electrolyte is needed in the corrosion reaction, the amount of time the steel stays wet will affect the rate of corrosion. Also, contaminants in the air, such as oxides or sulfur, accelerate corrosion. Thus areas with acid rain, coal-burning power plants, and other chemical plants may accelerate corrosion.

## Methods for Corrosion Resistance

Since steel contains three of the four elements needed for corrosion, protective coatings can be used to isolate the steel from moisture, the fourth element. There are three mechanisms by which coatings provide corrosion protection (Hare 1987).

1. *Barrier coatings* work solely by isolating the steel from the moisture. These coatings have low water and oxygen permeability.
2. *Inhibitive primer coatings* contain passivating pigments. They are low-solubility pigments that migrate to the steel surface when moisture passes through the film to passivate the steel surface.

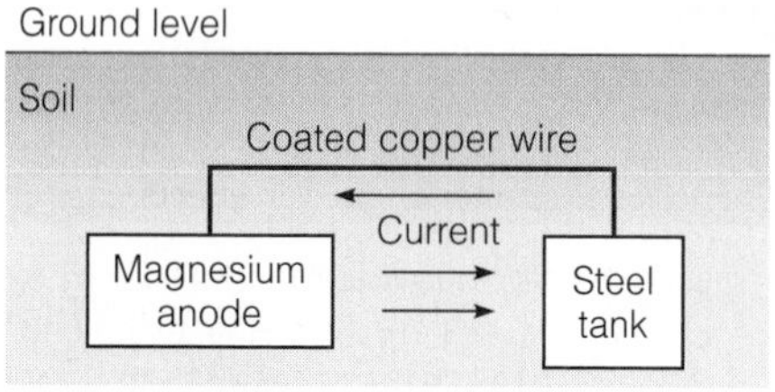

Ground level

**FIGURE 3.20** Cathodic protection of an underground steel tank using a magnesium sacrificial anode.

3. *Sacrificial primers (cathodic protection)* contain pigments such as elemental zinc. Since zinc is higher than iron in the galvanic series, when corrosion conditions exist the zinc gives up electrons to the steel, becomes the anode, and corrodes to protect the steel. There should be close contact between the steel and the sacrificial primer in order to have an effective corrosion protection.

Cathodic protection can take forms other than coating. For example, steel structures such as water heaters, underground tanks and pipes, and marine equipment, can be electrically connected to another metal that is more reactive in the particular environment, such as magnesium or zinc. Such reactive metal (sacrificial anode) experiences oxidation and gives up electrons to the steel, protecting the steel from corrosion. Figure 3.20 illustrates an underground steel tank that is electrically connected to a magnesium sacrificial anode (Fontana and Green 1978).

## SUMMARY

The history of civil engineering is closely tied to that of steel, and this will continue into the foreseeable future. With the development of modern production facilities, the availability of a wide variety of economical steel products is virtually assured. High strength, ductility, the ability to carry tensile as well as compressive loads, and the ability to join members either with welding or mechanical fastening are the primary positive attributes of steel as a structural material. The properties of steel can be tailored to meet the needs of specific applications through alloying and heat treatments. The primary shortcoming of steel is its tendency to corrode. When using steel in structures, the engineer should consider the means for protecting the steel from corrosion over the life of the structure.

## QUESTIONS AND PROBLEMS

**3.1.** What is the chemical composition of steel? What is the effect of carbon on the mechanical properties of steel?

**3.2.** Why does the iron-carbon phase diagram go only to 6.7% carbon?

**3.3.** Draw a simple iron-carbon phase diagram showing the liquid, liquid-solid, and solid phases.

**3.4.** What is the typical maximum percent of carbon in steel used for structures?

**3.5.** Briefly discuss four heat treatment methods to enhance the properties of steel. What are the advantages of each treatment?

**3.6.** Define alloy steels. Explain why alloys are added to steel.

**3.7.** Name three alloying agents and their principal effects.

**3.8.** Draw a typical stress-strain relationship for steel subjected to tension. On the graph show the modulus of elasticity, the yield strength, the ultimate stress, and the rupture stress.

**3.9.** A steel specimen is tested in tension. The specimen is 1.5 in. wide by 0.5 in. thick in the test region. By monitoring the load dial of the testing machine, it was found that the specimen yielded at a load of 37.5 kips and fractured at 52.5 kips.

   **a.** Determine the tensile stresses at yield and at fracture.

   **b.** Estimate how much increase in length would occur at 60% of the yield stress in a 2-in. gauge length.

**3.10.** A round steel alloy bar with a diameter of 0.5 in. and a gauge length of 2 in. is subjected to tension and showed the following results. Using a computer spreadsheet program plot the stress-strain relationship. From the graph determine the Young's modulus of the steel alloy and the deformation corresponding to a 8225-lb load.

| Load, lb | Deformation, $10^{-5}$ in. |
|---|---|
| 2,000 | 11.28 |
| 4,000 | 22.54 |
| 6,000 | 33.80 |
| 8,000 | 45.08 |
| 10,000 | 56.36 |
| 12,000 | 67.66 |

**3.11.** Testing a round steel alloy bar with a diameter of 15 mm and a gauge length of 250 mm produced the stress-strain relation below. Determine

   **a.** the elastic modulus

   **b.** the proportional limit

   **c.** the yield strength at a strain offset of 0.002

   **d.** the tensile strength

   **e.** the magnitude of the load required to produce an increase in length of 0.38 mm

   **f.** the final deformation if the specimen is unloaded after being strained by the amount specified in (e)

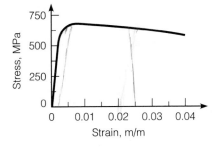

**3.12.** During the tension test on a steel rod within the elastic region the following data were measured:

Applied load          = 102 kN
Original diameter     = 25 mm
Current diameter      = 24.99325 mm
Original gauge length = 100 mm
Current gauge length  = 100.1 mm

Calculate the Young's modulus and Poisson's ratio.

**3.13.** A steel pipe having a length of 1 m, an outside diameter of 0.2 m, and a wall thickness of 10 mm is subjected to an axial compression of 200 kN. Assuming a modulus of elasticity of 200 GPa and a Poisson's ratio of 0.3, find

   **a.** the shortening of the pipe,

   **b.** the increase in the outside diameter, and

   **c.** the increase in the wall thickness.

**3.14.** A drill rod with a diameter of 12 mm is made of high-strength steel alloy with a shear modulus of 80 GPa. The rod is to be subjected to a torque $T$. What is the minimum required length $L$ of the rod so that the cross section at one end can be rotated 90° with respect to the other end without exceeding an allowable shear stress of 300 MPa?

**3.15.** What is the shear modulus of the material whose shear stress-strain relation is shown in Figure 3.12? Solve the problem using

   **a.** SI units

   **b.** U.S. customary units

**3.16.** A Charpy V Notch test was conducted for an ASTM A572 Grade 50 bridge steel. The average values of the test results at four different test temperatures were found as follows:

10 ft · lb at −50°F
15 ft · lb at 0°F
40 ft · lb at 40°F
60 ft · lb at 100°F

The bridge will be located in a region where specifications require a minimum of 25 ft · lb fracture toughness at 30°F for welded fracture-critical members. If the bridge contains a welded flange in a fracture critical member, does the steel have adequate Charpy V notch fracture toughness? Show your supporting calculations.

**3.17.** What are the typical uses of structural steel?

**3.18.** Why is reinforcing steel used in concrete? Discuss the typical properties of reinforcing steel.

**3.19.** Briefly define steel corrosion. What are the four elements necessary for corrosion to occur?

**3.20.** Discuss the main methods used to protect steel from corrosion.

**REFERENCES**

Budinski, K. G. 1996. *Engineering materials, properties and selection.* 5th ed. Upper Saddle River, NJ: Prentice-Hall.

Callister, W. D., Jr. 1985. *Materials science and engineering, an introduction.* New York: John Wiley.

Cordon, W. A. 1979. *Properties, evaluation, and control of engineering materials.* New York: McGraw-Hill.

Frank, K. H. and L. M. Smith. 1990. Highway materials engineering: steel, welding and coatings. Publication No. FHWA-HI-90-006. Washington, DC: Federal Highway Administration.

Fontana, M. G. and N. D. Greene. 1978. *Corrosion engineering.* 2nd ed. New York: McGraw-Hill.

Hare, C. H. 1987. Protective coatings for bridge steel. National Cooperative Highway Research Program, Synthesis of Highway Practice No. 136. Washington, DC: Transportation Research Board.

Somayaji, S. 1995. *Civil engineering materials.* Englewood Cliffs, NJ: Prentice-Hall.

# 4 Aluminum

Aluminum is the most plentiful metal on Earth, representing 8% of its crust. Although plentiful, aluminum exists primarily as oxides. The process to extract aluminum from oxide is very energy intensive. In fact, approximately 2% to 3% of the electricity used in the United States is consumed in aluminum production. This high energy requirement makes recycling of aluminum products economical. Of the 24 million tons of aluminum produced annually, approximately 75% is from ore reduction and 25% is from recycled materials.

The properties of pure aluminum are not suitable for structural applications. Some industrial applications require pure aluminum, but, otherwise, alloying elements are almost always added. These alloying elements, along with cold working and heat treatments, impart characteristics to the aluminum that make this product suitable for a wide range of applications. Here, the term *aluminum* is used to refer to both the pure element and to alloys.

In terms of the amount of metal produced, aluminum is second only to steel. About 25% of aluminum produced is used for containers and packaging, 20% for architectural applications, such as doors, windows, and siding, and 10% for electrical conductors. The balance is used for industrial goods, consumer products, aircraft, and highway vehicles.

Aluminum accounts for 80% of the structural weight of aircraft, and its use in the automobile and light truck industry has increased 300% since 1971 (Reynolds Metals Company 1996). However, use of aluminum for infrastructure applications has been limited. Of the approximately 600,000 bridges in the United States, only nine have primary structural members made of aluminum. Two reasons for the limited use of aluminum are the relatively high initial cost when compared to steel and the lack of performance information on aluminum structures.

Aluminum has many favorable characteristics and a wide variety of applications. The advantages of aluminum are that it (Budinski 1996)

**1.** has one-third the density of steel,

**2.** has good thermal and electrical conductivity,

**3.** has high strength-to-weight ratio,

**4.** can be given a hard surface by anodizing and hard coating,

**5.** has alloys that are weldable,

**6.** will not rust,

**7.** has high reflectivity,

**8.** can be die cast,

**9.** is easily machined,

**10.** has good formability,

**11.** is nonmagnetic, and

**12.** is nontoxic.

Aluminums high strength-to-weight ratio and its ability to resist corrosion are the primary factors that make aluminum an attractive structural engineering material. Although aluminum alloys can be formulated with strengths similar to steel products, the modulus of elasticity of aluminum is only about one-third that of steel. Thus the dimensions of structural elements must be increased to compensate for the lower modulus of aluminum.

## Aluminum Production

Aluminum production uses processes that were developed in the 1880s. Bayer developed the sodium aluminate leaching process to produce pure alumina ($Al_2O_3$). Hall and Héroult, working independently, developed an electrolytic process for reducing the alumina to pure aluminum. The essence of the aluminum production process is shown in Figure 4.1.

The production of aluminum starts with the mining of the aluminum ore, bauxite. Commercial grade bauxite contains between 45% to 60% alumina. The bauxite is crushed, washed to remove clay and silica materials, and is kiln dried to remove most of the water. The crushed bauxite is mixed with soda ash and lime and passed through a digester, pressure reducer, and settling tank to produce a concentrated solution of sodium aluminate. This step removes silica, iron oxide, and other impurities from the sodium aluminate solution. The solution is seeded with hydrated alumina crystals in precipitator towers. The seeds attract other alumina crystals and form groups that are heavy enough to settle out of solution. The alumina hydrate crystals are washed to remove remaining traces of impurities and are calcined in kilns to remove all water. The resulting alumina is ready to be reduced with the Hall-Héroult process. The alumina is melted in a cryolite bath (a molten salt of sodium-aluminum-floride). An electric current is passed between anodes and cathodes of carbon to separate the aluminum and oxygen molecules. The molten aluminum is collected at the cathode at the bottom of the bath. The molten aluminum, with better than 99% purity, is siphoned off to a crucible. It is then processed in a holding furnace. Hot gasses are passed through

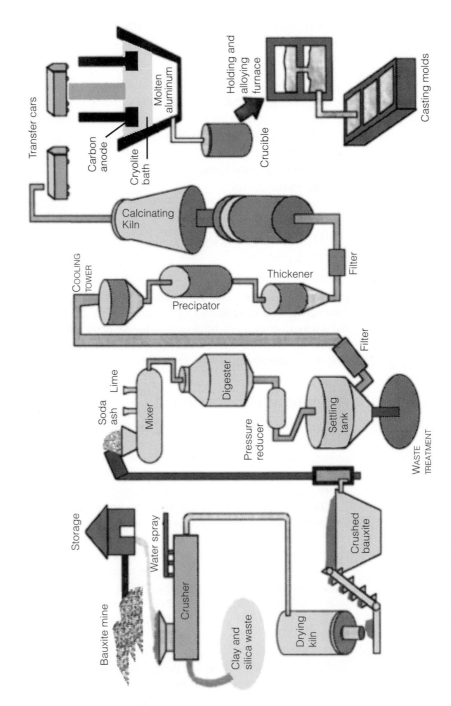

**FIGURE 4.1** Aluminum production process.

the molten material to further remove any remaining impurities. Alloying elements are then added.

The molten aluminum is either shipped to a foundry for casting into finished products or is cast into ingots. The ingots are formed by a direct-chill process that produces huge sheets for rolling mills, round log-like billets for extrusion presses, or square billets for production of wire, rod, and bar stock.

Final products are made by either casting, which is the oldest process, or deforming solid aluminum stock. Three forms of casting are used: die casting, permanent mold casting, and sand casting. The basic deformation processes are forging, impact extrusion, stamping, drawing, and drawing plus ironing. Many structural shapes are made with the extrusion process. Either cast or deformed products can be machined to produce the final shape and surface texture, and they can be heat treated to alter the mechanical behavior of the aluminum. Casting and forming methods are summarized in Table 4.1

**TABLE 4.1** Casting and Forming Methods for Aluminum Products*

| Casting Methods | |
| --- | --- |
| Sand casting | Sand with a binder is packed around a pattern. The pattern is removed and molten aluminum is poured in, reproducing the shape. Produces a rough texture that can be machined or otherwise surfaced if desired. Economical for low-volume production and for making very large parts. Also applicable when an internal void must be formed in the product. |
| Permanent mold casting | Molten aluminum is poured into a reusable metal mold. Economical for large volume production. |
| Die casting | Molten aluminum is forced into a permanent mold under high pressure. Suitable for mass production of precisely formed castings. |

| Forming Methods | |
| --- | --- |
| Extrusion | Aluminum heated to 425°C to 540°C (800°F to 1000°F) is forced through a die. Complex cross sections are possible, including incompletely or completely enclosed voids. A variety of architectural and structural members are formed by extrusion, including tubes, pipes, I-beams, and decorative components, such as window and door frames. |
| Rolling | Heated aluminum ingots are compressed and elongated with rollers producing plates more than 6 mm (0.25 in.) thick, sheets 0.15 mm to 6 mm (0.006 in. to 0.25 in.) thick, and foil less than 0.15 mm (0.006 in.). |
| Roll forming | Sheet aluminum is passed between a series of special rollers, usually in stages. Used for mass production of architectural products, such as moldings, gutters, downspouts, roofing, siding, and frames for windows and screens. |
| Brake forming | Sheet products are formed with a brake press. Uses simpler tooling than roll forming, but production rates are lower and the size of the product is limited. |
| Cutting operations | Outline shapes are produced by blanking and cutting. In blanking, a punch with the desired shape is pressed through a matching die. Used for mass production of flat shapes. Holes through a sheet are produced by piercing and perforating. Stacks of sheets can be trimmed or cut to an outline shape by a router or sheared in a guillotine-action shear. |
| Embossing | Sheet aluminum is pressed between mated rollers or dies, producing a raised pattern on one side and its negative indent on the other side. |
| Drawing | Sheet aluminum is drawn through the gap between two mated dies in a press. |
| Superplastic forming | Sheet aluminum is heated and forced over or into a mold by air pressure. Complex and deep contour shapes can be produced, but the process is slow. |

*From Reynolds Metals Company (1996).

When recycling aluminum, the scrap stock is melted in a furnace. The molten aluminum is purified and alloys are added. This process takes only about 5% of the electricity that is needed to produce aluminum from bauxite.

In addition to these conventional processes, very high-strength aluminum parts can be produced using powder metallurgy methods. A powdered aluminum alloy is compacted in a mold. The material is heated to a temperature that fuses the particles into a unified solid.

# Aluminum Metallurgy

Aluminum has a face center cubic (FCC) lattice structure. It is very malleable, with a typical elongation over a 50-mm- (2 in.) gauge length of over 40%. It has limited tensile strength, on the order of 28 MPa (4000 psi). The modulus of elasticity of aluminum is about 69 GPa (10,000 ksi). Commercially pure aluminum, more than 99% aluminum content, is limited to nonstructural applications, such as electrical conductors, chemical equipment, and sheet metal work.

Although the strength of pure aluminum is relatively low, aluminum alloys can be as much as 15 times stronger than pure aluminum through the addition of small amounts of alloying element, strain hardening by cold working, and heat treatment. The common alloying elements are copper, manganese, silicon, magnesium, and zinc. Cold working increases strength by causing a disruption of the slip planes in the material that resulted from the production process.

Figure 4.2 shows the two-phase diagram for aluminum and copper. This diagram is typical of the phase diagrams of other two-phase aluminum alloys. The alloying elements have low solubility in aluminum, and the solubility reduces as temperature drops.

As described previously, the properties of metals with this characteristic are very sensitive to heat treatments, which affect the grain size of the material and the distribution of the alloying element throughout the matrix of the lattice structures. Heat treatments typically used on aluminum alloys include annealing, hardening, aging, and stabilizing.

## Alloy Designation System

Aluminum classification starts by separating the product according to its production method, either casting or wrought methods. Aluminum alloys designed for casting are formulated to flow into the mold. Wrought aluminum alloys are used for products fabricated by deforming the aluminum into its final shape. The Aluminum Association has developed an aluminum alloy classification system shown in Table 4.2

The designation system for wrought alloys consists of a four-digit code. The first digit indicates the alloy series. The second digit, if different from 0, indicates a modification in the basic alloy. The third and forth digits identify the specific alloy in the series; these digits are arbitrarily assigned, except for the 1xxx series, where the final two digits indicate the minimum aluminum content. For the 1xxx series, the aluminum content is 99% plus the last two digits of the code expressed as a decimal fraction. For example, a 1060 contains a minimum aluminum content of 99.60%.

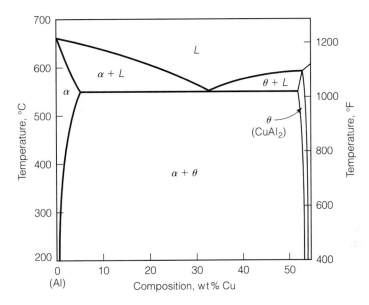

**FIGURE 4.2** Aluminum-copper phase diagram.

Cast alloys are assigned a three-digit number followed by one digit after the decimal point, as shown in Table 4.2. The first digit represents the alloy series. Note that series 3, 6, 8, and 9 have different meanings for cast versus wrought alloys. The second and third digits are arbitrarily assigned to identify specific alloys. The digit after the decimal indicates if the alloy composition is for the final casting (xxx.0) or for ingot (xxx.1 and xxx.2).

**TABLE 4.2** Designation System for Aluminum Alloys*

| Wrought Aluminum Alloys | | Cast Aluminum Alloys | |
|---|---|---|---|
| Alloy Series | Description/Major Alloying Elements | Alloy Series | Description/Major Alloying Elements |
| 1xxx | 99.00% minimum aluminum | 1xx.x | 99.00% minimum aluminum |
| 2xxx | Copper | 2xx.x | Copper |
| 3xxx | Manganese | 3xx.x | Silicon plus copper and/or magnesium |
| 4xxx | Silicon | 4xx.x | Silicon |
| 5xxx | Magnesium | 5xx.x | Magnesium |
| 6xxx | Magnesium and silicon | 6xx.x | Unused series |
| 7xxx | Zinc | 7xx.x | Zinc |
| 8xxx | Other element | 8xx.x | Tin |
| 9xxx | Unused series | 9xx.x | Other element |

*Aluminum Association (1993).

**TABLE 4.3**   Temper Designations for Aluminum Alloys

| Symbol | Meaning | Comment |
|---|---|---|
| F | As fabricated | No special control over thermal conditions or strain-hardening is employed. |
| O | Annealed | Wrought products, annealed to the lowest strength temper.<br>Cast products, annealed to improve ductility and dimensional stability.<br>The O may be followed by a digit other than zero, indicating a variation with special characteristics. |
| H | Strain hardened | Wrought products only. Strength is increased by strain-hardening, with or without supplemental thermal treatments. The H is always followed by two or more digits. The first digit indicates a specific combination of basic operations. The basic operations identified by the first digit are:<br>H1 strain-hardening only. Applies to products that are strain-hardened to obtain the desired strength, without supplementary thermal treatment.<br>H2 strain-hardened and partially annealed. Applies to products that are strain-hardened more than the desired final amount, and then reduced in strength to the desired level by partial annealing.<br>H3 strain-hardened and stabilized. Applies to products that are strain-hardened and whose mechanical properties are stabilized either by a low-temperature thermal treatment or as a result of heat introduced during fabrication. Stabilization usually improves ductility.<br>The second digit indicates the degree of strain-hardening. (Codes for the second digit are 2, quarter-hard; 4, half-hard; 8, full-hard; 9, extra-hard.) When used, the third digit indicates a variation of the two-digit temper. |
| W | Solution heat-treated | An unstable temper applicable only to alloys that spontaneously age at room temperature after solution heat-treatment. This designation is specific only when the period of natural aging is indicated, for example, W 1/2 hr. |
| T | Thermally treated to produce stable tempers other than F, O, or H | Applies to thermally treated products, with or without supplementary strain-hardening to produce stable tempers. The T is always followed by one or two digits:<br>T1 cooled from an elevated temperature-shaping process and naturally aged to a substantially stable condition. Products that are not cold-worked after cooling from an elevated-temperature shaping process or in which the effect of cold-working flattening or straightening may not be recognized are in mechanical property limits.<br>T2 cooled from an elevated temperature-shaping process, cold-worked, and naturally aged to a substantially stable condition. Products that are cold-worked to improve strength after cooling from an elevated-temperature shaping process or in which the effect of cold-work in flattening or straightening is recognized are in mechanical property limits.<br>T3 solution heat-treated, cold-worked, and naturally aged to a substantially stable condition. Products that are cold-worked to improve strength after solution heat-treatment or in which the effect of cold-work in flattening or straightening is recognized are in mechanical property limits.<br>T4 solution heat-treated and naturally aged to a substantially stable condition. Products that are not cold-worked after solution heat-treatment or in which the effect of cold-work in flattening or straightening is recognized are in mechanical property limits.<br>T5 cooled from an elevated-temperature shaping process, then artificially aged. Products that are not cold-worked after cooling from an elevated-temperature shaping process or in which the effect of cold-work in flattening or straightening may not be recognized are in mechanical property limits.<br>T6 solution heat-treated and then artificially aged. Products that are not cold-worked after solution heat-treatment or in which the effect of cold-work are in flattening or straightening may not be recognized in mechanical property limits.<br>T7 solution heat-treated and overaged/stabilized. Wrought products artificially aged after solution heat-treatment to carry them beyond a point of maximum strength to provide control of some significant characteristic. Cast products artificially aged after solution heat-treatment to provide dimensional and strength stability.<br>T8 solution heat-treated, cold-worked, and then artificially aged. Products cold-worked to improve strength or in which the effect of cold-work in flattening or straightening is recognized are in mechanical property limits.<br>T9 solution heat-treated, artificially aged, and then cold-worked. Products cold-worked to improve strength.<br>T10 cooled from an elevated temperature shaping process, cold-worked and then artificially aged. Products worked to improve strength, or in which the effect of cold-work in flattening and straightening is recognized are in mechanical property limits.<br>Additional digits can be appended to the above temper designations to indicate significant variations. |

### Temper Treatments

The mechanical properties of aluminum are greatly altered by both heat treatment and strain hardening. Therefore, specification of an aluminum material must include how the product was tempered. The processes described in Table 4.3 defines the types of tempering aluminum products undergo.

Aluminum alloys used for structural applications are classified as being either heat treatable or not. Non–heat treatable or "common" alloys contain elements that remain substantially in solid solution or that form insoluble constituents. Thus heat treatment does not influence their mechanical properties. The properties of these alloys is dependent on the amount of cold working introduced after annealing. Heat treatable or "strong" alloys contain elements, groups of elements, or constituents that have a considerable solid solubility at elevated temperatures and limited solubility at lower temperatures. The strength of these alloys is increased primarily by heat treatment.

## Aluminum Testing and Properties

Typical properties are provided in Tables 4.4 and 4.5 for non–heat treatable and heat treatable wrought aluminum alloys, respectively. Typical properties for cast aluminum alloys that may be used for structural applications are given in Table 4.6. These values are only an indication of the properties of cast aluminum alloys. Material properties of cast members can vary throughout the body of the casting due to differential cooling rates.

Tests performed on aluminum are similar to those described for steel. These typically include stress-strain tensile tests to determine elastic modulus, yield strength, ultimate strength, and percent elongation. In contrast to steel, aluminum alloys do not display an upper and lower yield point. Instead, the stress-strain curve is linear up to the proportional limit, and then is a smooth curve up to the ultimate strength. Yield strength is defined based on the 0.20% strain offset method, as shown on Figure 4.3.

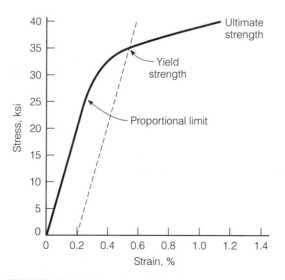

**FIGURE 4.3** Aluminum stress-strain diagram.

**TABLE 4.4**   Properties of Select Non–Heat Treatable Wrought Aluminum Alloys

| Alloy | | Ultimate ksi | Ultimate MPa | Yield ksi | Yield MPa | Elongation[a] (thickness) 1/16 in. | Elongation[a] (thickness) 1/2 in. | Hardness[b] | Shear Ultimate ksi | Shear Ultimate MPa | Fatigue[c] Endurance Limit ksi | Fatigue[c] Endurance Limit MPa | Nominal Chemical Composition |
|---|---|---|---|---|---|---|---|---|---|---|---|---|---|
| 1060 | O | 10 | 69 | 4 | 28 | 43 | | 19 | 7 | 48 | 3 | 21 | 99.6 Al |
| | H-12 | 12 | 83 | 11 | 76 | 16 | | 23 | 8 | 55 | 4 | 28 | |
| | H-14 | 14 | 97 | 13 | 90 | 12 | | 26 | 9 | 62 | 5 | 34 | |
| | H-16 | 16 | 110 | 15 | 103 | 5 | | 30 | 10 | 69 | 6.5 | 45 | |
| | H-18 | 19 | 131 | 18 | 124 | 6 | | 35 | 11 | 76 | 6.5 | 45 | |
| 1100 | O | 13 | 90 | 5 | 34 | 35 | 45 | 23 | 9 | 62 | 5 | 34 | 99.0 Al |
| | H-12 | 16 | 110 | 15 | 103 | 12 | 25 | 28 | 10 | 69 | 6 | 41 | |
| | H-14 | 18 | 124 | 17 | 117 | 9 | 50 | 32 | 11 | 76 | 7 | 48 | |
| | H-16 | 21 | 145 | 20 | 138 | 6 | 17 | 38 | 12 | 83 | 9 | 62 | |
| | H-18 | 24 | 165 | 22 | 152 | 5 | 15 | 44 | 13 | 90 | 9 | 62 | |
| 3003 | O | 16 | 110 | 6 | 41 | 30 | 40 | 28 | 11 | 76 | 7 | 48 | 1.2 Mn |
| | H-12 | 19 | 131 | 18 | 124 | 10 | 20 | 35 | 12 | 83 | 8 | 55 | |
| | H-14 | 22 | 152 | 21 | 145 | 8 | 16 | 40 | 14 | 97 | 9 | 62 | |
| | H-16 | 26 | 179 | 25 | 172 | 5 | 14 | 47 | 15 | 103 | 10 | 69 | |
| | H-18 | 29 | 200 | 27 | 186 | 4 | 10 | 55 | 16 | 110 | 10 | 69 | |
| 5005 | O | 18 | 124 | 6 | 41 | 25 | | 28 | 11 | 76 | | | 0.8 Mg |
| | H-12 | 20 | 138 | 19 | 131 | 10 | | | 14 | 97 | | | |
| | H-14 | 23 | 159 | 22 | 152 | 6 | | | 14 | 97 | | | |
| | H-16 | 26 | 1799 | 25 | 172 | 5 | | | 15 | 103 | | | |
| | H-18 | 29 | 200 | 28 | 193 | 4 | | | 16 | 110 | | | |
| | H-32 | 20 | 138 | 17 | 117 | 11 | | 36 | 14 | 97 | | | |
| | H-34 | 23 | 159 | 20 | 138 | 8 | | 41 | 14 | 97 | | | |
| | H-36 | 26 | 179 | 24 | 165 | 6 | | 46 | 15 | 103 | | | |
| | H-38 | 29 | 200 | 27 | 186 | 5 | | 51 | 16 | 110 | | | |
| 5086 | O | 38 | 262 | 17 | 117 | 22 | 30 | 60 | 23 | 159 | 21 | 145 | Mg |
| | H-32 | 42 | 290 | 30 | 207 | 12 | 16 | 72 | 25 | 172 | 22 | 152 | 0.45 Mn |
| | H-34 | 47 | 324 | 37 | 255 | 10 | 14 | 82 | 27 | 186 | 23 | 159 | |
| | H-111 | 40 | 276 | 27 | 186 | | 17 | 65 | 23 | 159 | 21 | 145 | |
| | H-112 | 39 | 269 | 19 | 131 | 14 | | 64 | 23 | 159 | 21 | 145 | |
| | H-116 | 42 | 290 | 30 | 207 | | 16 | 72 | 25 | 173 | 22 | 152 | |
| 5456 | O | 45 | 310 | 23 | 159 | 20 | 24 | 70 | 27 | 186 | 22 | 152 | 5.1 Mg |
| | H-111 | 47 | 324 | 33 | 228 | | 18 | 75 | 27 | 186 | 24 | 165 | 0.7 Mn |
| | H-112 | 45 | 310 | 24 | 156 | | 22 | 70 | 27 | 186 | | | 0.12 Cr |
| | H-116 | 51 | 352 | 37 | 255 | | 16 | 90 | 30 | 207 | 23 | 159 | |

[a]Percent elongation over 2 in.
[b]500,000,000 cycles of complete stress reversal using R. R. Moore type of machine and specimen.
[c]Brinell number, 500-kg load.

**TABLE 4.5**  Properties of Select Heat Treatable Wrought Aluminum Alloys

| Alloy | | Tension Ultimate ksi | Tension Ultimate MPa | Yield ksi | Yield MPa | Elongation[a] (thickness) $1/16$ in. | Elongation[a] (thickness) $1/2$ in. | Hard-ness[b] | Shear Ultimate ksi | Shear Ultimate MPa | Fatigue[c] Endurance Limit ksi | Fatigue[c] Endurance Limit MPa | Nominal Chemical Composition |
|---|---|---|---|---|---|---|---|---|---|---|---|---|---|
| 2014 | O | 27 | 186 | 14 | 97 | | 18 | 45 | 18 | 124 | 13 | 90 | 4.5 Cu, 0.8 Mn |
| | T4/T451 | 62 | 427 | 42 | 290 | | 50 | 105 | 38 | 262 | 20 | 138 | 0.8 Si, 0.4 Mg |
| | T6/T651 | 70 | 483 | 60 | 414 | | 13 | 135 | 42 | 290 | 18 | 124 | |
| 6053 | O | 16 | 110 | 8 | 55 | | 35 | 26 | 11 | 76 | 8 | 55 | 1.2 Mg, |
| | T6 | 37 | 255 | 32 | 221 | | 13 | 80 | 23 | 159 | 12 | 90 | 0.25 CR |
| 6061 | O | 18 | 124 | 8 | 55 | 25 | 30 | 30 | 12 | 83 | 9 | 62 | 1.0 Mg, 0.6 Si |
| | T4/T451 | 35 | 241 | 21 | 145 | 22 | 25 | 65 | 24 | 165 | 14 | 97 | 0.25 Cu, |
| | T6/T651 | 45 | 310 | 40 | 276 | 12 | 17 | 95 | 30 | 207 | 14 | 97 | 0.25 Cr |
| 6063 | O | 13 | 90 | 7 | 48 | | | 25 | 10 | 69 | 8 | 55 | 0.7 Mg |
| | T1 | 22 | 152 | 13 | 90 | 20 | 33 | 42 | 14 | 97 | 9 | 62 | 0.4 Si |
| | T4 | 25 | 172 | 13 | 90 | 22 | | | 16 | 110 | | | |
| | T5 | 27 | 186 | 21 | 145 | 12 | 22 | 60 | 17 | 117 | 10 | 69 | |
| | T6 | 35 | 241 | 31 | 214 | 12 | 18 | 73 | 22 | 152 | 10 | 69 | |
| | T83 | 37 | 255 | 35 | 241 | 9 | | 82 | 22 | 152 | | | |
| | T831 | 30 | 207 | 27 | 186 | 10 | | 70 | 18 | 124 | | | |
| | T832 | 42 | 290 | 39 | 269 | 12 | | 95 | 27 | 186 | | | |
| 7178 | O | 33 | 228 | 15 | 103 | 15 | 16 | 60 | 22 | 152 | | | 6.8 Zn, 2.0 Cu |
| | T6/T651 | 88 | 607 | 78 | 538 | 10 | 11 | 160 | 52 | 359 | 22 | 152 | 2.7 Mg, 0.3 Mn |
| | T76/T7651 | 83 | 572 | 73 | 503 | | 11 | | | | | | |

[a]Percent elongation over 2 in.
[b]500,000,000 cycles of complete stress reversal using R. R. Moore type of machine and specimen.
[c]Brinell number, 500-kg load.

As indicated earlier, the modulus of elasticity of aluminum alloys is on the order of 69 GPa (10,000 ksi) and is not very sensitive to types of alloys or temper treatments.

**SAMPLE PROBLEM 4.1**

An aluminum alloy rod with 10 mm diameter is subjected to 5 kN tensile load. After applying the load, the diameter was measured and found to be 9.997 mm. If the yield strength is 139 MPa, calculate the Poisson's ratio of the material.

*Solution:*

$$\sigma = \frac{5000}{\left(\frac{\pi \times 0.010^2}{4}\right)} = 63.694 \times 10^6 \text{ Pa} = 63.694 \text{ MPa}$$

It is clear that the applied stress is well below the yield stress and, as a result, the deformation is elastic.

**TABLE 4.6**  Typical Properties of Select Cast Aluminum Alloys

| Cast Alloy Designation | Tension | | | | | Hard-ness[b] | Shear Ultimate | | Fatigue[c] Endurance | |
| --- | --- | --- | --- | --- | --- | --- | --- | --- | --- | --- |
| | Ultimate | | Yield | | | | | | | |
| | ksi | MPa | ksi | MPa | Elongation[a] | | ksi | MPa | ksi | MPa |
| 356.0-T6[§] | 40 | 276 | 27 | 186 | 5 | 90 | 32 | 221 | 13 | 90 |
| 356.0-T7[§] | 33 | 228 | 24 | 165 | 5 | 70 | 25 | 172 | 11 | 76 |
| A356.0-T61[§] | 41 | 283 | 30 | 207 | 10 | 80 | | | | |
| A357.0-T6[§] | 50 | 345 | 40 | 276 | 10 | 85 | 43 | 296 | 16 | 110 |
| A444.0-T4[§] | 23 | 159 | 10 | 69 | 21 | 45 | | | | |
| 356.0-T6[‖] | 33 | 228 | 24 | 165 | 3.5 | 70 | 26 | 179 | 8.5 | 59 |
| 356.0-T7[‖] | 34 | 234 | 30 | 207 | 2.0 | 75 | 24 | 165 | 9.0 | 62 |
| Almag 35 535.0[‖] | 40 | 276 | 21 | 145 | 13 | 70 | 28 | 193 | 10 | 69 |

[a] Percent elongation over 2 in.
[b] 500,000,000 cycles of complete stress reversal using R. R. Moore type of machine and specimen.
[c] Brinell number, 500-kg load.
[§] Permanent mold
[‖] Sand casting

$$\text{Assume } E = 69 \text{ GPa}$$

$$\epsilon_{axial} = \frac{\sigma}{E} = \frac{63.694 \times 10^6}{69 \times 10^9} = 0.000923 \text{ m/m}$$

$$\Delta d = 9.997 - 10 = -0.003 \text{ m/m}$$

$$\epsilon_{laterial} = -\frac{0.003}{10} = -0.0003 \text{ m/m}$$

$$\nu = \frac{-\epsilon_{lateral}}{\epsilon_{axial}} = \frac{0.0003}{0.000923} = 0.33$$

Aluminum's coefficient of thermal expansion is 0.000023/°C (0.000013/°F), about twice as large as steel and concrete. Thus joints between aluminum and steel or concrete must be designed to accommodate the differential movement.     ◆

Strengths of aluminum are considerably affected by temperature as shown in Figure 4.4. At temperatures above 150°C (300°F) tensile strengths are reduced considerably. The temperature at which the reduction begins and the extent of the reduction depend on the alloy. At temperatures below room temperature, aluminum becomes stronger and tougher as the temperature decreases.

## Welding and Fastening

Aluminum pieces can be joined either by welding or by using fasteners. Welding requires that the tough oxide coating on aluminum be broken and kept from reforming during welding, so arc welding is generally performed in the presence of an inert gas that shields the weld from oxygen in the atmosphere. The two common processes by which aluminum is welded are gas metal arc welding, GMAW, and gas tungsten arc welding, GTAW. In the GMAW process, the filler wire also serves as the electrode. GTAW uses a tungsten electrode and a separate filler wire. Welding can alter the tempering of the aluminum in the area of the weld. For example, the tensile strength of

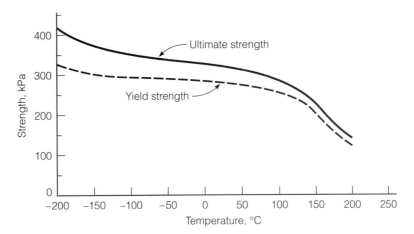

**FIGURE 4.4** Tensile strength of aluminum at different temperatures. Reprinted with permission. The Aluminum Association, 1987.

6061-T6 is 290 MPa (42 ksi), but the tensile strength of a weld in this alloy is only about 165 MPa (24 ksi). For design purposes it is assumed the weld affects an area of 25 mm (1 in.) on each side of the weld.

In addition to welding, either bolts or rivets can join aluminum pieces. Bolts can be either aluminum or steel. When steel bolts are used, they must be either galvanized, aluminized, cadmium plated, or made of stainless steel to prevent the development of galvanic corrosion. Rivet fasteners are made of aluminum and are cold driven. Both bolt and rivet joints are designed based on the shear strength of the fastener and the bearing strength of the material being fastened.

## Corrosion

Aluminum develops a thin oxidation layer immediately upon exposure to the atmosphere. This tough oxide film protects the surface from further oxidation. The alloying elements alter the corrosion resistance of the aluminum. The alloys used for airplanes are usually given extra protection by painting or "cladding" with a thin coat of a corrosion-resistant alloy. Painting is generally not needed for medium-strength alloys used for structural applications.

Galvanic corrosion occurs when aluminum is in contact with any of several metals in the presence of an electrical conductor, such as water. The best protection for this problem is to break the path of the galvanic cell by painting, using an insulator, or keeping the dissimilar metals dry.

**SUMMARY**

Although aluminum has many desirable attributes, its use as a structural material in civil engineering has been limited, primarily by economic considerations and a lack of performance information. Aluminum alloys and heat treatments provide products with a wide range of characteristics. The advantages of aluminum relative to steel include lightweight, high strength-to-weight ratio, and corrosion resistant.

**4.1.** Name the two primary factors that make aluminum an attractive structural engineering material.

**4.2.** Compare the strength and modulus of elasticity of aluminum alloys with those of steel.

**4.3.** An aluminum alloy specimen with a radius of 0.28 in. was subjected to tension until fracture and produced the following results.

| Stress, ksi | Strain, $10^{-3}$ in./in. |
|:---:|:---:|
| 8 | 0.6 |
| 17 | 1.5 |
| 27 | 2.4 |
| 35 | 3.2 |
| 43 | 4.0 |
| 50 | 4.6 |
| 58 | 5.2 |
| 62 | 5.8 |
| 64 | 6.2 |
| 65 | 6.5 |
| 67 | 7.3 |
| 68 | 8.1 |
| 70 | 9.7 |

    **a.** Using a spreadsheet program plot the stress-strain relationship.

    **b.** Calculate the modulus of elasticity of the aluminum alloy.

    **c.** Determine the proportional limit.

    **d.** What is the maximum load if the stress in the bar is not to exceed the proportional limit?

    **e.** Determine the 0.2% offset yield strength.

    **f.** Determine the tensile strength.

    **g.** Determine the percent of elongation at failure.

**4.4.** A 3003-H14 aluminum alloy rod with 0.5 in. diameter is subjected to 2000 lb tensile load. Calculate the resulting diameter of the rod. If the rod is subjected to a compressive load of 2000 lb, what will be the diameter of the rod? Assume that the modulus of elasticity is 10,000 ksi, Poisson's ratio is 0.33, and the yield strength is 21 ksi.

**4.5.** The stress-strain relation of an aluminum alloy bar having a length of 2 m and a diameter of 10 mm is expressed by the equation

$$\epsilon = \frac{\sigma}{70,000}\left[1 + \frac{3}{7}\left(\frac{\sigma}{270}\right)^9\right]$$

where $\sigma$ is in MPa. If the rod is axially loaded by a tensile force of 20 kN and then unloaded, what is the permanent deformation of the bar?

**4.6.** Discuss galvanic corrosion of aluminum. How can aluminum be protected from galvanic corrosion?

**REFERENCES**

Aluminum Association. 1987. *Structural design with aluminum.* Washington, DC: The Aluminum Association.

Aluminum Association. 1993. *Aluminum standards and data, 1993.* Washington, DC: The Aluminum Association.

Budinski, K. G. 1996. *Engineering materials, properties and selection.* 5th ed. Englewood Cliffs, NJ: Prentice-Hall.

Reynolds Metals Company. 1996. *Reynolds infrastructure.* Richmond VA: Reynolds Metals Company.

# 5 Aggregates

There are two main uses of aggregates in civil engineering: as an underlying material for foundations and pavements and as ingredients in portland cement and asphalt concretes. By dictionary definition, aggregates are a combination of distinct parts gathered into a mass or whole. Generally, in civil engineering the term *aggregate* means a mass of crushed stone, gravel, sand, etc., predominantly composed of individual particles, but in some cases including clays and silts. The largest particle size in aggregates may have a diameter as large as 150 mm (6 in.) and the smallest particle can be as fine as 5 to 10 microns.

## Aggregate Sources

*Natural* sources for aggregates include gravel pits, river run deposits, and rock quarries. Generally, *gravel* comes from pits and river deposits, whereas *crushed stones* are the result of processing rocks from quarries. Usually, gravel deposits must also be crushed to obtain the needed size distribution, shape, and texture.

*Manufactured* aggregates can use slag waste from steel mills and expanded shale and clays to produce lightweight aggregates. Heavy-weight concrete, used for radiation shields, can use steel slugs and bearings for the aggregate. Styrofoam beads can be used as an aggregate in lightweight concrete used for insulation.

## Geological Classification

All natural aggregates result from the breakdown of large rock masses. Geologists classify rocks into three basic types, *igneous, sedimentary,* and *metamorphic.* Volcanic action produces igneous rocks by hardening or crystallizing molten material, magma. The magma cools either at the earth's surface, when it is exposed to air or water, or within the crust of the earth. Cooling at the surface produces *extrusive* igneous rocks, while cooling underground produces *intrusive* igneous rocks. In general, the extrusive rocks cool much more rapidly than the intrusive rocks. Therefore, we would expect extrusive igneous rocks to have a fine grain size and potentially to include air voids and other inclusions. Intrusive igneous rocks have larger grain sizes and fewer flaws. Igneous rocks are classified based on grain size and composition. Coarse grains are larger than 2 mm and fine grains are less than 0.2 mm. Classification based on composition is a function of the silica content, specific gravity, color, and the presence of free quartz.

Sedimentary rocks coalesce from deposits of disintegrated existing rocks or inorganic remains of marine animals. Wind, water, glaciers, or direct chemical precipitation transport and deposit layers of material that become sedimentary rocks, resulting in a stratified structure. Natural cementing binds the particles together. Classification is based on the predominant mineral present: calcareous (limestone, chalk, etc.), siliceous (chert, sandstone, etc.), and argillaceous (shale, etc.).

Metamorphic rocks form from igneous or sedimentary rocks that are drawn back into the earth's crust and exposed to heat and pressure, reforming the grain structure. Metamorphic rocks generally have a crystalline structure with grain sizes ranging from fine to coarse.

All three classes of rock are used successfully in civil engineering applications. The suitability of aggregates from a given source must be evaluated by a combination of tests to check physical, chemical, and mechanical properties, and be supplemented by mineralogical examination. The best possible prediction of aggregate suitability for a given application is that based on historical performance in a similar design.

## Evaluation of Aggregate Sources

Civil engineers select aggregates for their ability to meet specific project requirements, rather than their geologic history. The physical and chemical properties of the rocks determine the acceptability of an aggregate source for a construction project. These characteristics vary within a quarry or gravel pit, making it necessary to continually sample and test the materials as the aggregates are being produced.

Due to the quantity of aggregates required for a typical civil engineering application, the cost and availability of the aggregates are important when selecting an aggregate source. Frequently, one of the primary challenges facing the materials engineer on a project is how to use the locally available material in the most cost effective manner.

Potential aggregate sources are usually evaluated for quality of the larger pieces, the nature and amount of fine material, and the gradation of the aggregate. The extent and quality of rock in the quarry is usually investigated by drilling cores and performing trial blasts (or shots) to evaluate how the rock breaks and by crushing some materials in the laboratory to evaluate grading, particle shape, soundness, durability, and amount of fine material. Cores are examined petrographically for general quality, suitability for various uses, and amount of deleterious materials. Potential sand and gravel pits are evaluated by collecting samples and performing sieve analysis tests. The amount of large gravel and cobble sizes determines the need for crushing, while the amount of fine material determines the need for washing. Petrographic examinations evaluate the nature of aggregate particles and the amount of deleterious material (Meininger and Nichols 1990).

Price and availability are universal criteria that apply to all uses of aggregates. However, the required aggregate characteristics depend on how they will be used in the structure; they may be used as base material, in asphalt concrete, or in portland cement concrete.

## Aggregate Uses

As mentioned, aggregates are primarily used as an underlying material for foundations and pavements and as ingredients in portland cement and asphalt concretes. Aggregate underlying materials, or base courses, can add stability to a structure, provide a drainage layer, and protect the structure from frost damage. Stability is a function of the interparticle friction between the aggregates and the amount of clay and silt "binder" material in the voids between the aggregate particles. However, increasing the clay and silt content will block the drainage paths between the aggregate particles, thereby inhibiting the ability of the material to act as a drainage layer.

In portland cement concrete, 60% to 75% of the volume and 79% to 85% of the weight is made up of aggregates. The aggregates act as a filler to reduce the amount of cement paste needed in the mix. In addition, aggregates have greater volume stability than the cement paste. Therefore, maximizing the amount of aggregate, to a certain extent, improves the quality and economy of the mix.

In asphalt concrete, aggregates constitute over 80% of the volume and 92% to 96% of the mass. The asphalt cement acts as a binder to hold the aggregates together, but does not have enough strength to lock the aggregate particles into position. As a result, the strength and stability of asphalt concrete depends mostly on interparticle friction between the aggregates and, to a limited extent, on the binder.

## Aggregate Properties

Aggregates' properties are defined by the characteristics of both the individual particles and the characteristics of the combined material. These properties can be further described by their physical, chemical, and mechanical characteristics, as shown in Table 5.1 (Meininger and Nichols, 1990). There are several individual particle characteristics that are important in determining if an aggregate source is suitable for a particular application. Other characteristics are measured for designing portland cement and asphalt concrete mixes (Goetz and Wood 1960).

### Particle Shape and Surface Texture

The shape of the individual aggregate particles, Figures 5.1 and 5.2, determines how the material will pack into a dense configuration and also determines the mobility of the stones within a mix. There are two considerations in the shape of the material: *angularity* and *flakiness*. Crushing rocks produces angular particles with sharp corners. Due to weathering, the corners of the aggregates break down, creating *subangular* particles. When the aggregates tumble while being transported in water, the corners can become completely *rounded*. Generally, angular aggregates produce bulk materials with higher stability than rounded aggregates. However, the angular aggregates will be more difficult to work into place than rounded aggregates, since its shape

**TABLE 5.1** Relative Importance of Basic Aggregate Properties for End Use[*,†]

| | Portland Cement Concrete | Asphalt Concrete | Base |
|---|---|---|---|
| *Physical* | | | |
| Particle shape (angularity) | M | V | V |
| Particle shape (flakiness, elongation) | M | M | M |
| Particle size—maximum | M | M | M |
| Particle size—distribution | M | V | M |
| Particle surface texture | M | V | V |
| Pore structure, porosity | V | M | U |
| Specific gravity, absorption | V | M | M |
| Soundness—weatherability | V | M | M |
| Unit weight, voids (loose, compacted) | V | M | M |
| Volumetric stability—thermal | M | U | U |
| Volumetric stability—wet/dry | M | U | M |
| Volumetric stability—freeze/thaw | V | M | M |
| Integrity during heating | U | M | U |
| Deleterious constituents | V | M | M |
| *Chemical* | | | |
| Solubility | M | U | U |
| Surface charge | U | V | U |
| Asphalt affinity | U | V | M |
| Reactivity to chemicals | V | U | U |
| Volume stability—chemical | V | M | M |
| Coatings | M | M | U |
| *Mechanical* | | | |
| Compressive strength | M | U | U |
| Toughness (impact resistance) | M | M | U |
| Abrasion resistance | M | M | M |
| Character of products of abrasion | M | M | U |
| Mass stability (stiffness, resilience) | U | V | V |
| Polishability | M | M | U |

*Meininger and Nichols (1990)
†V = Very important; M = Moderately important; U = Unimportant or importance unknown

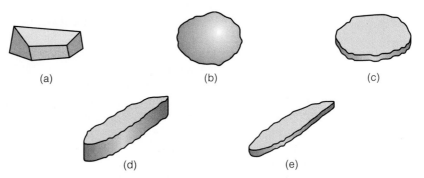

(a)  (b)  (c)

(d)  (e)

**FIGURE 5.1** Particle shapes: (a) angular, (b) rounded, (c) flaky, (d) elongated, (e) flaky and elongated.

**FIGURE 5.2** Angular and rounded aggregates.

makes it difficult for them to slide across each other. Flakiness describes the relationship between the smallest and largest dimensions of the aggregate.

The roughness of the aggregate surface plays an important role in the way the aggregate compacts and bonds with the binder material. Aggregates with a *rough* texture are more difficult to compact into a dense configuration than *smooth* aggregates. Rough texture generally improves bonding and increases interparticle friction. In general, natural gravel and sand have a smooth texture, whereas crushed aggregates have a rough texture.

For the purpose of preparing portland cement concrete, it is desirable to use rounded and smooth aggregate particles to improve the workability of fresh concrete during mixing. However, angular and rough particles are desirable for asphalt concrete and base courses in order to increase the stability of the materials in the field and to reduce rutting. Flaky and elongated aggregates are undesirable for asphalt concrete since they are difficult to compact during construction and easy to break.

Many specifications for aggregates used in asphalt concrete require a minimum percentage of aggregates with crushed faces as a surrogate *shape* and *texture requirement*. A crushed particle exhibits one or more mechanically induced fractured face and typically has a rough surface texture. To evaluate the angularity and surface texture of coarse aggregate, the percentages of particles with one and with two or more crushed faces are counted in a representative sample.

For fine aggregate, angularity and surface texture can be measured indirectly using the AASHTO TP33 Method A, Test Method for Uncompacted Void Content of Fine Aggregate. In this test a sample of fine aggregate is poured into a small cylinder by flowing it through a standard funnel, as shown in Figure 5.3. By determining the weight of the fine aggregate in the filled cylinder of known volume, the void content can be calculated as the difference between the cylinder volume and the fine aggregate volume collected in the cylinder. The volume of the fine aggregate is calculated by dividing the weight of the fine aggregate by its bulk density. The higher the amount of void content, the more angular and the rougher the surface texture of the fine aggregate.

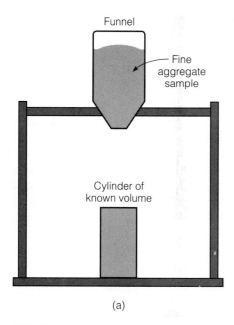

(a)

(b)

**FIGURE 5.3** Apparatus used to measure angularity and surface texture of fine aggregate.

## Soundness and Durability

The ability of aggregate to withstand weathering is defined as soundness or durability. Aggregates used in various civil engineering applications must be sound and durable, particularly if the structure is subjected to severe climatic conditions. Water freezing in the voids of aggregates generates stresses that can fracture the stones. The soundness test (ASTM C88) simulates weathering by soaking the aggregates in either a sodium sulfate or a magnesium sulfate solution. These sulfates cause crystals to grow in the aggregates, simulating the effect of freezing. The test starts with an oven-dry sample separated into different sized fractions. The sample is subjected to cycles of soaking in the sulfate for 16 hours followed by drying. Typically, the samples are subjected to five cycles. After this process, the aggregates are washed, dried, each size is weighed, and the weighted average percentage loss for the entire sample is computed. This result is compared to allowable limits to determine if the aggregate is acceptable. This is an empirical screening procedure for new aggregate sources when no service records are available.

The soundness by freeze thaw (AASHTO T103) and potential expansion from hydrated reactions (ASTM D4792) are alternative screening tests for evaluating soundness. The durability of aggregates in portland cement concrete can be tested by rapid freezing and thawing (ASTM C666), critical dilation by freezing (ASTM C671), and by frost resistance of coarse aggregates in air-entrained concrete by critical dilation (ASTM C682).

### Toughness, Hardness, and Abrasion Resistance

The ability of aggregates to resist the damaging effect of loads is related to the hardness of the aggregate particles and described as the toughness or abrasion resistance. The aggregate must resist crushing, degradation, and disintegration when stockpiled, mixed as either portland cement or asphalt concrete, placed and compacted, and exposed to loads.

The Los Angeles abrasion test (ASTM C131, C535) evaluates the aggregates' toughness and abrasion resistance. In this test, aggregates blended to a fixed size distribution are placed in a large steel drum with standard-sized steel balls that act as an abrasive charge. The drum is rotated, typically for 500 revolutions. The material is recovered from the machine and passed through a sieve that retained all the original material. The percentage weight loss is the LA abrasion number. This is an empirical test, that is, the test results do not have a scientific basis and are meaningful only when local experience defines the acceptance criteria.

### Absorption

Although aggregates are inert, they can capture water and asphalt binder in surface voids. The amount of water the aggregates absorb is important in the design of portland cement concrete since moisture captured in the aggregate voids is not available to improve the workability of the plastic concrete and to react with the cement. There is no specific level of aggregate absorption that is desirable for aggregates used in portland cement concrete, but aggregate absorption must be evaluated to determine the appropriate amount of water to mix into the concrete.

Absorption is also important for asphalt concrete since absorbed asphalt is not available to act as a binder. Thus highly absorptive aggregates require greater amounts of asphalt binder, making the mix less economical. On the other hand, some asphalt absorption is desired to promote bonding between the asphalt and the aggregate. Therefore, low-absorption aggregates are desirable for asphalt concrete.

Figure 5.4 demonstrates the four moisture condition states for an aggregate particle. *Bone dry* means the aggregate contains no moisture; this requires drying the aggregate in an oven to a constant mass. In an *air dry* condition, the aggregate may have some moisture but the saturation state is not quantified. In a *saturated surface–dry (SSD)* condition, the aggretate's voids are filled with moisture but the main

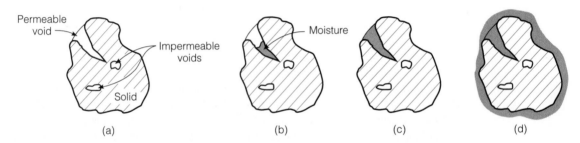

**FIGURE 5.4** Voids and moisture absorption of aggregates: (a) bone dry, (b) air dry, (c) saturated surface–dry (SSD), (d) moist.

surface area of the aggregate particles is dry. *Absorption* is defined as the moisture content in the SSD condition. Moist aggregates have a moisture content in excess of the SSD condition. Free moisture is the difference between the actual moisture content of the aggregate and the moisture content in the SSD condition.

## Specific Gravity

The weight-volume characteristics of aggregates are not an important indicator of aggregate quality, but they are important for concrete mix design. *Density*, the mass per unit volume, could be used for these calculations. However, *specific gravity* (Sp. Gr.), the mass of a material divided by the mass of an equal volume of distilled water, is more commonly used. Four types of specific gravity are defined based on how voids in the aggregate particles are considered. Three of these are widely accepted and used in portland cement and asphalt concrete mix design: *bulk-dry, bulk-saturated surface–dry,* and *apparent* specific gravity. These are defined as

$$\text{Bulk dry Sp. Gr.} = \frac{\text{Dry weight}}{(\text{Total particle volume})\gamma_w} = \frac{W_s}{(V_s + V_i + V_p)\gamma_w} \tag{5.1}$$

$$\text{Bulk SSD Sp. Gr.} = \frac{\text{SSD weight}}{(\text{Total particle volume})\gamma_w} = \frac{W_s + W_p}{(V_s + V_i + V_p)\gamma_w} \tag{5.2}$$

$$\text{Apparent Sp. Gr.} = \frac{\text{Dry weight}}{(\text{Volume not accessible to water})\gamma_w} = \frac{W_s}{(V_s + V_i)\gamma_w} \tag{5.3}$$

where
- $W_s$ = weight of solids
- $V_s$ = volume of solids
- $V_i$ = volume of water impermeable voids
- $V_p$ = volume of water permeable voids
- $W_p$ = weight of water in the permeable voids when the aggregate is in the SSD condition
- $\gamma_w$ = unit weight of water

Figure 5.5 shows that when aggregates are mixed with asphalt binder, only a portion of the water-permeable voids are filled with asphalt. Hence, a fourth type of specific gravity is defined, the *effective specific gravity.*

$$\text{Effective Sp. Gr.} = \frac{\text{Dry weight}}{(\text{Volume not accessible to asphalt})\gamma_w} = \frac{W_s}{(V_s + V_c)\gamma_w} \tag{5.4}$$

where $V_c$ is volume of voids not filled with asphalt cement.

At present, there is no standard method for determining the effective specific gravity of aggregates directly. The U.S. Corps of Engineers has defined a method for determining the effective specific gravity of aggregates that absorb more than 2.5% water.

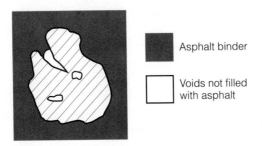

**FIGURE 5.5** Aggregate particle submerged in asphalt cement; not all voids are filled with asphalt.

The specific gravity and absorption of coarse aggregates are determined in accordance with ASTM C127. In this procedure a representative sample of the aggregate is soaked for 24 hours and weighed suspended in water. The sample is then dried to the SSD condition and weighed. Finally, the sample is dried to a constant weight and weighed. The specific gravity and absorption are determined by

$$\text{Bulk dry Sp. Gr. } = \frac{A}{B-C} \tag{5.5}$$

$$\text{Bulk SSD Sp. Gr. } = \frac{B}{B-C} \tag{5.6}$$

$$\text{Apparent Sp. Gr. } = \frac{A}{A-C} \tag{5.7}$$

$$\text{Absorption (\%) } = \frac{B-A}{A}(100) \tag{5.8}$$

where
  $A$ = dry weight
  $B$ = SSD weight
  $C$ = submerged weight

ASTM 128 defines the procedure for determining the specific gravity and absorption of fine aggregates. A representative sample is soaked in water for 24 hours and dried back to the SSD condition. A 500-g sample of the SSD material is placed in a *pycnometer*, a constant volume flask; water is added to the constant volume mark on the pycnometer and the weight is determined again. The sample is then dried and the weight is determined. The specific gravity and absorption are determined by

$$\text{Bulk dry Sp. Gr. } = \frac{A}{B+S-C} \tag{5.9}$$

$$\text{Bulk SSD Sp. Gr. } = \frac{S}{B+S-C} \tag{5.10}$$

$$\text{Apparent Sp. Gr. } = \frac{A}{B+A-C} \tag{5.11}$$

$$\text{Absorption} (\%) = \frac{S-A}{A}(100) \qquad\qquad (5.12)$$

where
   $A$ = dry weight
   $B$ = weight of the pycnometer filled with water
   $C$ = weight of the pycnometer filled with aggregate and water
   $S$ = saturated surface–dry weight of the sample

## Strength and Modulus

The strength of portland cement concrete and asphalt concrete cannot exceed that of the aggregates. It is difficult and rare to test the strength of aggregate particles. However, tests on the parent rock sample or a bulk aggregate sample provide an indirect estimate of these values. Aggregate strength is generally important in high-strength concrete and in the surface course on heavily traveled pavements. The tensile strength of aggregates ranges from 0.7 MPa to 16 MPa (100 psi to 2300 psi), while the compressive strength ranges from 35 MPa to 350 MPa (5000 psi to 50,000 psi) (Meininger and Nichols 1990; Barksdale 1991). Field service records are a good indication of the adequacy of the aggregate strength.

The modulus of elasticity of aggregates is not usually measured. However, new mechanistic-based methods of pavement design require an estimate of the modulus of aggregate bases. The response of bulk aggregates to stresses is nonlinear and depends on the confining pressure on the material. Since the modulus is used for pavement design, dynamic loads are used in a test to simulate the magnitude and duration of stresses in a pavement base caused by a moving truck. During the test, as the stresses are applied to the sample, the deformation response has two components, a recoverable or resilient deformation, and a permanent deformation. Only the resilient portion of the strain is used with the applied stress level to compute the modulus of the aggregate. Hence, the results are defined as the resilient modulus $M_R$.

In the resilient modulus test (AASHTO T292), a prepared cylindrical sample is placed in a triaxial cell, as shown in Figure 5.6. A specimen with large aggregates is typically 0.15 m (6 in.) in diameter by 0.30 m (12 in.) high, while soil samples are 71 mm (2.8 in.) in diameter by 142 mm (5.6 in.) high. The specimen is subjected to a specified confining pressure and a repeated axial load. Accurate transducers, such as LVDTs, measure the axial deformation. The test requires determining the modulus over a range of axial loads and confining pressures. The resilient modulus equals the repeated axial stress divided by the resilient strain for each combination of load level and confining pressure. The resilient modulus test requires measuring very small loads and deformations and is, therefore, difficult to perform. Currently, the test is mostly limited to research projects.

## Gradation and Maximum Size

Gradation describes the particle size distribution of the aggregate. The particle size distribution is an important attribute of the aggregates. Large aggregates are economically advantageous in portland cement and asphalt concrete, as they have less surface

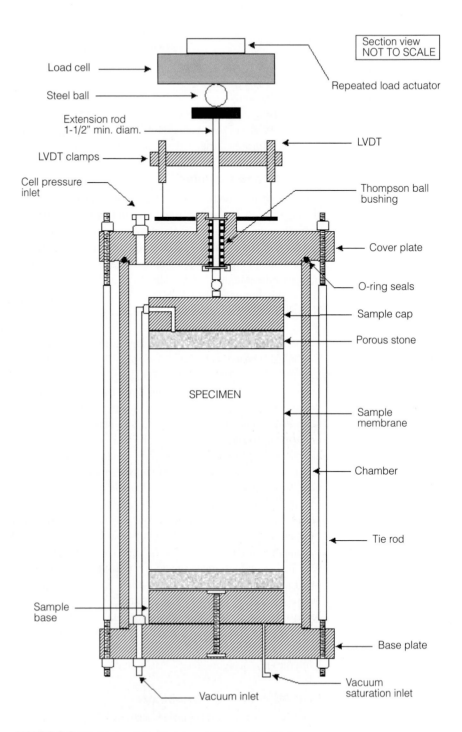

**FIGURE 5.6** Triaxial chamber with external LVDTs and load cell.

area and, therefore, require less binder. However, large aggregate mixes, whether asphalt or portland cement concrete, are harsher and more difficult to work into place. Hence construction considerations, such as equipment capability, dimensions of construction members, clearance between reinforcing steel, and layer thickness, limit the maximum aggregate size.

Two definitions are used to describe the maximum particle size in an aggregate blend

*Maximum aggregate size*—the smallest sieve size through which 100% of the aggregates sample particles pass.

*Nominal maximum aggregate size*—the largest sieve that retains any of the aggregate particles, but generally not more than 10%.

Some agencies define the maximum aggregate size as two sizes larger than the first sieve to retain more than 10% of the material, while the nominal maximum size is one size larger than the first sieve to retain more than 10% of the material (The Asphalt Institute 1995; McGennis et al. 1995).

**Sieve Analysis** Gradation is evaluated by passing the aggregates through a series of sieves, as shown in Figure 5.7 (ASTM C136, E11). The sieve retains particles larger than the opening, while smaller ones pass through. Metric sieve descriptions are based on the size of the openings measured in millimeters. Sieves smaller than 0.6 mm can be described in either millimeters of micrometers. In U.S. customary units, sieves with

**FIGURE 5.7** Sieve shakers for large samples of aggregate.

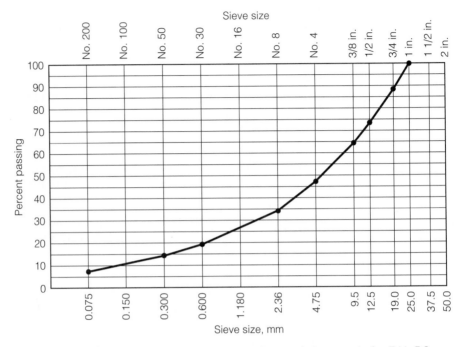

Sieve size

**FIGURE 5.8** Semi-log aggregate gradation chart showing a gradation example. See Table 5.2.

openings greater than 1/4 in. are designated by the size of the opening; the lengths of the sides of the square openings of a 2-in. sieve are 2 in. measured between the wires. This equals the diameter of a sphere that will exactly touch each side of the square at the midpoints. Sieves smaller than 1/4 in. are specified by the number of uniform openings per linear inch (a No. 8 sieve has 8 openings per inch, or 64 holes per square inch).

Gradation results are described by the cumulative percentage of aggregates that either pass through or are retained by a specific sieve size. Percentages are reported to the nearest whole number, except if the percentage passing the 0.075-mm (No. 200) sieve is less than 10%, it is reported to the nearest 0.1%. Gradation analysis results are generally plotted on a semi-log chart, as shown in Figures 5.8 and A.2.

Aggregates are usually classified by size as coarse aggregates, fine aggregates, and mineral fillers (fines). ASTM defines coarse aggregate as particles retained on the 4.75-mm (No. 4) sieve, fine aggregate as those passing the 4.75-mm sieve, and mineral filler as material mostly passing the 0.075-mm (No. 200) sieve.

**SAMPLE PROBLEM 5.1**

A sieve analysis test was performed on a sample of fine aggregate and produced the following results. Calculate the percent passing each sieve.

| Sieve, mm | 4.75 | 2.36 | 2.00 | 1.18 | 0.60 | 0.30 | 0.15 | 0.075 | pan |
|---|---|---|---|---|---|---|---|---|---|
| Amount retained, g | 0 | 33.2 | 56.9 | 83.1 | 151.4 | 40.4 | 72.0 | 58.3 | 15.6 |

## Solution:

| Sieve size | Amount Retained, g (a) | Cumulative Amount Retained, g (b) | Cumulative Percent Retained (c) = (b) × 100 / Total | Percent Passing (d) = 100 − (c) |
|---|---|---|---|---|
| 4.75 mm (No. 4) | 0 | 0 | 0 | 100 |
| 2.36 mm (No. 8) | 33.2 | 33.2 | 6 | 94 |
| 2.00 mm (No. 10) | 56.9 | 90.1 | 18 | 82 |
| 1.18 mm (No. 16) | 83.1 | 173.2 | 34 | 66 |
| 0.60 mm (No. 30) | 151.4 | 324.6 | 64 | 36 |
| 0.30 mm (No. 50) | 40.4 | 365.0 | 71 | 29 |
| 0.15 mm (No. 100) | 72.0 | 437.0 | 86 | 14 |
| 0.075 mm (No.200) | 58.3 | 495.3 | 96.9 | 3.1 |
| Pan | 15.6 | 510.9 | 100 | |
| Total | 510.9 | | | |

**Maximum Density Gradation** The density of an aggregate mix is a function of the size distribution of the aggregates. In 1907 Fuller established the relationship for determining the distribution of aggregates that provides the maximum density or minimum amount of voids

$$P_i = 100\left(\frac{d_i}{D}\right)^{0.45}$$ (5.13)

where
$P_i$ = percent passing a sieve of size $d_i$
$d_i$ = the sieve size in question
$D$ = maximum size of the aggregate

In the 1960s the Federal Highway Administration introduced the "0.45 power" gradation chart, Figures 5.9 and A.3, designed to produce a straight line for maximum density gradations (Federal Highway Administration 1988). Table 5.2 presents an example calculation of the particle size distribution required for maximum density. Note that the gradation in Table 5.2 is plotted on both gradation charts in Figures 5.8 and 5.9.

Frequently, a *dense* gradation but not necessarily the maximum possible density, is desired in many construction applications because of its high stability. Using a high-density gradation also means the aggregates occupy most of the volume of the material, limiting the binder content and thus reducing the cost. For example, aggregates for asphalt concrete must be dense, but must also have sufficient voids in the mineral aggregate to provide room for the binder plus room for voids in the mixture.

**Other Types of Gradation** In addition to maximum density (i.e., *well-graded*) aggregates can have other characteristic distributions, as shown in Figure 5.10. A *one-sized* distribution has the majority of aggregates passing one sieve and being retained on the next smaller sieve. Hence, the majority of the aggregates have essentially the same diameter; their gradation curve is nearly vertical. One-sized graded aggregates will

**Table 5.2**  Example Calculation of
Aggregate Distribution Required to
Achieve Maximum Density

| Sieve | $P_i = 100 \, (d_i \,/\, D)^{0.45}$ |
|---|---|
| 25 mm (1 in.) | 100 |
| 19 mm (3/4 in.) | 88 |
| 12.5 mm (1/2 in.) | 73 |
| 9.5 mm (3/8 in.) | 64 |
| 4.75 mm (No. 4) | 47 |
| 2.36 mm (No. 8) | 34 |
| 0.60 mm (No. 30) | 19 |
| 0.30 mm (No. 50) | 14 |
| 0.075 mm (No. 200) | 7.3 |

have good permeability, but poor stability and are used in such applications as chip
seals of pavements. *Gap*-graded aggregates are missing one or more sizes of material.
Their gradation curve has a near horizontal section indicating that nearly the same
portions of the aggregates pass two different sieve sizes. *Open*-graded aggregates are
missing small aggregate sizes that would block the voids between the larger aggregate.
Since there are a lot of voids, the material will be highly permeable but may not have
good stability.

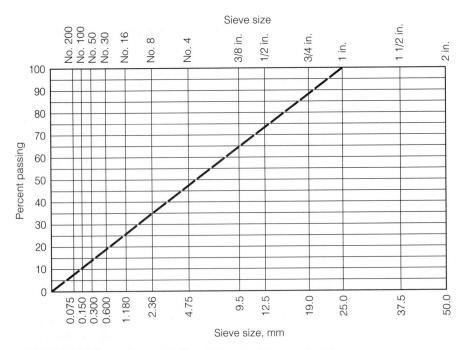

**FIGURE 5.9** Federal Highway Administration 0.45 power gradation chart showing the maximum
density gradation for a maximum size of 25 mm. See Table 5.2.

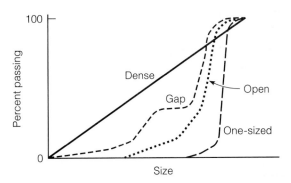

**FIGURE 5.10** Types of aggregate grain-size distributions plotted on a 0.45 gradation chart.

**TABLE 5.3** Effect of Amount of Fines on the Relative Properties of Aggregate Base Material

| Characteristic | No Fines (open or clean) | Well-Graded (dense) | Large Amount of Fines (dirty or rich) |
|---|---|---|---|
| Stability | Medium | Excellent | Poor |
| Density | Low | High | Low |
| Permeability | Permeable | Low | Impervious |
| Frost Susceptibility | No | Maybe | Yes |
| Handling | Difficult | Medium | Easy |
| Cohesion | Poor | Medium | Large |

As shown in Table 5.3, the amount of fines has a major effect on the characteristics of aggregate base materials. Aggregates with the percentage of fines equal to the amount required for maximum density have excellent stability and density but may have a problem with permeability, frost susceptibility, handling, and cohesion.

**Gradation Specifications** Gradation specifications define maximum and minimum cumulative percentages of material passing each sieve. Aggregates are commonly described as being either coarse or fine, depending on whether the material is predominantly retained on or passes through a 4.75-mm (No. 4) sieve.

Portland cement concrete requires separate specifications for coarse and fine aggregates. The ASTM C33 specifications for fine aggregates for concrete are given in Table 5.4. Table 5.5 shows the ASTM C33 gradation specifications for coarse concrete aggregates.

Generally, local agencies develop their own specifications for the gradation of aggregates for asphalt concrete. Table 5.6 gives the asphalt concrete aggregate specifications used by the Arizona Department of Transportation as an example (Arizona Department of Transportation 1990). These specifications define the range of allowable gradations for asphalt concrete for mix design purposes. Note the percent of material passing the 0.075-mm (No. 200) sieve, the fines or mineral filler, is carefully controlled for asphalt concrete due to its significance to the properties of the mix.

**TABLE 5.4** ASTM C33 Gradation Specifications for Fine Aggregates for Portland Cement Concrete*

| Sieve Size | Percent Passing |
|---|---|
| 9.5 mm (3/8 in.) | 100 |
| 4.75 mm (No. 4) | 95–100 |
| 2.36 mm (No. 8) | 80–100 |
| 1.18 mm (No. 16) | 50–85 |
| 0.60 mm (No. 30) | 25–60 |
| 0.30 mm (No. 50) | 10–30 |
| 0.15 mm (No. 100) | 2–10 |

*Copyright ASTM. Reprinted with permission.

Once aggregate gradation from asphalt concrete mix design is established for a project, the contractor must produce aggregates that fall within a narrow band around the single gradation line established for developing the mix design. For example, the Arizona Department of Transportation will only give the contractor full pay if the gradation of the aggregates are within the following limits with respect to the accepted mix design gradations.

| Sieve Size | Allowable Deviations for Full Pay |
|---|---|
| 9.5 mm (3/8 in.) and larger | ± 3% |
| 2.36 to 0.45 mm (No. 8 to No. 40) | ± 2% |
| 0.075 mm (No. 200) | ± 0.5% |

**Fineness Modulus** The *fineness modulus* is a measure of the fine aggregates' gradation and is used primarily for portland cement concrete mix design. It can also be used as a daily quality control check in the production of concrete. The fineness modulus is one-hundredth of the sum of the cumulative weight retained on the 0.15-mm, 0.3-mm, 0.6-mm, 1.18-mm, 2.36-mm, 4.75-mm, 9.5-mm, 19.0-mm, 37.5-mm, 75-mm, and 150-mm (No. 100, 50, 30, 16, 8, and 4 and 3/8 in., 3/4 in., 1-1/2 in., 3 in., and 6 in.) sieves. When the fineness modulus is determined for fine aggregates, sieves larger than 9.5 mm (3/8 in.) are not used. The fineness modulus should be in the range of 2.3 to 3.1, with a higher number being a coarser aggregate. Table 5.7 demonstrates the calculation of the fineness modulus.

**SAMPLE PROBLEM 5.2**

Calculate the fineness modulus of the sieve analysis results of sample problem 5.1.

*Solution:* According to the definition of fineness modulus, sieves 2.00 and 0.075 mm (No. 10 and 200) are not included.

$$\text{Fineness modulus} = \frac{6 + 34 + 64 + 71 + 86}{100} = 2.61$$

**TABLE 5.5** ASTM C33 Gradation Specifications for Coarse Aggregate for Portland Cement Concrete*†

| Size No. | Nominal Size | Sieve Size | | | | | | | | | | | | |
|---|---|---|---|---|---|---|---|---|---|---|---|---|---|---|
| | | 4 in. (100 mm) | 3-1/2 in. (90 mm) | 3 in. (75 mm) | 2-1/2 in. (63 mm) | 2 in. (50 mm) | 1-1/2 in. (37.5 mm) | 1 in. (25.0 mm) | 3/4 in. (19.0 mm) | 1/2 in. (12.5 mm) | 3/8 in. (9.5 mm) | No. 4 (4.75 mm) | No. 8 (2.36 mm) | No. 16 (1.18 mm) |
| 1 | 3-1/2 to 1-1/2 in. (90 to 37.5 mm) | 100 | 90 to 100 | — | 25 to 60 | — | 0 to 15 | — | 0 to 5 | — | — | — | — | — |
| 2 | 2-1/2 to 1-1/2 in. (63 to 37.5 mm) | — | — | 100 | 90 to 100 | 35 to 70 | 0 to 15 | — | 0 to 5 | — | — | — | — | — |
| 3 | 2 to 1 in. (50 to 25.0 mm) | — | — | — | 100 | 90 to 100 | 35 to 70 | 0 to 15 | — | 0 to 5 | — | — | — | — |
| 357 | 2 in. to No. 4 (50 to 4.75 mm) | — | — | — | 100 | 95 to 100 | — | 35 to 70 | — | 10 to 30 | — | 0 to 5 | — | — |
| 4 | 1-1/2 to 3/4 in. (37.5 to 19 mm) | — | — | — | — | 100 | 90 to 100 | 20 to 55 | 0 to 15 | — | 0 to 5 | — | — | — |
| 467 | 1-1/2 in. to No. 4 (37.5 to 4.75 mm) | — | — | — | — | 100 | 95 to 100 | — | 35 to 70 | — | 10 to 30 | 0 to 5 | — | — |
| 5 | 1 to 1/2 in. (25.0 to 12.5 mm) | — | — | — | — | — | 100 | 90 to 100 | 20 to 55 | 0 to 10 | 0 to 5 | — | — | — |
| 56 | 1 to 3/8 in. (25.0 to 9.5 mm) | — | — | — | — | — | 100 | 90 to 100 | 40 to 85 | 10 to 40 | 0 to 15 | 0 to 5 | — | — |
| 57 | 1 in. to No. 4 (25.0 to 4.75 mm) | — | — | — | — | — | 100 | 95 to 100 | — | 25 to 60 | — | 0 to 10 | 0 to 5 | — |
| 6 | 3/4 in. to 3/8 in. (19.0 to 9.5 mm) | — | — | — | — | — | — | 100 | 90 to 100 | 20 to 55 | 0 to 15 | 0 to 5 | — | — |
| 67 | 3/4 in. to No. 4 (19.0 to 4.75 mm) | — | — | — | — | — | — | 100 | 90 to 100 | — | 20 to 55 | 0 to 10 | 0 to 5 | — |
| 7 | 1/2 in. to No. 4 (12.5 to 4.75 mm) | — | — | — | — | — | — | — | 100 | 90 to 100 | 40 to 70 | 0 to 15 | 0 to 5 | — |
| 8 | 3/8 in. to No. 8 (9.5 to 2.36 mm) | — | — | — | — | — | — | — | — | 100 | 85 to 100 | 10 to 30 | 0 to 10 | 0 to 5 |

*Copyright ASTM. Reprinted with permission.
†Amounts passing through each laboratory sieve (square openings), wt./%.

**TABLE 5.6**   Example Aggregate Grading Requirements for Asphalt Concrete*

| Sieve size | 12.5 mm (1/2 in.) | | 19 mm (3/4 in.) | | Base | |
|---|---|---|---|---|---|---|
| | No Admixture | Admixture[†] | No Admixture | Admixture | No Admixture | Admixture |
| 31.75 mm (1-1/4 in.) | | | | | 100 | 100 |
| 25 mm (1 in.) | | | 100 | 100 | 90–100 | 90–100 |
| 19 mm (3/4 in.) | 100 | 100 | 90–100 | 90–100 | 85–95 | 85–95 |
| 12.5 mm (1/2 in.) | 90–100 | 90–100 | — | — | — | — |
| 9.5 mm (3/8 in.) | 70–85 | 70–85 | 65–80 | 65–80 | 60–75 | 60–75 |
| 2.36 mm (No. 8) | 43–51 | 44–52 | 40–49 | 41–50 | 35–45 | 35–46 |
| 0.45 mm (No. 40) | 12–20 | 13–21 | 12–20 | 13–21 | 10–18 | 11–19 |
| 0.075 mm (No. 200) | 2.0–5.0 | 3.0–5.0 | 2.0–5.0 | 3.0–6.5 | 1.5–4.0 | 2.0–5.5 |

*Arizona Department of Transportation (1990).
†Lime or portland cement used as an additive to reduce stripping.

**TABLE 5.7**   Example Calculation of Fineness Modulus

| Sieve Size | Percentage of Individual Fraction Retained, by weight | Cumulative Percentage Retained, by weight | Percentage Passing, by weight |
|---|---|---|---|
| 9.5 mm (3/8 in.) | 0 | 0 | 100 |
| 4.75 mm (No. 4) | 2 | 2 | 98 |
| 2.36 mm (No. 8) | 13 | 15 | 85 |
| 1.18 mm (No. 16) | 25 | 40 | 60 |
| 0.60 mm (No. 30) | 15 | 55 | 45 |
| 0.30 mm (No. 50) | 22 | 77 | 23 |
| 0.15 mm (No. 100) | 20 | 97 | 3 |
| Pan | 3 | — | 0 |
| Total | 100 | 286 | |
| **Fineness modulus = 286 / 100 = 2.86** | | | |

**TABLE 5.8**   Example of Aggregate Blending Analysis by Graphical Method*

| | Sieve size | | | | | | | | |
|---|---|---|---|---|---|---|---|---|---|
| | 19 mm (3/4 in.) | 12.5 mm (1/2 in.) | 9.5 mm (3/8 in.) | 4.75 mm (No. 4) | 2.36 mm (No. 8) | 0.60 mm (No. 30) | 0.30 mm (No. 50) | 0.15 mm (No. 100) | 0.075 mm (No. 200) |
| Specification | 100 | 80–100 | 70–90 | 50–70 | 35–50 | 18–29 | 13–23 | 8–16 | 4–10 |
| Aggregate A | 100 | 90 | 59 | 16 | 3 | 0 | 0 | 0 | 0 |
| Aggregate B | 100 | 100 | 100 | 96 | 82 | 51 | 36 | 21 | 9 |
| Blend | 100 | 95 | 80 | 56 | 43 | 26 | 18 | 11 | 4.5 |

*Numbers shown are percent passing each sieve.

**Blending Aggregates to Meet Specifications**   Generally, a single aggregate source is unlikely to meet gradation requirements for portland cement or asphalt concrete mixes. Thus, blending of aggregates from two or more sources would be required to satisfy the specifications. Figure 5.11 shows a graphical method for selecting the combination of two aggregates to meet a specification. Table 5.8 presents the data used for Figure 5.11. Determining a satisfactory aggregate blend with the graphical method entails the following steps (The Asphalt Institute 1995).

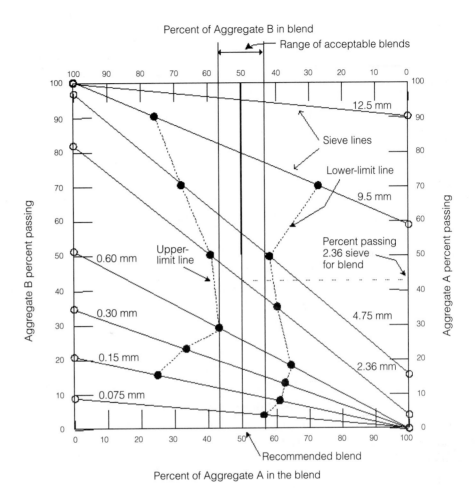

**FIGURE 5.11** Graphical method for determining aggregate blend to meet gradation requirements. See Table 5.8.

1. Plot the percentages passing through each sieve on the right axis for aggregate A and on the left axis for aggregate B, shown as open circles in Figure 5.11.

2. For each sieve size connect the left and right axes.

3. Plot the specification limits of each sieve on the corresponding sieve lines, that is, a mark is placed on the 9.5-mm (3/8 in.) sieve line corresponding to 70% and 90% on the vertical axis, shown as closed circles in Figure 5.11.

4. Connect the upper- and lower-limit points on each sieve line.

5. Draw vertical lines through the rightmost point of the upper-limit line and the leftmost point of the lower-limit line. If the upper- and lower-limit lines overlap, no combination of the aggregates will meet specifications.

6. Any vertical line drawn between these two vertical lines identifies an aggregate blend that will meet the specification. The intersection with the upper axis defines the percentage of aggregate B required for the blend. The projection to the lower axis defines the percentage of aggregate A required.

**7.** Projecting intersections of the blend line and the sieve lines horizontally gives an estimate of the gradation of the blended aggregate. Figure 5.11 shows that a 50-50 blend of aggregates A and B will result in a blend with 43% passing through the 2.36-mm (No. 8) sieve. The gradation of the blend is shown in the last line of Table 5.8.

When more than two aggregates are required, the graphical procedure can be repeated in an iterative manner. However, a trial and error process is generally used to determine the proportions. The basic equation for blending is

$$P_i = Aa + Bb + Cc + \cdots \tag{5.14}$$

where

$P_i$ = percent blend material passing sieve size $i$

$A, B, C, \ldots$ = percent of aggregates A, B, C,... passing sieve $i$

$a, b, c, \ldots$ = decimal fractions by weight of aggregates A, B, and C used in the blend, where the total is 1.00.

Table 5.9 demonstrates these calculations for two aggregate sources. The table shows the required specification range and the desired (or target) gradation, usually the midpoint of the specification. A trial percentage of each aggregate source is assumed and multiplied by the percentage passing each sieve. These gradations are added to get the composite percentage passing each sieve for the blend. The gradation of the blend is compared to the specification range to determine if the blend is acceptable. With practice, blends of four aggregates can be readily resolved. These calculations are easily performed by a spreadsheet computer program.

**Properties of Blended Aggregates** When two or more aggregates from different sources are blended, some of the properties of the blend can be calculated from the properties of individual components. With the exception of specific gravity and density, the properties of the blend are the simple weighted average of the properties of the components as demonstrated in Equation 5.15. This equation applies to properties such as angularity, absorption, strength, and modulus.

$$X = P_1X_1 + P_2X_2 + P_3X_3 + \cdots \tag{5.15}$$

**TABLE 5.9**   Example of Aggregate Blending Analysis by Iterative Method*

| | Sieve size | | | | | | |
|---|---|---|---|---|---|---|---|
| | 12.5 mm (1/2 in.) | 9.5 mm (3/8 in.) | 4.75 mm (No. 4) | 2.00 mm (No. 10) | 0.425 mm (No. 40) | 0.180 mm (No. 80) | 0.075 mm (No. 200) |
| Specification | 100 | 95–100 | 70–85 | 55–70 | 20–40 | 10–20 | 4–8 |
| Target gradation | 100 | 98 | 77.5 | 62.5 | 30 | 15 | 6 |
| % Aggregate A (A) | 100 | 100 | 98 | 90 | 71 | 42 | 19 |
| % Aggregate B (B) | 100 | 94 | 70 | 49 | 14 | 2 | 1 |
| 30% A (a) | 30 | 30 | 29.4 | 27 | 21.3 | 12.6 | 5.7 |
| 70% B (b) | 70 | 65.8 | 49 | 34.3 | 9.8 | 1.4 | 0.7 |
| Blend ($P_i$) | 100 | 96 | 78 | 61 | 31 | 14 | 6.4 |

*Numbers shown are percent passing each sieve.

where

$X$ = composite property of the blend

$X_1, X_2, X_3$ = properties of fractions 1, 2, and 3

$P_1, P_2, P_3$ = decimal fractions by weight of aggregates 1, 2, and 3 used in the blend, where the total is 1.00.

**SAMPLE PROBLEM 5.3**

Aggregates from two sources having crushed faces of 40% and 90% were blended at a ratio of 30:70 by weight, respectively. What is the percent of crushed faces of the aggregate blend?

*Solution:*

Crushed faces of the blend = $(0.3)\,(40) + (0.7)\,(90) = 75\%$ ◆

Asphalt concrete mix design requires that the engineer knows the composite specific gravity of all aggregates in the mix. The composite specific gravity of a mix of different aggregates is obtained by using Equation 5.16.

$$G = \frac{1}{\dfrac{P_1}{G_1} + \dfrac{P_2}{G_2} + \dfrac{P_3}{G_3} + \cdots} \tag{5.16}$$

where

$G$ = composite specific gravity

$G_1, G_2, G_3$ = specific gravities of fractions 1, 2, and 3

$P_1, P_2, P_3$ = decimal fractions by weight of aggregates 1, 2, and 3 used in the blend, where the total is 1.00.

Note that Equation 5.16 is used only to obtain the combined specific gravity and density of the blend, whereas Equation 5.15 is used to obtain other combined properties.

**SAMPLE PROBLEM 5.4**

Aggregates from three sources having bulk specific gravities of 2.753, 2.649, and 2.689 were blended at a ratio of 70:20:10 by weight, respectively. What is the bulk specific gravity of the aggregate blend?

*Solution:*

$$G = \frac{1}{\dfrac{0.7}{2.753} + \dfrac{0.2}{2.649} + \dfrac{0.1}{2.689}} = 2.725$$ ◆

## Deleterious Substances in Aggregate

A deleterious substance is any material that adversely affects the quality of portland cement or asphalt concrete made with the aggregate. Table 5.10 identifies the main deleterious substances in aggregates and their effects on portland cement concrete. In asphalt concrete, deleterious substances are clay lumps, soft or friable particles, and coatings. These substances decrease the adhesion between asphalt and aggregate particles.

**TABLE 5.10**  Deleterious Substances and Their Affects on Portland Cement Concrete

| Substance | Harmful Effect |
| --- | --- |
| Organic impurities | Delay settling and hardening, may reduce strength gain, may cause deterioration |
| Materials less than 0.075 mm (No. 200) | Weaken bond, may increase water requirements |
| Coal, lignite, or other low-density materials | Reduce durability, may cause popouts or stains |
| Clay lumps and friable particles | Popouts, reduce durability and wear resistance |
| Soft particles | Reduce durability and wear resistance, popouts |

## Alkali-Aggregate Reactivity

Some aggregates react with portland cement, harming the concrete structure. The most common reaction, particularly in humid and warm climates, is between the active silica constituents of an aggregate and the alkalis in cement (sodium oxide, $Na_2O$, and potassium oxide, $K_2O$). The alkali-silica reaction results in excessive expansion, cracking, or popouts in concrete. Other constituents in the aggregate, such as carbonates, can also react with the alkali in the cement; however, their reaction is less harmful. The alkali-aggregate reactivity is affected by amount, type, and particle size of the reactive material, as well as the soluble alkali and water content of the concrete.

The best way to evaluate the potential for alkali-aggregate reactivity is by reviewing the field service history. For aggregates without field service history, several laboratory tests are available to check the potential alkali-aggregate reactivity. The ASTM C227 test can be used to determine the potentially expansive alkali-aggregate reactivity of cement-aggregate combinations. In this test a mortar bar is stored under a prescribed temperature and moisture conditions and its expansion is determined. The quick chemical test (ASTM C289) can be used to identify potentially reactive siliceous aggregates. ASTM C586 is used to determine potentially expansive carbonate rock aggregates (alkali-carbonate reactivity).

If alkali-reactive aggregate must be used, the reactivity can be minimized by limiting the alkali content of the cement (e.g., using Type II cement). The reactivity can also be reduced by keeping the concrete structure as dry as possible. Pozzolanic admixtures, such as fly ash, reduce the alkali-aggregate reactivity. Pozzolans, however, may increase reactivity of high-water soluble alkali contents. Finally, replacing about 30% of a reactive sand-gravel aggregate with crushed limestone (limestone sweetening) can minimize the alkali reactivity.

## Affinity for Asphalt

*Stripping,* or moisture-induced damage, is a separation of the asphalt film from the aggregate through the action of water, reducing the durability of the asphalt concrete and resulting in pavement failure. The mechanisms causing stripping are complex and not fully understood. One important factor is the relative affinity of the aggregate for either water or asphalt. *Hydrophilic* (water-loving) aggregates, such as silicates, have a greater affinity for water than for asphalt. They are usually acidic in nature and have a negative surface charge. Conversely, *hydrophobic* (water-repelling) aggregates have a greater affinity for asphalt than water. These aggregates, such as limestone, are basic in nature and have a positive surface charge. Hydrophilic aggregates are more susceptible to stripping than hydrophobic aggregates. Other stripping factors include porosity, absorption, and the existence of coatings and other deleterious substances.

Since stripping is the result of a compatibility problem between the asphalt and the aggregate, tests for stripping potential are performed on the asphalt concrete mix. Early compatibility tests submerged the sample in either room-temperature water (ASTM D1664) or boiling water (ASTM D3625); after a period of time, the technician observes the percentage of particles stripped from the asphalt. More recent procedures subject asphalt concrete to cycles of freeze-thaw conditioning. The strength or modulus of the specimens is measured and compared to the values of unconditioned specimens (ASTM D1075).

## Handling Aggregates

Aggregates must be handled and stockpiled in such a way to minimize segregation, degradation, and contamination. If aggregates roll down the slope of the stockpile, the different sizes will segregate, with large stones at the bottom and small ones at the top. Building stockpiles in thin layers circumvents this problem. The drop height should be limited to avoid breakage, especially for large aggregates. Vibration and jiggling on a conveyor belt tends to work fine material downward while coarse particles rise. Segregation can be minimized by moving the material on the belt frequently (up and down, side to side, in and out) or by installing a baffle plate, rubber sleeve, or paddle wheel at the end of the belt to remix coarse and fine particles. Rounded aggregates segregate more than crushed aggregates. Also, large aggregates segregate more readily than smaller aggregates. Therefore, different sizes should be stockpiled and batched separately. Stockpiles should be separated by dividers to avoid mixing and contamination (Meininger and Nichols 1990).

## Sampling Aggregates

In order for any of the tests described in this chapter to be valid, the sample of material being tested must represent the whole population of materials that is being quantified with the test. This is a particularly difficult problem with aggregates due to potential segregation problems. Samples of aggregates can be collected from any location in the production process, that is, from the stockpile, conveyor belts, or from bins within the mixing machinery. Usually, the best location for sampling the aggregate is on the conveyor belt that feeds the mixing plant. However, since the aggregate segregates on the belt, the entire width of the belt should be sampled at several locations or times throughout the production process. The samples would then be mixed to represent the entire lot of material.

Sampling from stockpiles must be performed carefully. Typically, aggregate samples are taken from the top, middle, and bottom of the stockpile and then combined. Before taking the samples, discard the 75-mm to 150-mm (3 in. to 6 in.) material at the surface. A board is used to prevent rolling of coarse aggregates during

**FIGURE 5.12** Aggregate sample splitter.

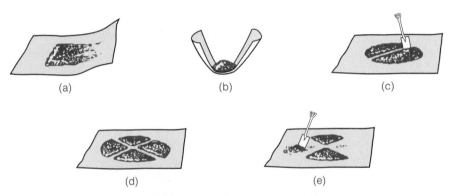

**FIGURE 5.13** Steps for reducing the sample size by quartering: (a) mixing by rolling on blanket, (b) forming a cone after mixing, (c) flattening the cone and quartering, (d) finishing quartering, (e) retaining opposite quarters (the other two quarters are rejected). (ASTM C702). Copyright ASTM. Reprinted with permission.

sampling. Samples are collected using a square shovel and are placed in sample bags or containers and labeled.

Sampling tubes 1.8 m (6 ft) long and 30 mm (1.25 in.) in diameter are used to sample fine aggregate stockpiles. At least five samples should be collected from random locations in the stockpile. These samples are then combined before laboratory testing.

Field sample sizes are governed by the nominal maximum size of aggregate particles (ASTM D75). Larger-sized aggregates require larger samples to minimize segregation errors. Field samples are typically larger than the samples needed for testing. Therefore, field samples must be reduced using sample splitters (Figure 5.12) or by quartering (Figure 5.13) (ASTM C702).

**SUMMARY**

Aggregates are widely used as a base material for foundations and as an ingredient in portland cement concrete and asphalt concrete. While the geological classification of aggregates gives insight into the properties of the material, the suitability of a specific source of aggregates for a particular application requires testing and evaluation. The most significant attributes of aggregates include the gradation, specific gravity, shape and texture, and soundness. When used in concrete, the compatibility of the aggregate and the binder must be evaluated.

**QUESTIONS AND PROBLEMS**

**5.1.** What are the three mineralogical or geological classifications of rocks and how are they formed?

**5.2.** Discuss five different desirable characteristics of aggregate used in portland cement concrete.

**5.3.** Discuss five different desirable characteristics of aggregate used in asphalt concrete.

**5.4.** Use the following information to determine the total and free moisture contents in percent.

Mass of wet sand   = 627.3 g

Mass of dry sand   = 590.1 g

Absorption         = 1.5%

**5.5.** Calculate the sieve analysis of the following aggregate and plot on a semi-log gradation paper.

| Sieve size | Amount Retained, g | Cumulative Amount Retained, g | Cumulative Percent Retained | Percent Passing |
|---|---|---|---|---|
| 25 mm (1 in.) | 0 | | | |
| 9.5 mm (3/8 in.) | 35.2 | | | |
| 4.75 mm (No. 4) | 299.6 | | | |
| 2.00 mm (No. 10) | 149.7 | | | |
| 0.425 mm (No. 40) | 125.8 | | | |
| 0.075 mm (No.200) | 60.4 | | | |
| Pan | 7.3 | | | |

**5.6.** Draw a graph to show the cumulative percent passing through the sieve versus sieve size for well-graded, gap-graded, open-graded, and one-sized aggregates.

**5.7.** The table below shows the grain size distributions of aggregates A, B, and C. The three aggregates must be blended at a ratio of 15:25:60 by weight, respectively. Using a spreadsheet program, determine the grain size distribution of the blend.

Percent Passing

| | Size | | | | | | | | |
|---|---|---|---|---|---|---|---|---|---|
| | 25 mm | 19 mm | 12.5 mm | 9.5 mm | 4.75 mm | 1.18 mm | 0.60 mm | 0.30 mm | 0.15 mm |
| Aggregate A | 100 | 100 | 100 | 83 | 67 | 49 | 37 | 25 | 18 |
| Aggregate B | 100 | 100 | 74 | 51 | 32 | 24 | 19 | 13 | 7 |
| Aggregate C | 100 | 82 | 66 | 42 | 27 | 14 | 5 | 0 | 0 |

**5.8.** The table below shows the grain size distribution for two aggregates and the specification limits for an asphalt concrete. Determine the blend proportion required to meet the specification and the gradations of the blend. On a semi-log gradation graph, plot the gradations of aggregate A, aggregate B, the selected blend, and the specification limits.

Percent Passing

| | Size | | | | | | | | |
|---|---|---|---|---|---|---|---|---|---|
| | 19 mm (3/4 in.) | 12.5 mm (1/2 in.) | 9.5 mm (3/8 in.) | 4.75 mm (No. 4) | 2.36 mm (No. 8) | 0.60 mm (No. 30) | 0.30 mm (No. 50) | 0.15 mm (No. 100) | 0.075 mm (No. 200) |
| Spec. limits | 100 | 80–100 | 70–90 | 50–70 | 35–50 | 18–29 | 13–23 | 8–16 | 4–10 |
| Aggregate A | 100 | 85 | 55 | 20 | 2 | 0 | 0 | 0 | 0 |
| Aggregate B | 100 | 100 | 100 | 85 | 67 | 45 | 32 | 19 | 11 |

**5.9.** Define the fineness modulus of aggregate. What is it used for?

**5.10.** Calculate the fineness modulus of aggregate B in problem 5.8. (Note that the percent passing the 1.18-mm (No. 16) sieve is not given and must be estimated.)

**5.11.** A portland cement concrete mix requires mixing well-graded sand having a gradation following the midpoint of the ASTM gradation band (Table 5.4) and well-graded gravel having a gradation following the midpoint of size number 467 of the ASTM gradation band (Table 5.5) at a ratio of 2:3 by weight. On a 0.45 power gradation chart, plot the gradations of the sand, gravel, and the blend. Is the gradation of the blend still well graded? If not, what would you call it?

**5.12.** Discuss the effect of the amount of material passing the 0.075-mm (N0. 200) sieve on the stability, drainage, and frost susceptibility of aggregate base courses.

**5.13.** Aggregates from three sources having the properties shown in the table below were blended at a ratio of 60:30:10 by weight. Determine the properties of the aggregate blend.

| Property | Aggregate 1 | Aggregate 2 | Sand |
|---|---|---|---|
| Coarse aggregate angularity, crushed faces | 100 | 87 | N/A |
| Bulk specific gravity | 2.631 | 2.711 | 2.614 |
| Apparent specific gravity | 2.732 | 2.765 | 2.712 |

**5.14.** Three aggregates are blended by weight in the following percentages.

| | |
|---|---|
| 50% Crushed limestone | Bulk dry Sp. Gr. = 2.702 |
| 30% Blast furnace slag | Bulk dry Sp. Gr. = 2.331 |
| 20% Natural sand | Bulk dry Sp. Gr. = 2.609 |

What is the specific gravity of the blended aggregates?

**5.15.** Define the alkali-aggregate reactivity. What is its effect on portland cement concrete?

**5.16.** What are the typical deleterious substances in aggregates that affect portland cement concrete? Discuss these effects.

**REFERENCES**

Arizona Department of Transportation. 1990. *Standard specifications for roads and bridge construction.* Phoenix, AZ: Arizona Department of Transportation.

The Asphalt Institute. 1995. *Mix design methods for asphalt concrete and other hot-mix types.* 6th ed. Manual Series No. 2 (MS-2). Lexington, KY: The Asphalt Institute.

Barksdale, R. D., ed. 1991. *Aggregate handbook.* Washington, DC: National Stone Association.

Federal Highway Administration. 1988. *Asphalt concrete mix design and field control.* Technical Advisory T 5040.27. Washington, DC: Federal Highway Administration.

Goetz, W. H. and L. E. Wood. 1960. *Bituminous materials and mixtures.* In *Highway engineering handbook,* Section 18. New York: McGraw-Hill.

McGennis, R. B., et al. 1995. *Background of Superpave asphalt mixture design and analysis.* Publication no. FHWA-SA-95-003. Washington, DC: Federal Highway Administration.

Meininger, R. C. and F. P. Nichols. 1990. *Highway materials engineering, aggregates and unbound bases.* Publication no. FHWA-HI-90-007, NHI Course No. 13123. Washington, DC: Federal Highway Administration.

# 6 Portland Cement

Portland cement concrete is the most widely used manufactured construction material in the world. The importance of concrete in our daily lives cannot be overstated. It is used in structures such as buildings, bridges, tunnels, dams, factories, pavements, and playgrounds. Portland cement concrete consists of portland cement, aggregates, water, air voids, and, in many cases, admixtures. This chapter covers the topics of portland cement, mixing water, and admixtures; Chapter 7 describes portland cement concrete.

There are many types of concrete based on different cements. However, portland cement concrete is so prevalent that unless otherwise identified, the term concrete is always assumed to mean portland cement concrete. Portland cement was patented by Joseph Aspdin in 1824 and was named after the limestone cliffs on the Isle of Portland in England (Kosmatka and Panarese 1988).

Portland cement is an instant glue (just add water) that bonds aggregates together to make portland cement concrete. Materials specialists concerned with the selection, specification, and quality control of civil engineering projects should understand the production, chemical composition, hydration rates, and physical properties of portland cement.

## Portland Cement Production

Production of portland cement starts with two basic raw ingredients, a calcareous material and an argillaceous material. The calcareous material is a calcium oxide such as limestone, chalk, or oyster shells. The argillaceous material is a combination of silica and alumina that can be obtained from clay, shale, and blast furnace slag. As shown in Figure 6.1, these materials are crushed then stored in silos. The raw materials, in the

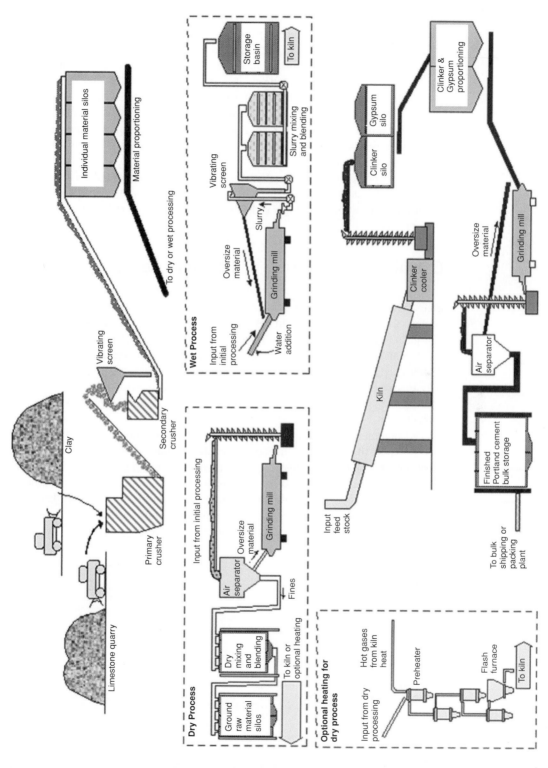

**FIGURE 6.1** Steps in the manufacture of Portland cement.

desired proportions, are passed through a grinding mill using either a wet or dry process. Modern dry process cement plants use a heat recovery cycle to preheat the ground material, or feed stock, with the exhaust gas from the kiln. In addition, some plants use a flash furnace to further heat the feed stock. Both the preheater and flash furnace improve the energy efficiency of cement production. The ground material is stored until it can be sent to the kiln. In the kiln the raw materials are melted at temperatures of 1400°C to 1650°C (2500°F to 3000°F) changing the raw materials into cement *clinker*. The clinker is cooled. A small amount of gypsum is added during this operation to regulate the set time of the cement, and the clinker and gypsum are pulverized into a fine powder.

The finished product may be stored and transported in either bulk or sacks. In the United States, a standard *sack* of cement is 94 lb, which is approximately equal to 1 ft$^3$ of loose cement when freshly packed. The cement can be stored for long periods of time, provided it is kept dry.

## Chemical Composition of Portland Cement

The raw materials used to manufacture portland cement are lime, silica, alumina, and iron oxide. These raw materials interact in the kiln, forming complex chemical compounds. *Calcination* in the kiln restructures the molecular composition producing four main compounds, as shown in Table 6.1.

$C_3S$ and $C_2S$, when hydrated, provide the desired characteristics of the concrete. Alumina and iron, which produce $C_3A$ and $C_4AF$, are included with the other raw materials to reduce the temperature required to produce $C_3S$ from 2000°C to 1350°C (3500°F to 2500°F). This saves energy and reduces the cost of producing the portland cement.

In addition to these main compounds, there are minor compounds, such as magnesium oxide, titanium oxide, manganese oxide, sodium oxide, and potassium oxide. These minor compounds represent a few percent by weight of cement. The term minor compounds refers to their quantity and not to their importance. In fact, two of the minor compounds, sodium oxide ($Na_2O$) and potassium oxide ($K_2O$), are known as alkalis. These alkalis react with some aggregates causing disintegration of concrete and affecting the rate of strength development, as discussed in Chapter 5.

**TABLE 6.1**   Main Compounds of Portland Cement

| Compound | Chemical Formula | Common Formula* | Usual Range, wt. % |
|---|---|---|---|
| Tricalcium silicate | 3CaO SiO$_2$ | C$_3$S | 45–60 |
| Dicalcium silicate | 2CaO SiO$_2$ | C$_2$S | 15–30 |
| Tricalcium aluminate | 3CaO Al$_2$O$_3$ | C$_3$A | 6–12 |
| Tetracalcium aluminoferrite | 4CaO Al$_2$O$_3$ Fe$_2$O$_3$ | C$_4$AF | 6–8 |

*The cement industry commonly uses shorthand notation for chemical formulas: C = calcium oxide, S = silicon dioxide, A = aluminum oxide, and F = iron oxide.

# Fineness of Portland Cement

Fineness of cement particles is an important property that must be carefully controlled. Since hydration starts at the surface of cement particles, the finer the cement particles, the larger the surface area and the faster the hydration. Therefore, finer material results in faster strength development and a greater initial heat of hydration. Increasing fineness beyond the requirements for a type of cement increases production costs and can be detrimental to the quality of the concrete.

The maximum size of the cement particles is 0.09 mm (0.0035 in.), 85% to 95% of the particles are smaller than 0.045 mm (0.0018 in.), and the average diameter is 0.01 mm (0.0004 in.). (For reference, a number 200 sieve passes material smaller than 0.075 mm.) A kilogram of portland cement has approximately 7 trillion particles with a total surface area of about 300 $m^2$ to 400 $m^2$ (1500 $ft^2$ to 2000 $ft^2$ per pound). The total surface area per unit weight is a function of the size of the particles and is more readily measured. Thus particle size specifications are defined in terms of the surface area per unit weight.

**FIGURE 6.2** Blaine air permeability apparatus.

Fineness of cement is usually measured indirectly by measuring the surface area with the Blaine air permeability apparatus (ASTM C204) or the Wagner turbidimeter apparatus (ASTM C115). In the Blaine test (Figure 6.2) the surface area of the cement particles in $cm^2/g$ is determined by measuring the air permeability of a cement sample and relating it to the air permeability of a standard material. The Wagner turbidimeter determines the surface area by measuring the rate of sedimentation of cement suspended in kerosene. The finer the cement particles, the slower the sedimentation. Both the Blaine and Wagner tests are indirect measurements of surface area and use somewhat different measurement principles. Therefore, tests on a single sample of cement will produce different results. Fineness can also be measured by determining the percent passing the 0.045 mm sieve (No. 325) (ASTM C430).

# Specific Gravity of Portland Cement

The specific gravity of cement is needed for mixture proportioning calculations. The specific gravity of portland cement (without voids between particles) is about 3.15 and can be determined according to ASTM C188. The density of the bulk cement (including voids between particles) varies considerably, depending on how it is handled and stored. For example, vibration during transportation of bulk cement consolidates the cement and increases its bulk density. Thus, cement quantities are specified and measured by weight rather than volume.

# Hydration of Portland Cement

Hydration is the chemical reaction between the cement particles and water. The features of this reaction are the change in matter, the change in energy level, and the rate of reaction. The primary chemical reactions are shown in Table 6.2. Since portland cement is composed of several compounds, many reactions are occurring concurrently.

**TABLE 6.2**   Primary Chemical Reactions During Cement Hydration

| | | | | | | | |
|---|---|---|---|---|---|---|---|
| $2(3CaO \cdot SiO_2)$ Tricalcium silicate | $+$ | $6H_2O$ Water | | | $=$ | $3CaO \cdot 2SiO_2 \cdot 3H_2O$ Calcium silicate hydrates | $+$ $3Ca(OH)_2$ Calcium hydroxide |
| $2(2CaO \cdot SiO_2)$ Dicalcium silicate | $+$ | $4H_2O$ Water | | | $=$ | $3CaO \cdot 2SiO_2 \cdot 3H_2O$ Calcium silicate hydrates | $+$ $Ca(OH)_2$ Calcium hydroxide |
| $3CaO \cdot Al_2O_3$ Tricalcium aluminate | $+$ | $12H_2O$ Water | $+$ | $Ca(OH)_2$ Calcium hydroxide | $=$ | $3CaO \cdot Al_2O_3 \cdot Ca(OH)_2 \cdot 12H_2O$ Calcium aluminate hydrate | |
| $4CaO \cdot Al_2O_3 \cdot Fe_2O_3$ Tetracalcium aluminoferrite | $+$ | $10H_2O$ Water | $+$ | $2Ca(OH)_2$ Calcium hydroxide | $=$ | $6CaO \cdot Al_2O_3 \cdot Fe_2O_3 \cdot 12H_2O$ Calcium aluminoferrite hydrate | |
| $3CaO \cdot Al_2O_3$ Tricalcium aluminate | $+$ | $10H_2O$ Water | $+$ | $CaSO_4 \cdot 2H_2O$ Gypsum | $=$ | $3CaO \cdot Al_2O_3 \cdot CaSO_4 \cdot 12H_2O$ Calcium monosulfoaluminate hydrate | |

The hydration process occurs through two mechanisms, through-solution and topochemical. The through-solution process involves (Mehta and Monteiro 1993)

**1.** dissolution of anhydrous compounds into constituents
**2.** formation of hydrates in solution
**3.** precipitation of hydrates from the supersaturated solution

The through-solution mechanism dominates the early stages of hydration. Topochemical hydration is a solid-state chemical reaction occurring at the surface of the cement particles.

The aluminates hydrate much faster than the silicates. The reaction of tricalcium aluminate with water is immediate and liberates large amounts of heat. Gypsum is used to slow down the rate of aluminate hydration. The gypsum goes into the solution quickly, producing sulfate ions that suppress the solubility of the aluminates. The balance of aluminate to sulfate determines the rate of *setting* (solidification). Cement paste that sets at a normal rate requires low concentrations of both aluminate and sulfate ions. The cement paste will remain workable for about 45 minutes; thereafter, the paste starts to stiffen as crystals displace the water in the pores. The paste begins to solidify within 2 to 4 hours after the water is added to the cement. If there is an excess of both aluminate and sulfate ions, the workability stage may only last for 10 minutes and setting may occur in 1 to 2 hours. If the availability of aluminate ions is high, and sulfates are low, either a *quick set* (10 to 45 minutes) or *flash set* (less than 10 minutes) can occur. Finally, if the aluminate ions availability is low and the sulfate ions availability is high, the gypsum can recrystalize in the pores within 10 minutes, producing a flash set. Flash set is associated with large heat evolution and poor ultimate strength (Mehta and Monteiro 1993).

Calcium silicates combine with water to form calcium-silicate-hydrate, C-S-H. The crystals begin to form a few hours after the water and cement are mixed and can be continuously developed as long as there are unreacted cement particles and free water. C-S-H is not a well-defined compound. The calcium to silicate ratio varies between 1.5 and 2.0 and the structurally combined water content is more variable.

As shown in Table 6.2, the silicate hydration produces both C-S-H and calcium hydroxide. Complete hydration of $C_3S$ produces 61% C-S-H and 39% calcium hydroxide; hydration of $C_2S$ results in 82% C-S-H and 18% calcium hydroxide. Since C-S-H is what makes the hydrated cement paste strong, the ultimate strength of the concrete is enhanced by increasing the content of $C_2S$ relative to the amount of $C_3S$. Furthermore, calcium hydroxide is susceptible to attack by sulfate and acidic waters. Increasing the proportion of $C_2S$ relative to $C_3S$ reduces the quantity of calcium hydroxide and, therefore, improves the durability of the concrete.

$C_3S$ hydrates more rapidly than $C_2S$, contributing to the final set time and early strength gain of the cement paste. The rate of hydration is accelerated by sulfate ions in solution. Thus a secondary effect of the addition of gypsum to cement is to increase the rate of development of the C-S-H.

## Structure Development in Cement Paste

The sequential development of the structure in a cement paste is summarized in Figure 6.3. The process begins immediately after water is added to the cement [Figure 6.3(a)]. In less than 10 minutes, the water becomes highly alkaline. As the cement particles hydrate, the volume of the cement particle reduces, increasing the space between the particles. During the early stages of hydration, weak bonds can form, particularly from the hydrated $C_3A$ [Figure 6.3(b)]. Further hydration stiffens the mix and begins locking the structure of the material in place [Figure 6.3(c)]. Final set occurs when the C-S-H phase has developed a rigid structure and all components of the paste lock into place and the spacing between grains increases as the grains are consumed by hydration [Figure 6.3(d)]. The cement paste continues hardening and gains strength as hydration continues [Figure 6.3(e)]. Hardening develops rapidly at early ages and continues, as long as unhydrated cement particles and free water exist. However, the rate of hardening decreases with time.

## Evaluation of Hydration Progress

Several methods are available to evaluate the progress of cement hydration in hardened concrete. These include measuring (Neville 1981)

1. the heat of hydration,
2. the amount of calcium hydroxide in the paste developed due to hydration,
3. the specific gravity of the paste,
4. the amount of chemically combined water,
5. the amount of unhydrated cement paste using X ray quantitative analysis, or
6. the strength of the hydrated paste, an indirect measurement.

# Voids in Hydrated Cement

Due to the random growth of the crystals and the different types of crystals, voids are left in the paste structure as the cement hydrates. Concrete strength, durability, and volume stability are greatly influenced by voids. Two types of voids are formed during hydration, the interlayer hydration space and capillary voids.

The C-S-H phase is initially formed. $C_3A$ forms a gel fastest.

The volume of cement grain decreases as a gel forms at the surface. Cement grains are still able to move independently, but as hydration grows, weak interlocking begins. Part of the cement is in a thixotropic state; vibration can break the weak bonds.

The initial set occurs with the development of a weak skeleton in which cement grains are held in place.

Final set occurs as the skeleton becomes rigid, cement particles are locked in place, and spacing between cement grains increases due to the volume reduction of the grains.

Spaces between the cement grains are filled with hydration products as cement paste develops strength and durability.

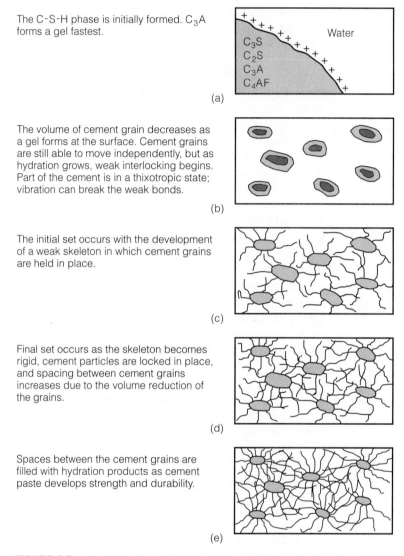

**FIGURE 6.3** Development of structure in the cement paste: (a) initial C-S-H phase, (b) forming of gels, (c) initial set—development of weak skeleton, (d) final set—development of rigid skeleton, (e) hardening. (Hover and Phillco 1990)

*Interlayer hydration space* occurs between the layers in the C-S-H. The space thickness is between 0.5 nm and 2.5 nm, which is too small to affect the strength. It can, however, contribute 28% to the porosity of the paste. Water in the interparticle space is strongly held by hydrogen bonds, but can be removed when humidity is less than 11%, resulting in considerable shrinkage.

*Capillary voids* are the result of the hydrated cement paste having a lower bulk specific gravity than the cement particles. The amount and size of capillary voids depends on the initial separation of the cement particles, which is largely controlled by the ratio of water to cement paste. For a highly hydrated cement paste where a minimum amount of water was used, the capillary voids will be on the order of 10 nm to 50 nm. A poorly hydrated cement produced with excess water can have capillary voids on the order of 3 $\mu$m to 5 $\mu$m. Capillary voids greater than 50 nm decrease strength and increase permeability. Removal of water from capillary voids greater than 50 nm does not cause shrinkage, whereas removal of water from the smaller voids causes shrinkage.

In addition to the interlayer space and capillary voids, air can be trapped in the cement paste during mixing. The *trapped air* reduces strength and increases permeability. However, well-distributed, minute air bubbles can greatly increase the durability of the cement paste. Hence, as described later in this chapter, admixtures are widely used to *entrain air* into the cement paste.

## Properties of Hydrated Cement

The proper hyudration of portland cement is a fundamental quality control issue for cement producers. While specifications control the quality of the portland cement, they do not guarantee the quality of the concrete made with the cement. Mix design, quality control, and the characteristics of the mixing water and aggregates also influence the quality of the concrete. Properties of the hydrated cement are evaluated with either *cement paste* (water and cement) or *mortar* (paste and sand).

### Setting

*Setting* refers to the stiffening of the cement paste or the change from a plastic state to a solid state. Although with setting comes some strength, it should be distinguished from *hardening*, which refers to the strength gain in a set cement paste. Setting is usually described by two levels: initial set and final set. The definitions of the initial and final sets are arbitrary, based on measurements by either the Vicat apparatus (ASTM C191) or the Gillmore needles (ASTM C266).

The Vicat test (Figure 6.4) requires that a sample of cement paste be prepared using the amount of water required for normal consistency according to a specified procedure. The 1 mm (0.04 in.) diameter needle is allowed to penetrate the paste for 30 seconds and the amount of penetration is measured. The penetration process is repeated every 15 minutes (every 10 minutes for Type III cement) until a penetration of 25 mm (1 in.) or less is obtained. By interpolation, the time when a penetration of 25 mm occurs is determined and recorded as the initial set time. The final set time is when the needle does not penetrate visibly into the paste.

Similar to the Vicat test, the Gillmore test (Figure 6.5) requires that a sample of cement paste of normal consistency is prepared. A pat with a flat top is molded and the initial Gillmore needle is applied lightly to its surface. The application process is repeated until the pat bears the force of the needle without appreciable indentation, and the elapsed time is recorded as the initial set time. This process is then repeated with the final Gillmore needle and the final set time is recorded. Due to the differences in the test apparatuses and procedures, the Vicat and Gillmore tests produce different results for a single sample of material.

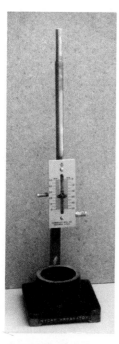

**FIGURE 6.4** Vicat set time apparatus.

**FIGURE 6.5** Gillmore set time apparatus.

The initial set time must allow for handling and placing the concrete before stiffening. The maximum final set time is specified and measured to ensure normal hydration. During cement manufacturing, gypsum is added to regulate the setting time. Other factors that affect the set time include the fineness of the cement, the water-cement ratio, and the use of admixtures.

If the cement is exposed to humidity during storage, a *false set* might occur in which the cement stiffens within a few minutes of being mixed, without the evolution of much heat. To resolve this problem, the cement paste can be vigorously remixed, without adding water, in order to restore plasticity of the paste and to allow it to set in a normal manner without losing strength. A false set is different than a quick set and a flash set mentioned earlier; a false set can be remedied by remixing, whereas a quick set and a flash set cannot be remedied.

### Soundness

Soundness of the cement paste refers to its ability to retain its volume after setting. Expansion after setting, caused by delayed or slow hydration or other reactions, could result if the cement is unsound. The autoclave expansion test (Figure 6.6) (ASTM C151) is used to check the soundness of the cement paste. In this test, cement paste bars are subjected to heat and high pressure, and the amount of expansion is measured. ASTM C150 limits autoclave expansion to 0.8%.

### Compressive Strength

Compressive strength of mortar is measured by preparing 50-mm (2 in.) cubes and subjecting them to compression according to ASTM C109. The mortar is prepared with cement, water, and standard sand (ASTM C778). Minimum compressive strength values are specified by ASTM C150 for different cement types at different ages. The compressive strength of mortar cubes is proportional to the compressive strength of

**FIGURE 6.6** Cement autoclave expansion apparatus.

concrete cylinders. However, the compressive strength of the concrete cannot be predicted accurately from mortar cube strength since the concrete strength is also affected by the aggregate characteristics, the concrete mixing, and the construction procedures.

## Water-Cement Ratio

In 1918 Abrams found that the ratio of the weight of water to the weight of cement, *water-cement ratio,* influences all the desirable qualities of concrete. For fully compacted concrete made with sound and clean aggregates, strength and other desirable properties are improved by reducing the weight of water used per unit weight of cement. This concept is frequently identified as Abrams' law.

Hydration requires approximately 0.22 kg to 0.25 kg of water per 1 kg of cement. Concrete mixes generally require excess moisture, beyond the hydration needs, for workability. Excess water causes the development of capillary voids in the concrete. These voids increase the porosity and permeability of the concrete and reduce strength. Figure 6.7 shows a typical relationship between the water-cement ratio and compressive strength. It is easy to see that increasing the water-cement ratio decreases the compressive strength of the concrete for various curing times. A low

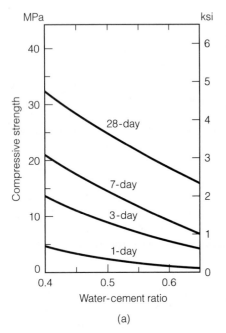
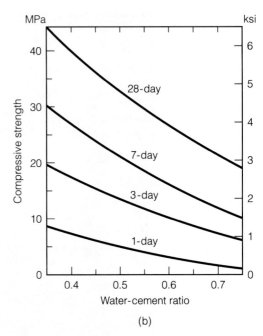

**FIGURE 6.7** Typical age-strength relationships of concrete based on compression tests of 0.15 × 0.30 m (6 × 12 in.) cylinders, using Type I portland cement and moist-curing at 21°C (70°F): (a) air-entrained concrete, (b) non–air-entrained concrete. (Kosmatka and Panarese 1988)

water-cement ratio also increases resistance to weathering, provides a good bond between successive concrete layers, provides a good bond between concrete and steel reinforcement, and limits volume change due to wetting and drying. Air-entrained concrete includes an air entraining agent, an admixture, which is used to increase the concrete's resistance to freezing and thawing, as discussed later in this chapter. Curing maintains satisfactory moisture content and temperature in the hardened concrete for a definite period of time to allow for hydration (see Chapter 7).

## Types of Portland Cement

Different concrete applications require cements with different properties. Some applications require rapid strength gain to expedite the construction process. Other applications require a low heat of hydration to control volume change and associated shrinkage cracking. In some cases, the concrete is exposed to sulfates ($SO_4$), which can deteriorate normal portland cement concrete. Fortunately, these situations can be accommodated by varying the raw materials used to produce the cement, thereby altering the ratios of the four main compounds of portland cement listed in Table 6.1. The rate of hydration can also be altered by varying the fineness of the cement produced in the final grinding mill. Cement is classified to five standard types, as well as other special types.

## Standard Portland Cement Types

Table 6.3 describes the five standard types of portland cement (Types I through V) specified by ASTM C150. In addition to these five types, air entrainers can be added to Type I, II, and III cements during manufacturing, producing Types IA, IIA, and IIIA, which produce better resistance to freeze and thaw than do non–air-entrained cements. The use of air-entrained cements (Types IA, IIA, and IIIA) has diminished, due to improved availability and reliability of the air entrainer admixtures that can be added during concrete mixing. The uses and effects of air entrainers are described in the section on admixtures. The ASTM specifications of the standard cement types are shown in Table 6.4.

The allowable maximum compound compositions are given in Table 6.5, along with the required Blaine fineness (controls particle size). Note the chemical composition of Type I and III cements are almost identical; the primary difference is the much greater surface area of the Type III cement. The $C_3A$ content of Type II and V cements are lower than Type I to improve sulfate resistance. $C_3S$ and $C_3A$ are limited in Type IV cement to limit the rate of hydration.

The existence of an ASTM specification for a type of cement does not assure its availability. Type I cement is widely available and represents most of the United States' cement production. Type II is the second most available type. Cements can be manufactured that meet all the requirements of both Types I and II; these are labeled Type I/II. Type III cement represents about 4% of U.S. production. Due to the stricter grinding requirements for Type III, it is more expensive than Type I. The strength gain of Type I cement can be accelerated by increasing the cement content per unit volume of concrete, so the selection of Type III becomes a question of economics and availability. Type IV can be manufactured on demand. As discussed later, adding fly ash to Type I or II portland cement reduces the heat of hydration, producing the benefits of Type IV, but at a lower cost. Type V cement is produced only in locations with a severe sulfate problem.

**TABLE 6.3** Types and Applications of Standard Portland Cement

| Type | Name | Application |
|------|------|-------------|
| I | Normal | General concrete work when the special properties of other types are not needed. Suitable for floors, reinforced concrete structures, pavements, etc. |
| II | Moderate sulfate resistance | Protection against moderate sulfate exposure, 0.1%–0.2% weight water-soluble sulfate in soil or 150 ppm–1500 ppm sulfate in water (seawater). Can be specified with a moderate heat of hydration making it suitable for large piers, heavy abutments, and retaining walls. The moderate heat of hydration is also beneficial when placing concrete in warm weather. |
| III | High early strength | Used for fast track construction when forms need to be removed as soon as possible or structure needs to be put in service as soon as possible. In cold weather, reduces time required for controlled curing. |
| IV | Low heat of hydration | Used when mass of structure, requires careful control of the heat of hydration, such as in large dams. |
| V | High sulfate resistance | Protection from severe sulfate exposure, 0.2%–2.0% weight water-soluble sulfate in soils or 1500 ppm–10,800 ppm sulfate in water. |

**TABLE 6.4**   Specifications of Standard Properties of Portland Cement (ASTM C150)*

| | I | IA | II | IIA | III | IIIA | IV | V |
|---|---|---|---|---|---|---|---|---|
| | | | | *Cement Type* | | | | |
| Air content of mortar,[†] vol. % | | | | | | | | |
| max. | 12 | 22 | 12 | 22 | 12 | 22 | 12 | 12 |
| min. | — | 16 | — | 16 | — | 16 | — | — |
| Fineness,[‡] specific surface, m²/kg (alternative methods) | | | | | | | | |
| Turbidimeter test, min. | 160 | 160 | 160 | 160 | — | — | 160 | 160 |
| Air permeability test, min. | 280 | 280 | 280 | 280 | — | — | 280 | 280 |
| Autoclave expansion, max., % | 0.80 | 0.80 | 0.80 | 0.80 | 0.80 | 0.80 | 0.80 | 0.80 |
| Minimum compressive strength , psi (MPa)** | | | | | | | | |
| 1 day | — | — | — | — | 1800 | 1450 | — | — |
| | — | — | — | — | (12.4) | (10.0) | — | — |
| 3 days | 1800 | 1450 | 1500 | 1200 | 3500 | 2800 | — | 1200 |
| | (12.4) | (10.0) | (10.3)** | (8.3) | (24.1) | (19.3) | — | (8.3) |
| | | | 1000[††] | 800[††] | | | | |
| | | | (6.9)[††] | (5.5)[††] | | | | |
| 7 days | 2800 | 2250 | 2500 | 2000 | — | — | 1000 | 2200 |
| | (19.3) | (15.5) | (17.2) | (13.8) | — | — | (6.9) | (15.2) |
| | | | 1700[††] | 1350[††] | | | | |
| | | | (11.7)[††] | (9.3)[††] | | | | |
| 28 days | — | — | — | — | — | — | 2500 | 3000 |
| | — | — | — | — | — | — | (17.2) | (20.7) |
| Time of setting (alternative methods):[‡‡] | | | | | | | | |
| Gillmore test: | | | | | | | | |
| Initial set, minutes, not less than | 60 | 60 | 60 | 60 | 60 | 60 | 60 | 60 |
| Final set, minutes, not more than | 600 | 600 | 600 | 600 | 600 | 600 | 600 | 600 |
| Vicat test: | | | | | | | | |
| Initial set, minutes, not less than | 45 | 45 | 45 | 45 | 45 | 45 | 45 | 45 |
| Final set, minutes, not more than | 375 | 375 | 375 | 375 | 375 | 375 | 375 | 375 |

*Copyright ASTM. Reprinted with permission.
[†]Compliance with the requirements of this specification does not necessarily ensure that the desired air content will be obtained in concrete.
[‡]Either of the two alternative fineness methods may be used at the option of the testing laboratory. However, when the sample fails to meet the requirements of the air permeability test, the turbidimeter test shall be used and the requirements in this table for the turbidimetric method shall govern.
**The strength at any specified test age shall be not less than that attained at any previous specified test age.
[††]When the optional heat of hydration or the chemical limit on the sum of the tricalcium silicate and tricalcium aluminate is specified.
[‡‡]The purchaser should specify the type of setting time test required. In case the purchaser does not so specify, the requirements of the Vicat test only shall govern.

**TABLE 6.5**   Required Chemical Composition and Fineness for Portland Cement (ASTM C150)*

| Type | $C_3S$ | $C_2S$ | $C_3A$ | $C_4AF$ | Blaine Fineness, m²/kg |
|---|---|---|---|---|---|
| | *Maximum Compound Composition, %* | | | | |
| I | 55 | 19 | 10 | 7 | 370 |
| II | 51 | 24 | 6 | 11 | 370 |
| III | 56 | 19 | 10 | 7 | 540 |
| IV | 28 | 49 | 4 | 12 | 380 |
| V | 38 | 43 | 4 | 9 | 380 |

*Copyright ASTM. Reprinted with permission.

## Other Cement Types

Other than the five standard types of portland cement, several hydraulic cements are manufactured in the United States, including:

> white portland cement
> blended hydraulic cements
> > portland blast-furnace slag cement (Type IS)
> > portland-pozzolan cement (Type IP and Type P)
> > slag cement (Type S)
> > pozzolan-modified portland cement (Type I(PM))
> > slag modified portland cement (Type I(SM))
> masonry cements
> expansive cements (Type K)
> specialty cements

In general, these cements have limited applications. Civil and construction engineers should be aware of their existence, but should study them further before using them.

## Mixing Water

Any potable water is suitable for making concrete. However, some nonpotable water may also be suitable. Frequently, material suppliers will use unprocessed surface or well water if it can be obtained at a lower cost than processed water. However, impurities in the mixing water can affect concrete set time, strength, and long-term durability. In addition, chloride ions in the mixing water can accelerate corrosion of reinforcing steel.

The acceptance criteria for questionable water are specified in ASTM C94. After 7 days, the compressive strength of mortar cubes made with the questionable water should not be less than 90% of the strength of cubes made with potable or distilled water (ASTM C109). Also, the set time of cement paste made with the questionable water should, as measured using the Vicat apparatus (ASTM C191), not be 1 hour less than or 1-1/2 hours more than the set time of paste made with potable or distilled water.

**TABLE 6.6**  Chemical Limits for Wash Water Used as Mixing Water (ASTM C94)[*]

| Chemical | Maximum Concentration, ppm | Test Method |
|---|---|---|
| Chloride, as Cl | | ASTM D512 |
|    Prestressed concrete or concrete in bridge decks | 500 | |
|    Other reinforced concrete in moist environments or containing aluminum embedments or dissimilar metals or with stay-in-place galvanized metal forms | 1000 | |
| Sulfate, as $SO_4$ | 3000 | ASTM D516 |
| Alkali, as ($Na_2O + 0.658K_2O$) | 600 | |
| Total solids | 50,000 | AASHTO T26 |

*Copyright ASTM. Reprinted with permission.

**TABLE 6.7**  Summary of Effects of Water Impurities on Concrete Quality

| Impurity | Effect |
|---|---|
| Alkali carbonate and bicarbonate | Can retard or accelerate strength test setting and 28-day strength when total dissolved salts exceed 1000 ppm. Can also aggravate alkali-aggregate reaction. |
| Chloride | Corrosion of reinforcing steel is primary concern. Chloride can enter the mix through admixtures, aggregates, cement, and mixing water, so limits are expressed in terms of total free chloride ions. ACI limits water-soluble ion content based on type of reinforcement:<br>  Prestressed concrete                                     0.06%<br>  Reinforced concrete exposed to chloride in service       0.15%<br>  Reinforced concrete protected from moisture              1.00%<br>  Other reinforced concrete                                0.30% |
| Sulfate | Can cause expansive reaction and deterioration |
| Other salts | Not harmful when concentrations limited to:<br>  Calcium bicarbonate       400 ppm<br>  Magnesium bicarbonate     400 ppm<br>  Magnesium sulfate       25,000 ppm<br>  Magnesium chloride      40,000 ppm<br>  Iron salts              40,000 ppm<br>  Sodium sulfide             100 ppm |
| Seawater | Do not use for reinforced concrete. Can accelerate strength gain but reduces ultimate strength. Can aggravate alkali reactions. |
| Acid water | Limit concentrations of hydrochloric, sulfuric, and other inorganic acids to less than 10,000 ppm. |
| Alkaline water | Possible increase in alkali-aggregate reactivity. Sodium hydroxide may introduce quick set at concentrations higher than 0.5%, strength may be lowered. Potassium hydroxide in concentrations over 1.2% may reduce 28-day strength of some cements. |
| Sanitary sewage water | Dilute to reduce organic matter to less than 20 ppm. |
| Sugar | Concentrations over 500 ppm can retard setting time and alter strength development. Sucrose in the range of 0.03% to approximately 0.15% usually retards setting; concentrations over 0.25% by weight of cement can accelerate strength gain but substantially reduce 28-day strength. |
| Oils | Mineral oil (petroleum) in excess of 2.5% by weight of mix may reduce strength by 20%. |
| Algae | Can reduce hydration and entrain air. Do not use water containing algae. |

Other adverse effects caused by excessive impurities in mixing water include efflorescence (white stains forming on the concrete surface due to the formation of calcium carbonate), staining, corrosion of reinforcing steel, volume instability, and reduced durability. Therefore, in addition to the compressive strength and set time, there are maximum chemical limits that should not be exceeded in the mixing water, as shown in Table 6.6. Several tests are available to evaluate the chemical impurities of questionable water. Over 100 different compounds and ions can exist in the mixing water and can affect concrete quality; the more important effects are described in the Table 6.7.

## Admixtures for Concrete

Admixtures are ingredients other than portland cement, water, and aggregates that may be added to concrete to impart a specific quality to either the plastic (fresh) mix or the hardened concrete (ASTM C494). Some admixtures are charged into the mix as solutions. In such cases the liquid should be considered part of the mixing water. If admixtures cannot be added in solution, they are either weighed or measured by volume as recommended by the manufacturer. Admixtures are classified by chemical and functional physical characteristics (Hewlett 1978). These are:

1. air entrainers
2. water reducers
3. high-range water reducers—superplasticizers
4. retarders
5. accelerators
6. fine minerals
7. specialty admixtures

The Portland Cement Association (PCA) identifies four major reasons for using admixtures (Kasmatka and Panarese 1988). These are

1. to reduce cost of concrete construction,
2. to achieve certain properties in concrete more effectively than by other means,
3. to ensure quality of concrete during the stages of mixing, transporting, placing, and curing in adverse weather conditions, and
4. to overcome certain emergencies during concrete operations.

## Air Entrainers

Air entrainers produce tiny air bubbles in the hardened concrete to provide space for water to expand upon freezing. As moisture within the concrete pore structure freezes, three mechanisms contribute to the development of internal stresses in the concrete:

1. Critical saturation—Upon freezing, water expands in volume by 9%. If the percent saturation exceeds 91.7%, the volume increase generates stress in the concrete.
2. Hydraulic pressure—Freezing water draws unfrozen water to it. The unfrozen water moving throughout the concrete pores generates stress depending on length of flow path, rate of freezing, permeability, and concentration of salt in pores.
3. Osmotic pressure—Water moves from the gel to capillaries to satisfy thermo-dynamic equilibrium and to equalize alkali concentrations. Voids permit water to flow from the interlayer hydration space and capillaries into the air voids where it has room to freeze without damaging the parts.

Internal stresses reduce the durability of hardened concrete, especially when cycles of freeze and thaw are repeated many times. The impact of each of these mechanisms is mitigated by providing a network of tiny air voids in the hardened concrete using air entrainers. In the late 1930s, the introduction of air entrainment in concrete represented a major advance in concrete technology. Air entrainment is recommended for all concrete exposed to freezing.

All concrete contains entrapped air voids, which have diameters of 1 mm or larger and which represent approximately 0.2% to 3% of the concrete volume. Entrained air voids have diameters that range from 0.01 mm to 1 mm, with the majority being less than 0.1 mm. The entrained air voids are not connected and have a total volume between 1% and 7.5% of the concrete volume. Concrete mixed for severe frost conditions should contain approximately 14 billion bubbles per cubic meter. Frost resistance improves with decreasing void size, and small voids reduce strength less than large ones. The fineness of air voids is measured by the specific surface index, equal to the total surface area of voids in a unit volume of paste. The specific surface index should exceed 23,600 $m^2/m^3$ (600 in.$^2$/in.$^3$) for frost resistance.

In addition to improving durability, air entrainment provides other important benefits to both freshly mixed and hardened concrete. Air entrainment improves concrete's resistance to several destructive factors, including freeze-thaw cycles, de-icers and salts, sulfates, and alkali-silica reactivity. Air entrainment also increases the workability of fresh concrete. Air entrainment decreases the strength of concrete, as shown in Figure 6.7; however, this effect can be reduced for moderate-strength concrete by lowering the water-cement ratio and increasing the cement factor. High strength is difficult to attain with air-entrained concrete.

Air entraining admixtures are available from several manufacturers and can be composed of a variety of materials, such as

> salts of wood resins (Vinsol resin)
> synthetic detergents
> salts of sulfonated lignin (by-product of paper production)
> salts of petroleum acids
> salts of proteinaceous material
> fatty and resinous acids
> alkylbenzene sulfonates
> salts of sulfonated hydrocarbons

Air entrainers are usually liquid and should meet the specifications of ASTM C260. The agents enhance air entrainment by lowering the surface tension of the mixing water. Anionic air-entrainers are hydrophobic (water hating). The negative charge of the agent is attracted to the positive charge of the cement particle. The hydrophobic agent forms tough, elastic, air-filled bubbles. Mixing disperses the air bubbles throughout the paste and the sand particles form a grid that holds the air bubbles in place. Other types of air-entrainers have different mechanisms but produce similar results.

## Water Reducers

Workability of fresh or plastic concrete requires more water than is needed for hydration. The excess water, beyond the hydration requirements, is detrimental to all desirable properties of hardened concrete. Thus water reducer admixtures were developed to minimize the amount of water required for workability. Water reducers increase the mobility of the cement particles in the plastic mix, allowing workability to be achieved at lower water contents. Hewlett (1978) demonstrated that water reducers can actually be used to accomplish three different objectives, as shown in Table 6.8.

> **1.** Adding a water reducer without altering the other quantities in the mix increases the slump, which is a measure of concrete consistency and an indicator of workability, as discussed in Chapter 7.
> **2.** The strength of the mix can be increased by using the water reducer by lowering the quantity of water and keeping the cement content constant.
> **3.** The cost of the mix, which is primarily determined by the amount of cement, can be reduced. In this case, the water reducer allows decreasing the amount

**TABLE 6.8**  Effects of Water Reducer

| | Cement Content, kg/m$^3$ | Water-Cement Ratio | Slump, mm | Compressive Strength, MPa | |
|---|---|---|---|---|---|
| | | | | 7-day | 28-day |
| Base mix | 300 | 0.62 | 50 | 25 | 37 |
| Improve consistency | 300 | 0.62 | 100 | 26 | 38 |
| Increase strength | 300 | 0.56 | 50 | 34 | 46 |
| Reduce cost | 270 | 0.62 | 50 | 25.5 | 37.5 |

of water. The amount of cement is then reduced to keep the water-cement ratio equal to the original mix. Thus the quality of the mix, as measured by compressive strength, is kept constant, although the amount of cement is decreased.

## Superplasticizers

Superplasticizers, or high-range water reducers, can either greatly increase the flow of the fresh concrete or reduce the amount of water required for a given consistency. For example, adding a superplasticizer to a concrete with a 75-mm (3 in.) slump can increase the slump to 230 mm (9 in.), or the original slump can be maintained by reducing the water content 12% to 30%. Reducing the amount of mixing water reduces the water-cement ratio, which in turn increases the strength of hardened concrete. In fact, the use of superplasticizers has resulted in a major breakthrough in the concrete industry; now, high-strength concrete that was not previously attainable can be produced. Superplasticizers can be used in the following cases:

1. a low water-cement ratio is beneficial (e.g., high-strength concrete, early-strength gain, and reduced porosity)
2. placing thin sections
3. placing concrete around tightly spaced reinforcing steel
4. placing cement underwater
5. placing concrete by pumping
6. consolidating the concrete is difficult

When superplasticizers are used, the fresh concrete stays workable for a short time, 30 min to 60 min, and is followed by rapid loss in workability. Thus, superplasticizers are usually added at the job site. The setting time varies with the type of agents, the amount used, and the interactions with other admixtures used in the concrete.

## Retarders

Some construction conditions require that the time between mixing and placing or finishing the concrete be increased. In such cases retarders can be used to delay the initial set of concrete. Retarders are used for several reasons, such as

1. offsetting the effect of hot weather,
2. allowing for unusual placement or long haul distances, and
3. providing time for special finishes (e.g., exposed aggregate).

Retarders can reduce the strength of concrete at early ages, 1 to 3 days. In addition, some retarders entrain air and improve workability. Other retarders increase the time required for the initial set but reduce the time between the initial and final set. The properties of retarders vary with the materials used in the mix and with job conditions. Thus the use and effect of retarders must be evaluated experimentally during the mix design process.

## Accelerators

Accelerators are used to develop early strength of concrete at a faster rate than that developed in normal concrete. The ultimate strength, however, of high early strength concrete is about the same is that of normal concrete. Accelerators are used to

1. reduce the amount of time before finishing operations begin,
2. reduce curing time,
3. increase rate of strength gain, and
4. plug leaks under hydraulic pressure efficiently.

The first three reasons are particularly applicable to concrete work placed during cold temperatures. The increased strength gained helps to protect the concrete from freezing and the rapid rate of hydration generates heat that can reduce the risk of freezing.

Calcium chloride, $CaCl_2$, is the most widely used accelerator (ASTM D98). Both initial and final set times are reduced with calcium chloride. The initial set time of 3 hours for a typical concrete can be reduced to 1.5 hours by adding an amount of calcium chloride equal to 1% of the cement weight; 2% reduces the initial set time to 1 hour. Typical final set times are 6 hours, 3 hours, and 2 hours for 0%, 1%, and 2% calcium chloride. Figure 6.8 shows that strength development is also affected by $CaCl_2$ for plain portland cement concrete (PCC) and portland cement concrete with 2% calcium chloride. Concrete with $CaCl_2$ develops higher early strength compared to plain concrete cured at the same temperature (Hewlett 1978).

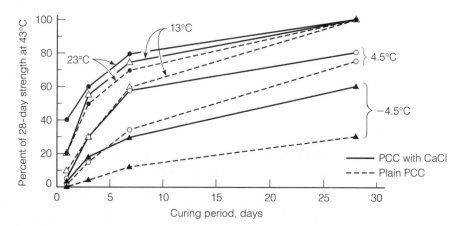

**FIGURE 6.8** Effect of $CaCl_2$ on strength development at different curing temperatures.

The PCA recommends against using calcium chloride when

1. concrete is prestressed;
2. concrete contains embedded aluminum such as conduits, especially if the aluminum is in contact with steel;
3. concrete is subjected to alkali-aggregate reaction;
4. concrete is in contact with water of soils containing sulfates;
5. concrete is placed during hot weather; and
6. mass applications of concrete.

The American Concrete Institute (ACI) recommends the following limits to water-soluble chloride-ion content based on percent weight of cement (American Concrete Institute 1986).

| Member Type | Chloride Ion Limit, % |
|---|---|
| Prestressed concrete | 0.06 |
| Reinforced concrete subjected to chloride in service | 0.15 |
| Reinforced concrete protected from moisture | 1.00 |
| Other reinforced concrete | 0.30 |

Several alternatives to the use of calcium chloride are available. These include

1. using high early strength (Type III) cement,
2. increasing cement content,
3. curing at higher temperatures, and
4. using non–calcium chloride accelerators such as triethanolamine, sodium thiocyanate, calcium formate, calcium nitrite, or calcium nitrate.

## Fine Minerals

Powdered or pulverized mineral admixtures are siliceous materials added to concrete in relatively large amounts (20% to 100% of the cement weight) to improve the characteristics of both plastic and hardened concrete. Mineral admixtures are frequently the waste from a production process. Use of these by-products provides an environmental benefit by reducing waste disposal. The PCA classifies mineral admixtures by chemical and physical characteristics as: (1) cementitious, (2) pozzolanic, (3) pozzolanic-cementitious, and (4) nominally inert materials (Kosmatka and Panarese 1988).

Cementitious minerals, such as blast furnace slag, natural cement, and hydraulic hydrated lime, have hydraulic cementing properties. Iron blast-furnace slag primarily consists of silicates and aluminosilicates of calcium. The molten slag is quenched in water and ground to less than 0.045 mm with a Blaine fineness of 400 $m^2$/kg to 600 $m^2$/kg. The rough-angular slag hydrates and sets in the presence of NaOH and CaOH (both produced by hydrating portland cement).

A pozzolan is a siliceous and aluminous material which, in itself, possesses little or no cementitious value but will, in finely divided form and in the presence of moisture, react chemically with calcium hydroxide at ordinary temperatures to form compounds possessing cementitious properties (ASTM C595). Naturally occurring pozzolans, such as fine volcanic ash, combined with burned lime, were used about 2000 years ago for building construction and pozzolan continues to be used today. As shown in Table 6.2, calcium hydroxide is one of the products generated by the hydration of $C_3S$ and $C_2S$. In fact, up to 15% of the weight of the portland cement is hydrated lime. Adding a pozzolan to portland cement generates an opportunity to convert this free lime to a cementitious material.

Fly ash and natural pozzolans are classified (ASTM C618) as follows:

*Class N*—Raw or calcined natural pozzolans, including diatomaceous earths, opaline cherts and shales, ruffs and volcanic ashes or pumicites, and some calcined clays and shales.
*Class F*—Fly ash with pozzolan properties.
*Class C*—Fly ash with pozzolan and cementitious properties.

While natural and other synthetic pozzolans exist, fly ash is the most commonly used pozzolan in civil engineering structures and is, therefore, the focus of this discussion.

Combusting pulverized coal in an electric power plant burns off the carbon and most volatile materials. However, depending on the source and type of coal, a significant amount of impurities pass through the combustion chamber. The carbon contents of common coals are the following:

| Coal Type | Carbon Content, % |
|---|---|
| Lignite | 70 |
| Subbituminous | 75 |
| Bituminous | 85 |
| Anthracite | 94 |
| Graphite | 100 |

The noncarbon percentages are impurities (e.g., clay, feldspar, quartz, and shale), which fuse as they pass through the combustion chamber. Exhaust gas carries the fused material, fly ash, out of the combustion chamber. The fly ash cools into spheres, which may be solid, hollow (cenospheres), or hollow and filled with other spheres (plerospheres). Particle diameters range from 1 $\mu$m to more than 0.1 mm, with an average of 0.015 mm to 0.020 mm, and are 70% to 90% smaller than 0.045 mm. Fly ash is primarily a silica glass composed of:

| Chemical | Content, % |
|---|---|
| Silica ($SiO_2$) | 40–90 |
| Alumina ($Al_2O_3$) | 20–60 |
| Iron oxide ($Fe_2O_3$) | 5–25 |
| Lime ($CaO$) | 1–30 |

Class F fly ash usually has less than 5% CaO but may contain up to 10%. Class C fly ash has 15% to 30% CaO.

Silica fume, or microsilica, is a by-product of silicon or ferrosilicon alloy production and occurs when high-purity quartz and coal are reduced in an electric arc furnace. Silica fume is a silicon dioxide in noncrystalline form with spherical shape. Silica fume has an average diameter of about 0.1 $\mu$m and a maximum size of 1.0 $\mu$m. The surface area is about 20,000 m$^2$/kg (about twice the surface area of tobacco smoke).

The spherical shape of fly ash increases the workability of the fresh concrete. In addition, both fly ash and silica fume extend the hydration process, allowing a greater strength development and reduced porosity. Studies have shown that concrete containing more than 20% pozzolan by weight of cement has a much smaller pore size distribution than portland cement concrete without fly ash. The lower heat of hydration reduces the early strength of the concrete. The extended reaction permits a continuous gaining of strength beyond what can be accomplished with plain portland cement.

Nominally inert materials, finely divided quartz, limestone, marble, etc., can be used to improve workability; however, these materials have no cementitious value.

Tables 6.9 and 6.10 summarize the effects of mineral admixtures on fresh and hardened concrete. These summaries are based on general trends and should be verified experimentally for specific materials and construction conditions.

**TABLE 6.9**  Effect of Mineral Admixtures on Freshly Mixed Concrete

| Quality Measure | Effect |
| --- | --- |
| Water requirements | Fly ash reduces water requirements. Silica fume increases water requirements. |
| Air content | Fly ash and silica fume reduce air content; compensate by increasing air entrainer. |
| Workability | Fly ash, ground slag, and inert minerals generally increase workability. Silica fume reduces workability; compensate by using superplasticizer. |
| Hydration | Fly ash reduces heat of hydration. Silica fume may not affect, but superplasticizer used with silica fume can increase heat. |
| Set time | Fly ash, natural pozzolans and blast furnace slag increase set time; can compensate by using accelerator. |

**TABLE 6.10**  Effect of Mineral Admixtures on Hardened Concrete

| Quality Measure | Effect |
| --- | --- |
| Strength | Fly ash increases ultimate strength but reduces rate of strength gain. Silica fume has less effect on rate of strength gain than pozzolans. |
| Drying shrinkage and creep | Low concentrations usually have little effect. High concentrations of ground slag or fly ash may increase shrinkage. Silica fume may reduce shrinkage. |
| Permeability and absorption | Generally reduced permeability and absorption. Silica fume is especially effective. |
| Alkali-aggregate reactivity | Generally reduced reactivity, extent of improvement depends on type of admixture. |
| Sulfate resistance | Improved due to reduced permeability. |

## Specialty Admixtures

In addition to the previously mentioned admixtures, several admixtures are available to improve concrete quality in particular ways. The civil engineer should be aware of these admixtures, but will need to study their application in detail, as well as their cost, before using them. Examples of specialty admixtures include:

workability agents
corrosion inhibitors
damp proofing agents
permeability reducing agents
fungicidal, germicidal, and insecticidal admixtures
pumping aids
bonding agents
grouting agents
gas forming agents
coloring agents

## SUMMARY

The development of portland cement as the binder material for concrete is one of the most important innovations of civil engineering. It is extremely difficult to find civil engineering projects that do not include some component constructed with portland cement concrete. The properties of portland cement are governed by the chemical composition and the fineness of the particles. These control the rate of hydration and the ultimate strength of the concrete. Abrams' discovery of the importance of the water to cement ratio as the factor that controls the quality of concrete is perhaps the single most important advance in concrete technology. Second to this development was the introduction of air entrainment. The subsequent development of additional admixtures for concrete has improved the workability, set time, strength and economy of concrete construction.

## QUESTIONS AND PROBLEMS

**6.1.** What ingredients are used for the production of portland cement?

**6.2.** What is the role of gypsum in the production of portland cement?

**6.3.** What is a typical value for the fineness of portland cement?

**6.4.** What are the primary chemical reactions during the hydration of portland cement?

**6.5.** Define the C-S-H phase of cement paste.

**6.6.** What are the four main chemical compounds in portland cement?

**6.7.** What chemical compounds contribute to early strength gain?

**6.8.** Define:

   **a.** interlayer hydration space

   **b.** capillary voids

   **c.** entrained air

   **d.** entrapped air

**6.9.** Discuss the effect of water to cement ratio on the quality of hardened concrete. Explain why this effect happens.

**6.10.** What are the five primary types and functions of portland cement? Which can be replaced with the substitution of pozzolan for some of the cement?

**6.11.** Why isn't pozzolan used with Type III cement?

**6.12.** In order to evaluate the suitability of nonpotable water available at the job site for mixing concrete, six standard mortar cubes were made using that water and six others using potable water. The cubes were tested for compressive strength after 7 days of curing and produced the following loads to failure (in pounds).

| Cubes Made with Nonpotable Water | Cubes Made with Potable Water |
|---|---|
| 17,810 | 16,730 |
| 15,110 | 18,870 |
| 14,200 | 15,230 |
| 18,290 | 17,470 |
| 14,650 | 16,990 |
| 16,430 | 17,850 |

**a.** Based on these results only, would you accept that water for mixing concrete according to ASTM C94?

**b.** According to ASTM C94, are there other tests to be performed to evaluate the suitability of that water? Discuss briefly.

**6.13.** State five types of admixtures and discuss their applications.

**6.14.** If you were a materials engineer in Minnesota (cold climate) and could only use one type of admixture, which would you select? Explain.

**6.15.** Under what condition is an air entraining agent is needed? Why? Discuss how the air entraining agent performs its function.

**6.16.** Why is a superplasticizer used? How does it perform its function?

**6.17.** A materials engineer is working in a research project to evaluate the effect of one type of admixture on the compressive strength of concrete. He tested 10 mortar cubes made without admixture and 10 others with admixture after 28 days of curing. The compressive strengths of cubes in MPa without admixture were 25.1 in MPa, 24.4 in MPa, 25.8 in MPa, 25.2 in MPa, 23.9 in MPa, 24.7 in MPa, 24.3 in MPa, 26.0 in MPa, 23.8 in MPa, and 24.6 in MPa. The compressive strengths of cubes with admixture were 25.3 in MPa, 26.8 in MPa, 26.5 in MPa, 24.5 in MPa, 27.2 in MPa, 24.8 in MPa, 24.1 in MPa, 25.9 in MPa, 25.3 in MPa, and 25.0 in MPa. Using the statistical $t$-test, does this admixture show an increase of the compressive strength of the cement mortar at a level of significance of 0.05?

**REFERENCES**

American Concrete Institute, 1986. *Building code requirements for reinforced concrete.* ACI Committee 318 Report, ACI 318-83. Farmington Hills, MI: American Concrete Institute.

Hewlett, P. C. 1978. *Concrete admixtures: use and applications.* ed. M. R. Rixom. London: The Construction Press.

Hover, K. and R. E. Phillco. 1990. *Highway materials engineering, concrete.* Publication No. FHWA-H1-90-009, NHI Course No. 13123. Washington, D.C.: Federal Highway Administration.

Kosmatka, S. H. and W. C. Panarese. 1988. *Design and control of concrete mixtures.* 13th ed. Skokie, IL: Portland Cement Association.

Mehta, P. K. and P. J. M. Monteiro. 1993. *Concrete structure, properties, and materials.* 2nd ed. Englewood Cliffs, NJ: Prentice-Hall.

Neville, A. M. 1981. *Properties of concrete.* 3rd ed. London: Pitman Books Ltd.

# 7 Portland Cement Concrete

Civil and construction engineers are directly responsible for the quality control of portland cement concrete and the proportions of the components used in it. The quality of the concrete is governed by the chemical composition of the portland cement, hydration and development of the microstructure, admixtures, and aggregate characteristics. The quality is strongly affected by placement, consolidation, and curing, as well.

How a concrete structure performs throughout its service life is largely affected by the methods of mixing, transporting, placing, and curing the concrete in the field. In fact, the ingredients of a "good" concrete may be the same as those of a "bad" concrete. The difference, however, is often the expertise of the engineer and technicians who are handling the concrete during construction.

Because of the advances made in concrete technology in the past few decades, concrete can be used in many more applications. Civil and construction engineers should be aware of the alternatives to conventional concrete, such as lightweight concrete, high-strength concrete, polymer concrete, fiber-reinforced concrete, and roller-compacted concrete. Before using these alternatives to conventional concrete, the engineer needs to study them, and their costs, in detail. This chapter covers basic principles of conventional portland cement concrete, its proportioning, mixing and handling, curing, and testing. Alternatives to conventional concrete that increase the applications and improve the performance of concrete are also introduced.

## Proportioning of Concrete Mixes

The properties of concrete depend on the mix proportions and the placing and curing methods. Designers generally specify or assume a certain strength or modulus of elasticity of the concrete when determining structural dimensions. The materials engineer is responsible for assuring that the concrete is properly proportioned, mixed, placed, and cured to have the properties specified by the designer.

The proportioning of the concrete mix affects its properties in both the plastic and solid states. During the plastic state, the materials engineer is concerned with the workability and finishing characteristics of the concrete. Properties of the hardened concrete important to the materials engineer are the strength, modulus of elasticity, durability, and porosity. Strength is generally the controlling design factor. Unless otherwise specified, concrete strength, $f'_c$, refers to the average compressive strength of three tests. Each test is the average result of two 0.15 m × 0.30 m (6 in. × 12 in.) cylinders tested in compression after curing for 28 days.

The PCA specifies three qualities required of properly proportioned concrete mixtures (Kosmatka and Panarese 1988):

1. acceptable workability of freshly mixed concrete
2. durability, strength, and uniform appearance of hardened concrete
3. economy

In order to achieve these characteristics, the materials engineer must determine the proportions of cement, water, fine and coarse aggregates, and the use of admixtures. Several mix design methods have been developed over the years, ranging from an arbitrary volume method (1:2:3 cement:sand:coarse aggregate) to the weight and absolute-volume methods prescribed by the American Concrete Institute's Committee 211. The weight method provides relatively simple techniques for estimating mix proportions using an assumed or known unit weight of concrete. The absolute volume method uses the specific gravity of each ingredient to calculate the unit volume each will occupy in a unit volume of concrete. The absolute volume method is more accurate than the weight method. The mix design process for the weight and absolute volume methods differs only in how the amount of fine aggregates is determined.

## Basic Steps for Weight and Absolute Volume Methods

The basic steps required for determining mix design proportions for both weight and absolute volume methods are as follows (Kosmatka and Panarese 1988):

1. Evaluate strength requirements.
2. Determine the water-cement ratio required.
3. Evaluate coarse aggregate requirements.
   - maximum aggregate size of the coarse aggregate
   - quantity of the coarse aggregate
4. Determine air entrainment requirements.
5. Evaluate workability requirements of the plastic concrete.
6. Estimate the water content requirements of the mix.
7. Determine cement content and type needed.
8. Evaluate the need and application rate of admixtures.
9. Evaluate fine aggregate requirements.
10. Determine moisture corrections.
11. Make and test trial mixes.

Most concrete supply companies have a wealth of experience about how their materials perform in a variety of applications. This experience, accompanied with reliable test data on the relationship between strength and water-cement ratio, is the most dependable method for selecting mix proportions. However, understanding the basic principles of mixture design and the proper selection of materials and mixture characteristics is as important as the actual calculation. Therefore, the PCA procedure provides guidelines and can be adjusted to match the experience obtained from local conditions. The PCA mix design steps are discussed next.

**1. Strength Requirements** Variations in materials, and batching and mixing of concrete results in deviations in the strength of the concrete produced by a plant. Generally, the structural design engineer does not consider this variability when determining the size of the structural members. If the materials engineer provides a material with an average strength equal to the strength specified by the designer, then half of the concrete will be weaker than the specified strength. Obviously this is undesirable. To compensate for the variance in concrete strength, the materials engineer designs the concrete to have an average strength greater than the strength specified by the structural engineer.

In order to compute the strength requirements for concrete mix design, three quantities must be known:

1. the specified compressive strength $f'_c$,
2. the variability or standard deviation, $s$, of the concrete, and
3. the allowable risk of making concrete with an unacceptable strength.

The standard deviation in the strength is determined for a plant by making batches of concrete, testing the strength for many samples, and computing the standard deviation using Equation 1.15 in Chapter 1. The allowable risk has been established by the American Concrete Institute (ACI). One of the risk rules states that there should be less than 10% chance that the strength of a concrete mix is less than the specified strength. Assuming the concrete strength has a normal distribution, the implication of the ACI rule is that 10% of the area of the distribution must be to the left of $f'_c$, as shown in Figure 7.1 on page 168. Using a table of standard $z$ values for a normal distribution curve, we can determine that 90% of the area under the curve will be to the right of $f'_c$ if the average strength is 1.34 standard deviations from $f'_c$. In other words, the required average strength, $f'_{cr}$, for this criterion can be calculated as

$$f'_{cr} = f'_c + 1.34s \qquad (7.1)$$

where

$f'_{cr}$ = required average compressive strength, MPa or psi
$f'_c$ = specified compressive strength, MPa or psi
$s$ = standard deviation, MPa or psi

For mixes with a large standard deviation in strength, the ACI has another risk criterion that requires

$$f'_{cr} = f'_c + 2.33s - 3.45 \qquad (7.2)$$

The required average compressive strength $f'_{cr}$ is determined as the larger value obtained from Equations 7.1 and 7.2.

Note that if the U.S. customary units are used, $f'_{cr}$, $f'_c$, and $s$ are recorded in psi and the constant 3.45 in Equation 7.2 should be changed to 500.

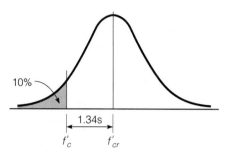

**FIGURE 7.1** Use of normal distribution and risk criteria to estimate average required concrete strength.

The standard deviation should be determined from at least 30 strength tests. If the standard deviation is computed from 15 to 30 samples, then the standard deviation is multiplied by the following factor, $F$, to determine the modified standard deviation $s'$.

| Number of Tests | Modification Factor $F$ |
|---|---|
| 15 | 1.16 |
| 20 | 1.08 |
| 25 | 1.03 |
| 30 or more | 1.00 |

Linear interpolation is used for an intermediate number of tests, and $s'$ is used in place of $s$ in Equations 7.1 and 7.2.

If fewer than 15 tests are available, the following adjustments are made to the specified strength instead of using Equations 7.1 and 7.2.

| Specified Compressive Strength $f'_c$, MPa (psi) | Required Average Compressive Strength $f'_{cr}$, MPa (psi) |
|---|---|
| < 20.7 (< 3000) | $f'_c + 6.9$ (1000) |
| 20.7 to 34.5 (3000 to 5000) | $f'_c + 8.3$ (1200) |
| > 34.5 (> 5000) | $f'_c + 9.7$ (1400) |

These estimates are very conservative and should not be used for large projects since the concrete will be over-designed and, therefore, not economical.

**SAMPLE PROBLEM 7.1**

The design engineer specifies a concrete strength of 31.0 MPa (4500 psi). Determine the required average compressive strength for

    **a.**  a new plant where $s$ is unknown,
    **b.**  a plant where $s = 3.6$ MPa (520 psi) for 17 test results,
    **c.**  a plant with extensive history of producing concrete with
        $s = 2.4$ MPa (350 psi),
    **d.**  a plant with extensive history of producing concrete with
        $s = 3.8$ MPa (550 psi).

**Solution:**

    **a.** $f'_{cr} = f'_c + 8.3 = 31.0 + 8.3 = 39.3$ MPa (5700 psi)

    **b.** Need to interpolate modification factor

$$F = 1.16 - \left(\frac{1.16 - 1.08}{20 - 15}\right)(17 - 15) \cong 1.13$$

Multiply standard deviation by the modification factor

$$s' = (s)(F) = 3.6\,(1.13) = 4.1 \text{ MPa (590 psi)}$$

Determine maximum from Equations 7.1 and 7.2

$$f'_{cr} = 31.0 + 1.34\,(4.1) = 36.5 \text{ MPa (5300 psi)}$$

$$f'_{cr} = 31.0 + 2.33\,(4.1) - 3.45 = 37.1 \text{ MPa (5390 psi)}$$

Use $f'_{cr} = 37.1$ MPa (5390 psi)

    **c.** Determine maximum from Equations 7.1 and 7.2

$$f'_{cr} = 31.0 + 1.34\,(2.4) = 34.2 \text{ MPa (4970 psi)}$$

$$f'_{cr} = 31.0 + 2.33\,(2.4) - 3.45 = 33.1 \text{ MPa (4810 psi)}$$

Use $f'_{cr} = 34.2$ MPa (4970 psi)

    **d.** Determine maximum from Equations 7.1 and 7.2

$$f'_{cr} = 31.0 + 1.34\,(3.8) = 36.1 \text{ MPa (5240 psi)}$$

$$f'_{cr} = 31.0 + 2.33\,(3.8) - 3.45 = 36.4 \text{ MPa (5280 psi)}$$

Use $f'_{cr} = 36.4$ MPa (5280 psi)     ◆

**2. Water-Cement Ratio Requirements** The next step is to determine the water-cement ratio needed to produce the required strength. Historical records are used to plot a strength versus water-cement ratio curve such as that seen in Figure 7.2. If historical data are not available, three trial batches are made at different water-cement

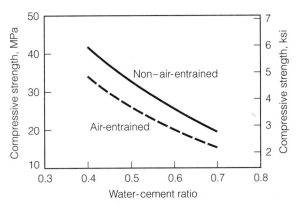

**FIGURE 7.2** Example trial-mixture or field-data strength curves.

**TABLE 7.1**  Typical Relationship between Water-Cement Ratio and Compressive Strength of Concrete*

| Average compressive strength at 28 Days, $f'_{cr}$, MPa (psi)[†] | Water-Cement Ratio, by wt | |
|---|---|---|
| | Non–air-entrained Concrete | Air-entrained Concrete |
| 41.4 (6000) | 0.41 | — |
| 34.5 (5000) | 0.48 | 0.40 |
| 27.6 (4000) | 0.57 | 0.48 |
| 20.7 (3000) | 0.68 | 0.59 |
| 13.8 (2000) | 0.82 | 0.74 |

*American Concrete Institute (1985). Authorized reprint.
†Values are estimated average strengths for concrete containing not more than the percentage of air shown in Table 7.6. For a constant water-cement ratio, the strength of concrete is reduced as the air content is increased. Strength is based on 0.15 m by 0.30 m (6 in. by 12 in.) cylinders moist-cured 28 days at 23 ± 1.7°C (73.4 ± 3°F) in accordance with Section 9b of ASTM C31. The relationship assumes maximum size of aggregate about 19 mm to 25 mm (3/4 in. to 1 in.).

ratios to establish a curve similar to Figure 7.2. Table 7.1 can be used for estimating the water-cement ratios for the trial mixes when no other data are available. The required average compressive strength is used with the strength versus water-cement relationship to determine the water-cement ratio required for the strength requirements of the project.

For small projects of noncritical applications, Table 7.2 can be used in lieu of trial mixes with the permission of the project engineer. Table 7.2 is conservative with respect to the strength versus water-cement ratio relationship. This results in higher cement factors and greater average strengths than would be required if a mix design is performed. This table is not intended for use in designing trial batches; use Table 7.1 for trial batch design.

The water-cement ratio required for strength is checked against the maximum allowable water-cement ratio for the exposure conditions. Tables 7.3 and 7.4 provide guidance on the maximum allowable water-cement ratio for exposure conditions. Generally, more severe exposure conditions require lower water-cement ratios. The

**TABLE 7.2**  Maximum Permissible Water-Cement Ratios for Concrete When Strength Data from Field Experience or Trial Mixtures Are Not Available*

| Specified 28-Day Compressive Strength $f'_c$, MPa (psi) | Water-Cement Ratio, by wt | |
|---|---|---|
| | Non–air-entrained Concrete | Air-entrained Concrete |
| 17.2 (2500) | 0.67 | 0.54 |
| 20.7 (3000) | 0.58 | 0.46 |
| 24.1 (3500) | 0.51 | 0.40 |
| 27.6 (4000) | 0.44 | 0.35 |
| 31.0 (4500) | 0.38 | † |
| 34.5 (5000) | † | † |

*American Concrete Institute (1986b). Authorized reprint.
†For strength above 31.0 MPa (4500 psi) (non–air-entrained concrete) and 27.6 MPa (4000 psi) (air-entrained concrete), concrete proportions shall be established from field data or trial mixtures.

**TABLE 7.3**  Maximum Water-Cement Ratios for Various Exposure Conditions*

| Exposure Condition | Maximum Water-cement Ratio for Normal-Weight Concrete, by weight |
|---|---|
| Concrete protected from exposure to freezing and thawing or application of de-icer chemicals | Select water-cement ratio on basis of strength, workability, and finishing needs |
| Concrete intended to be watertight, exposed to: | |
|   **a.** Fresh water | 0.50 |
|   **b.** Brackish water or seawater | 0.45 |
| Concrete exposed to freezing and thawing in a moist condition (air-entrained concrete) | |
|   **a.** Curbs, gutters, guardrails, or this sections | 0.45 |
|   **b.** Other elements | 0.50 |
|   **c.** In presence of de-icing chemicals | 0.45 |
| For corrosion protection for reinforced concrete exposed to de-icing salts, brackish water, seawater, or spray from these sources | 0.40[†] |

*American Concrete Institute (1986b). Authorized reprint.
†If minimum concrete cover required by ACI 318, Section 7.7 is increased by 12 mm (0.5 in.) the water-cement ratio may be increased to 0.45 for normal-weight concrete.

**TABLE 7.4**  Requirements for Concrete Exposed to Sulfate-Containing Solutions*

| Sulfate Exposure | Water-Soluble Sulfate in Soil, % by wt | Sulfate in Water, ppm | Cement Type[†] | Normal-Weight Concrete<br>Maximum Water-Cement Ratio, by wt | Lightweight Concrete<br>Minimum Compressive Strength, MPa (psi) |
|---|---|---|---|---|---|
| Negligible | 0.00–0.10 | 0–150 | — | — | — |
| Moderate[‡] | 0.10–0.20 | 150–1500 | II, IP(MS), IS(MS), P(MS), I(PM)(MS), I(SM)(MS) | 0.50 | 25.9 (3750) |
| Severe | 0.20–2.00 | 1500–10,000 | V | 0.45 | 29.3 (4250) |
| Very severe | Over 2.00 | Over 10,000 | V plus pozzolan** | 0.45 | 29.3 (4250) |

*American Concrete Institute (1986b). Authorized reprint.
†Cement Types II and V are specified in ASTM C150; remaining types, blended cements, are in ASTM C595.
‡Seawater.
**Pozzolan (ASTM C618 or silica fume) that has been determined by test or service record to improve sulfate resistance when used in concrete containing Type V cement.

minimum of the water-cement ratio for strength and exposure is selected for proportioning the concrete.

If a pozzolan is used in the concrete, the water-cement plus pozzolan ratio by weight may be used instead of the traditional water-cement ratio. In other words, the weight of the water is divided by the sum of the weights of cement plus pozzolan.

**3. Coarse Aggregate Requirements** The next step is to determine the suitable aggregate characteristics for the project. In general, large-dense graded aggregates provide the most economical mix. Large aggregates minimize the amount of water required and, therefore, reduce the amount of cement required per cubic meter of mix. Round aggregates require less water than angular aggregates for an equal workability.

The maximum allowable aggregate size is limited by the dimensions of the structure and the capabilities of the construction equipment. The largest maximum aggregate size practical under job conditions that satisfies the size limits in the table should be used.

| Situation | Maximum Aggregate Size |
|---|---|
| Form dimensions | 1/5 of minimum clear distance |
| Clear space between reinforcement or prestressing tendons | 3/4 of minimum clear space |
| Clear space between reinforcement and form | 3/4 of minimum clear space |
| Unreinforced slab | 1/3 of thickness |

**SAMPLE PROBLEM 7.2**

A structure is to be built with concrete with a minimum dimension of 0.2 m, minimum space between rebars of 40 mm, and minimum cover over rebars of 40 mm. Two types of aggregate are locally available with maximum sizes of 19 mm and 25 mm, respectively. If both types of aggregate have essentially the same cost, which one is more suitable for this structure?

*Solution:*

$$25 \text{ mm} < \left(\tfrac{1}{5}\right) (200 \text{ mm}) \text{ minimum dimensions.}$$

$$25 \text{ mm} < \left(\tfrac{3}{4}\right) (40 \text{ mm}) \text{ rebar spacing.}$$

$$25 \text{ mm} < \left(\tfrac{3}{4}\right) (40 \text{ mm}) \text{ rebar cover.}$$

Therefore, both sizes satisfy the dimension requirements. However, 25 mm aggregate is more suitable because it will produce more economical concrete mix.     ◆

The gradation of the fine aggregates is defined by the fineness modulus. The desirable fineness modulus depends on the coarse aggregate size and the quantity of cement paste. A low fineness modulus is desired for mixes with low cement content to promote workability.

Once the fineness modulus of the fine aggregate and the maximum size of the coarse aggregate are determined, the volume of coarse aggregate per unit volume of concrete is determined using Table 7.5. For example, if the fineness modulus of the fine aggregate is 2.60 and the maximum aggregate size is 25 mm (1 in.), the coarse aggregate will have a volume of 0.69 m$^3$/m$^3$ (yd$^3$/yd$^3$) of concrete. Table 7.5 is based on the unit weight of aggregates in a dry-rodded condition (ASTM C29). The values given are based on experience in producing an average degree of workability. The volume of coarse aggregate can be increased by 10% when less workability is required, such as in pavement construction. The volume of coarse aggregate should be reduced by 10% to increase workability, for example to allow placement by pumping.

**TABLE 7.5** Volume of Coarse Aggregate per Unit of Volume of Concrete for Different Fineness Moduli of Fine Aggregate*

| Maximum Size of Aggregate, mm (in.) | Fineness Modulus | | | |
|---|---|---|---|---|
| | 2.40 | 2.60 | 2.80 | 3.00 |
| 9.5 (3/8) | 0.50 | 0.48 | 0.46 | 0.44 |
| 12.5 (1/2) | 0.59 | 0.57 | 0.55 | 0.53 |
| 19 (3/4) | 0.66 | 0.64 | 0.62 | 0.60 |
| 25 (1) | 0.71 | 0.69 | 0.67 | 0.65 |
| 37.5 (1-1/2) | 0.75 | 0.73 | 0.71 | 0.69 |
| 50 (2) | 0.78 | 0.76 | 0.74 | 0.72 |
| 75 (3) | 0.82 | 0.80 | 0.78 | 0.76 |
| 150 (6) | 0.87 | 0.85 | 0.83 | 0.81 |

*American Concrete Institute (1985). Authorized reprint.

**4. Air Entrainment Requirements** Next the need for air entrainment is evaluated. Air entrainment is required whenever concrete is exposed to freeze-thaw conditions and de-icing salts. Air entrainment is also used for workability in some situations. The amount of air required varies based on exposure conditions and is affected by the size of the aggregates. The exposure levels are defined as:

*Mild-exposure*—Indoor or outdoor service where concrete is not exposed to freezing and de-icing salts. Air entrainment may be used to improve workability.

*Moderate exposure*—Some freezing exposure occurs but concrete not exposed to moisture or free water for long periods prior to freezing. Concrete not exposed to de-icing salts. Examples include exterior beams, columns, walls, etc., not exposed to wet soil.

*Severe exposure*—Concrete exposed to de-icing salts, saturation, or free water. Examples include pavements, bridge decks, curbs, gutters, canal linings, etc.

Table 7.6 presents the recommended air contents for different combinations of exposure conditions and maximum aggregate sizes. The values shown in Table 7.6 are the entrapped air for non–air-entrained concrete and the entrapped plus entrained air in case of air-entrained concrete. The recommended air content decreases with increasing maximum aggregate size.

**5. Workability Requirements** The next step in the mix design is to determine the workability requirements for the project. Workability is defined as the ease of placing, consolidating, and finishing freshly mixed concrete. Concrete should be workable but should not segregate or excessively bleed (migration of water to the top surface of concrete). The slump test, Figure 7.3, is an indicator of workability when evaluating similar mixtures. This test consists of filling a truncated cone with concrete, removing the cone, then measuring the distance the concrete slumps (ASTM C143). The slump is increased by adding water, air entrainer, water reducer, superplasticizer, or by using round aggregates. Table 7.7 provides recommendations for the slump of concrete used in different types of projects. For batch adjustments, slump increases about 25 mm (1 in.) for each 6 kg of water added per m$^3$ (10 lb per cubic yard) of concrete.

**TABLE 7.6**   Approximate Target Air Requirements for Maximum Sizes of Aggregates*

| | Maximum Aggregate Size | | | | | | | |
|---|---|---|---|---|---|---|---|---|
| | 9.5 mm (3/8 in.) | 12.5 mm (1/2 in.) | 19 mm (3/4 in.) | 25 mm (1 in.) | 37.5 mm (1-1/2 in.) | 50 mm (2 in.) | 75 mm (3 in.) | 150 mm (6 in.) |
| Non–air-entrained concrete, approximate entrapped air, % | 3 | 2.5 | 2 | 1.5 | 1 | 0.5 | 0.3 | 0.2 |
| Air-entrained concrete, recommended air content, for level of exposure, %[†] | | | | | | | | |
| Mild exposure | 4.5 | 4.0 | 2.5 | 3.0 | 2.5 | 2.0 | 1.5 | 1.0 |
| Moderate exposure | 6.0 | 5.5 | 5.0 | 4.5 | 4.5 | 4.0 | 3.5 | 3.0 |
| Severe exposure | 7.5 | 7.0 | 6.0 | 6.0 | 5.5 | 5.0 | 4.5 | 4.0 |

*American Concrete Institute (1985; 1986b). Authorized reprint.
†The air content in job specifications should be specified to be delivered within –1 to +2 percentage points of the table target value for moderate and severe exposures.

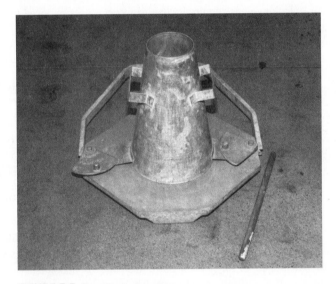

**FIGURE 7.3** Slump test apparatus.

**TABLE 7.7**   Recommended Slumps for Various Types of Construction*

| | Slump, mm (in.) | |
|---|---|---|
| Concrete Construction | Maximum[†] | Minimum |
| Reinforced foundation walls and footings | 75 (3) | 25 (1) |
| Plain footings, caissons, and substructure walls | 75 (3) | 25 (1) |
| Beams and reinforced walls | 100 (4) | 25 (1) |
| Building columns | 100 (4) | 25 (1) |
| Pavements and slabs | 75 (3) | 25 (1) |
| Mass concrete | 50 (2) | 25 (1) |

*American Concrete Institute (1985). Authorized reprint.
†May be increased 25 mm (1 in.) for consolidation by hand methods, such as rodding and spading.

**TABLE 7.8**  Approximate Mixing Water for Different Slumps and
Maximum Aggregate Sizes in kg/m$^3$ (lb/yd$^3$)*

| Slump, mm (in.) | Maximum Aggregate Size, mm (in.)† | | | | | | | |
|---|---|---|---|---|---|---|---|---|
| | 9.5 (3/8) | 12.5 (1/2) | 19 (3/4) | 25 (1) | 37.5 (1-1/2) | 50 (2) | 75 (3) | 150 (6) |
| *Non–Air-Entrained Concrete* | | | | | | | | |
| 25 to 50 (1 to 2) | 208 (350) | 211 (355) | 187 (315) | 178 (300) | 163 (275) | 154 (260) | 130 (220) | 113 (190) |
| 75 to 100 (3 to 4) | 228 (385) | 216 (365) | 202 (340) | 190 (321) | 178 (300) | 160 (285) | 145 (245) | 125 (210) |
| 150 to 175 (6 to 7) | 243 (410) | 228 (385) | 213 (360) | 202 (340) | 187 (315) | 178 (300) | 160 (270) | — |
| *Air-Entrained Concrete* | | | | | | | | |
| 25 to 50 (1 to 2) | 181 (305) | 175 (295) | 166 (280) | 160 (270) | 148 (250) | 142 (240) | 122 (205) | 107 (180) |
| 75 to 100 (3 to 4) | 202 (340) | 193 (325) | 181 (305) | 175 (295) | 163 (275) | 157 (265) | 133 (225) | 119 (200) |
| 150 to 175 (6 to 7) | 216 (365) | 205 (345) | 193 (325) | 184 (310) | 172 (290) | 166 (280) | 154 (260) | — |

*American Concrete Institute (1985; 1986). Authorized reprint.
†These quantities of mixing water are for use in computing cement factors for trial batches. They are maximums for reasonably well-shaped angular coarse aggregates graded within limits of accepted specifications.

**6. Water Content Requirements**  The water content required for a given slump depends on the maximum size and shape of the aggregates and whether an air entrainer is used. Table 7.8 gives the approximate mixing water requirements for angular coarse aggregates (crushed stone). The recommendations in Table 7.8 are reduced for other aggregate shapes.

| Aggregate Shape | Reduction in Water Content, kg/m$^3$ (lb/yd$^3$) |
|---|---|
| Sub-angular | 12 (20) |
| Gravel with crushed particles | 21 (35) |
| Round gravel | 27 (45) |

These recommendations are approximate and should be verified with trial batches for local materials.

**7. Cement Content Requirements**  With the water-cement ratio and the required amount of water estimated, the cement required for the mix is determined by dividing the weight of the water by the water-cement ratio. PCA recommends a minimum cement content of 334 kg/m$^3$ (564 lb/yd$^3$) for concrete exposed to severe freeze-thaw, de-icers, and sulfate exposures, and not less than 385 kg/m$^3$ (650 lb/yd$^3$) for concrete placed under water. In addition, Table 7.9 shows the minimum cement requirements for proper placing, finishing, abrasion resistance, and durability in flatwork, such as slabs.

**TABLE 7.9**  Minimum Cement Requirements for Normal-Weight Concrete Used in Flatwork*

| Maximum Size of Aggregate, mm (in.) | Cement, kg/m$^3$ (lb/yd$^3$) |
|---|---|
| 39.5 (1-1/2) | 279 (470) |
| 25.0 (1) | 308 (520) |
| 19.0 (3/4) | 320 (540) |
| 12.5 (1/2) | 350 (590) |
| 9.5 (3/8) | 361 (610) |

*American Concrete Institute (1980). Authorized reprint.

**8. Admixture Requirements**  If one or more admixtures are used to add a specific quality in the concrete (as discussed in Chapter 6), their quantities should be considered in the mix proportioning. Admixture manufacturers provide specific information on the quantity of admixture required to achieve the desired results.

**9. Fine Aggregate Requirements**  At this point, water, cement, and coarse aggregate weights per cubic meter (cubic yard) are known and the volume of air is estimated. The only remaining factor is the amount of fine aggregates needed. The weight mix design method uses Table 7.10 to estimate the total weight of a "typical" freshly mixed concrete for different maximum aggregate sizes. The weight of the fine aggregates is determined by subtracting the weight of the other ingredients from the total weight. Since Table 7.10 is based on a "typical" mix, the weight-based mix design method is only approximate.

In the absolute volume method of mix design, the component weight and the specific gravity are used to determine the volumes of the water, coarse aggregate, and cement. These volumes, along with the volume of the air, are subtracted from a unit volume of concrete to determine the volume of the fine aggregate required. The volume of the fine aggregate is then converted to a weight using the unit weight. Generally, the bulk SSD specific gravity of aggregates is used for the weight-volume conversions of both fine and coarse aggregates.

**10. Moisture Corrections**  Mix designs assume that water used to hydrate the cement is the free water in excess of the moisture content of the aggregates at the SSD

**TABLE 7.10**  Estimate of Mass (Weight) of Freshly Mixed Concrete

| Maximum Aggregate Size, mm (in.) | Non–air-entrained Concrete, kg/m$^3$ (lb/yd$^3$) | Air-entrained Concrete, kg/m$^3$ (lb/yd$^3$) |
|---|---|---|
| 9.5 (3/8) | 2276 (3840) | 2187 (3690) |
| 12.5 (1/2) | 2305 (3890) | 2228 (3760) |
| 19.0 (3/4) | 2347 (3960) | 2276 (3840) |
| 25.0 (1) | 2376 (4010) | 2311 (3900) |
| 37.5 (1-1/2) | 2412 (4070) | 2347 (3960) |
| 50.0 (2) | 2441 (4120) | 2370 (4000) |
| 75.0 (3) | 2465 (4160) | 2394 (4040) |
| 150 (6) | 2507 (4230) | 2441 (4120) |

condition (absorption) as discussed in Chapter 5. Therefore, the final step in the mix design process is to adjust the weight of water and aggregates to account for the existing moisture content of the aggregates. If the moisture content of the aggregates is more than the SSD moisture content, the weight of mixing water is reduced by an amount equal to the free weight of the moisture on the aggregates. Similarly, if the moisture content is below the SSD moisture content, the mixing water must be increased.

**11. Trial Mixes** After computing the required amount of each ingredient, a trial batch is mixed to check the mix design. Three 0.15 m × 0.30 m (6 in. × 12 in.) cylinders are made, cured for 28 days, and tested for compressive strength. In addition, the air content and slump of fresh concrete are measured. If the slump, air content or compressive strength does not meet the requirements, the mixture is adjusted and other trial mixes are made until the design requirements are satisfied.

Additional trial batches could be made by slightly varying the material quantities in order to determine the most workable and economical mix.

**SAMPLE PROBLEM 7.3**

Design a concrete mix for the following conditions and constraints using the absolute volume method.

*Design Environment*
Bridge pier exposed to freezing and subjected to de-icing chemicals
Required design strength = 24.1 MPa (3500 psi)
Minimum dimension = 0.3 m (12 in.)
Minimum space between rebars = 50 mm (2 in.)
Minimum cover over rebars = 40 mm (1.5 in.)
Standard deviation of compressive strength of 2.4 MPa (350 psi) is expected
    (more than 30 samples)
Only air entrainer is allowed

*Available Materials*
Cement
    Select Type V due to exposure
Air Entrainer
    Manufacture specification 6.3 ml/1% air/100 kg cement (0.1 fl oz/1% air/100 lb cement)
Coarse aggregate
    25 mm (1 in.) maximum size, river gravel (round)
    Bulk oven dry specific gravity = 2.621, Absorption = 0.4%
    Oven dry-rodded density = 1681 kg/m$^3$ (105 pcf)
    Moisture content = 3%
Fine aggregate
    Natural sand
    Bulk oven-dry specific gravity = 2.572, Absorption = 0.8%
    Moisture content = 4%
    Fineness modulus = 2.60

### Solution

**1.** Strength Requirements

$s = 2.4$ MPa (350 psi) (enough samples so that no correction is needed)

$f'_{cr} = f'_c + 1.34\,s = 24.1 + 1.34\,(2.4) = 27.3$ MPa (3960 psi)

$f'_{cr} = f'_c + 2.33\,s - 3.45 = 24.1 + 2.33\,(2.4) - 3.45 = 26.2$ MPa (3810 psi)

**$f'_{cr} = 27.3$ MPa (3960 psi)**

**2.** Water-Cement Ratio

Strength requirement (Table 7.1),
water-cement ratio = 0.48 by interpolation

Exposure requirement (Tables 7.3 and 7.4),
maximum water-cement ratio = 0.45

**Water-cement ratio = 0.45**

**3.** Coarse Aggregate Requirements

$25$ mm $< \left(\frac{1}{5}\right)$ (300 mm) minimum dimensions

$25$ mm $< \left(\frac{3}{4}\right)$ (50 mm) rebar spacing

$25$ mm $< \left(\frac{3}{4}\right)$ (40 mm) rebar cover

**Aggregate size Okay for dimensions**

(Table 7.5)
25 mm maximum size coarse aggregate and 2.60 FM fine aggregate

Coarse aggregate factor = 0.69

Oven-dry weight of coarse aggregate =
(1681) (0.69) = 1160 kg/m$^3$ (1956 lb/yd$^3$)

**Coarse aggregate = 1160 kg/m$^3$ (1956 lb/yd$^3$)**

**4.** Air Content

(Table 7.6)
Severe exposure, target air content = 6.0%

Job range = 5% to 8% base

**Design on 7%**

**5.** Workability

(Table 7.7)
Pier best fits the column requirement in the table

Slump range = 25 to 100 mm (1 to 4 in.)

**Use 75 mm (3 in.)**

**6.** Water Content

(Table 7.8)
25 mm aggregate with air entrainment and 75 mm slump

Water = 175 kg/m$^3$ (295 lb/yd$^3$) for angular aggregates. Since we have round coarse aggregates, reduce by 27 kg/m$^3$ (45 lb/yd$^3$)

**Required water = 148 kg/m$^3$ (250 lb/yd$^3$)**

**7.** Cement Content

Water-cement ratio = 0.45, water = 148 kg/m$^3$ (250 lb/yd$^3$)

Cement = 148 / 0.45 = 329 kg/m$^3$ (556 lb/yd$^3$)

Increase for minimum criterion of 334 kg/m$^3$ (564 lb/yd$^3$) for exposure

**Cement = 334 kg/m$^3$ (564 lb/yd$^3$)**

**8.** Admixture

7% air, cement = 334 kg/m$^3$ (564 lb/yd$^3$)

Admixture = (6.3) (7) (334 / 100) = 147 ml/m$^3$ (3.9 fl oz/yd$^3$)

**Admixture = 147 ml/m$^3$ (3.9 fl oz/yd$^3$)**

**9.** Fine Aggregate Requirements

Find fine aggregate content; use the absolute volume method.

Water volume = 148 / (1 × 1000) = 0.148 m$^3$/m$^3$ (4.006 ft$^3$/yd$^3$)

Cement volume = 334 / (3.15 × 1000) = 0.106 m$^3$/m$^3$ (2.869 ft$^3$/yd$^3$)

Air volume = 0.07 m$^3$/m$^3$ (0.07 × 27 = 1.890 ft$^3$/yd$^3$)

Coarse aggregate volume = 1160 / (2.621 × 1000)
= 0.443 m$^3$/m$^3$ (11.960 ft$^3$/yd$^3$)
Subtotal volume = 0.767 m$^3$/m$^3$ (20.725 ft$^3$/yd$^3$)

Fine aggregate volume = 1 − 0.767 = 0.233 m$^3$/m$^3$ (27 − 20.725 = 6.275 ft$^3$/yd$^3$)

Fine aggregate weight = (0.233) (2.572) (1000) = 599 kg/m$^3$ (1007 lb/yd$^3$)

**Fine aggregate = 599 kg/m$^3$ (1007 lb/yd$^3$)**

**10.** Moisture Corrections

**Coarse aggregate:** Need 1160 kg/m$^3$ (1956 lb/yd$^3$) in SSD condition, so increase by 3% for excess moisture

Moist coarse aggregate = (1160) (1.03) = 1195 kg/m$^3$ (2015 lb/yd$^3$)

**Fine aggregate:** Need 599 kg/m$^3$ (1007 lb/yd$^3$) in SSD condition, so increase 4% for excess moisture

Fine aggregate in moist condition = (599) (1.04)
= 623 kg/m$^3$ (1047 lb/yd$^3$)

**Water:** Reduce for free water on aggregates
= 148 − 1160 (0.03 − 0.004) − 599 (0.04 − 0.008)
= 99 kg/m$^3$ (176 lb/yd$^3$)

Summary

| | Batch Ingredients Required | |
|---|---|---|
| | 1 m$^3$ PCC | 1 yd$^3$ PCC |
| Water | 99 kg | 176 lb |
| Cement | 334 kg | 564 lb |
| Fine aggregate | 623 kg | 1047 lb |
| Coarse aggregate | 1195 kg | 2015 lb |
| Admixture | 147 ml | 3.9 fl oz |

## Mixing Concrete for Small Jobs

The mix design process applies to large jobs. For small jobs, where a large design effort is not economical (e.g., jobs requiring less than one cubic meter of concrete), Tables 7.11 and 7.12 can be used as a guide. The values in these tables may need to be adjusted to obtain a workable mix using the locally available aggregates. Recommendations related to exposure conditions discussed earlier should be followed.

Tables 7.11 and 7.12 are used for proportioning concrete mixes by weight and volume, respectively. The tables provide ratios of components with a sum of one unit. Therefore, the required total weight or volume of the concrete mix can be multiplied by the given ratios to obtain the weight or volume of each component. Note that for proportioning by volume, the combined volume is approximately two-thirds of the sum of the original bulk volumes of the components since water and fine materials fill the voids between coarse materials.

◆**SAMPLE PROBLEM 7.4**    Determine the required volumes of ingredients to make a 0.5-m³ batch of air-entrained concrete mix with a maximum gravel size of 19 mm (3/4 in.).

**Solution:** Sum of the original bulk volumes of the components = $0.5 \times 1.5 = 0.75$ m³
From Table 7.12:

**TABLE 7.11**   Proportions of Concrete for Small Jobs, by weight*

| Maximum Size of Coarse Aggregate, mm (in.) | Air-Entrained Concrete | | | | Non–air-Entrained Concrete | | | |
|---|---|---|---|---|---|---|---|---|
| | Cement | Wet Fine Aggregate | Wet Coarse Aggregate† | Water | Cement | Wet Fine Aggregate | Wet Coarse Aggregate† | Water |
| 9.5 (3/8) | 0.210 | 0.384 | 0.333 | 0.073 | 0.200 | 0.407 | 0.317 | 0.076 |
| 12.5 (1/2) | 0.195 | 0.333 | 0.399 | 0.073 | 0.185 | 0.363 | 0.377 | 0.075 |
| 19 (3/4) | 0.176 | 0.296 | 0.458 | 0.070 | 0.170 | 0.320 | 0.442 | 0.068 |
| 25 (1) | 0.169 | 0.275 | 0.493 | 0.063 | 0.161 | 0.302 | 0.470 | 0.067 |
| 37.5 (1-1/2) | 0.159 | 0.262 | 0.517 | 0.062 | 0.153 | 0.287 | 0.500 | 0.060 |

*Portland Cement Association (1980). Used by permission.
†If crushed stone is used, decrease coarse aggregate by 2 kg and increase fine aggregate by 2 kg for each cubic meter of concrete (or decrease coarse aggregate by 3 lb and increase fine aggregate by 3 lb for each cubic foot of concrete).

**TABLE 7.12**   Proportions of Concrete for Small Jobs, by volume*

| Maximum Size of Coarse Aggregate, mm (in.) | Air-Entrained Concrete | | | | Non-air-Entrained Concrete | | | |
|---|---|---|---|---|---|---|---|---|
| | Cement | Wet Fine Aggregate | Wet Coarse Aggregate† | Water | Cement | Wet Fine Aggregate | Wet Coarse Aggregate† | Water |
| 9.5 (3/8) | 0.190 | 0.429 | 0.286 | 0.095 | 0.182 | 0.455 | 0.272 | 0.091 |
| 12.5 (1/2) | 0.174 | 0.391 | 0.348 | 0.087 | 0.167 | 0.417 | 0.333 | 0.083 |
| 19 (3/4) | 0.160 | 0.360 | 0.400 | 0.080 | 0.153 | 0.385 | 0.385 | 0.077 |
| 25 (1) | 0.154 | 0.346 | 0.423 | 0.077 | 0.148 | 0.370 | 0.408 | 0.074 |
| 37.5 (1-1/2) | 0.148 | 0.333 | 0.445 | 0.074 | 0.143 | 0.357 | 0.429 | 0.071 |

*Portland Cement Association (1980). Used by permission.
†The combined volume is approximately two-thirds of the sum of the original bulk volumes.

Volume of cement = $0.75 \times 0.160 = 0.12$ m$^3$

Volume of wet fine aggregate = $0.75 \times 0.360 = 0.27$ m$^3$

Volume of wet coarse aggregate = $0.75 \times 0.400 = 0.3$ m$^3$

Volume of water = $0.75 \times 0.080 = 0.06$ m$^3$

# Mixing and Handling Fresh Concrete

The proper batching, mixing, and handling of fresh concrete are important prerequisites for strong and durable concrete structures. Next we will discuss the basic steps and precautions to be followed in mixing and handling fresh concrete (Mehta and Monteiro 1993; American Concrete Institute 1982; American Concrete Institute 1983).

Batching is measuring and introducing the concrete ingredients into the mixer. Batching by weight is more accurate than batching by volume since weight batching avoids the problem created by bulking of damp sand. Water and liquid admixtures, however, can be measured accurately either by weight or volume. On the other hand, batching by volume is commonly used with continuous mixers and when hand mixing.

Concrete should be mixed thoroughly either in a mixer or by hand until it becomes uniform in appearance. Hand mixing is usually limited to small jobs or where mechanical mixers are not available. Mechanical mixers include on-site mixers and central mixers in ready mix plants. The capacity of these mixers varies from 1.5 m$^3$ to 9 m$^3$ (2 yd$^3$ to 12 yd$^3$). Mixers also vary in type such as tilting, nontilting, and pan-type mixers. Most of the mixers are batch mixers, although some mixers are continuous.

Mixing time and number of revolutions vary with the size and type of the mixer. Specifications usually require a minimum of 1 minute of mixing for stationary mixers of up to 0.75 m$^3$ (1 yd$^3$) of capacity, with an increase of 15 seconds for each additional 0.75 m$^3$ of capacity. Mixers are usually charged with 10% of the water, followed by uniform additions of solids and 80% of the water. Finally, the remainder of the water is added to the mixer.

## Ready Mixed Concrete

Ready mixed concrete is mixed in a central plant, and delivered to the job site in mixing trucks ready for placing (Figure 7.4). Three mixing methods can be used for ready mixed concrete.

1. Central-mixed concrete is mixed completely in a stationary mixer and delivered in an agitator truck (2 rpm to 6 rpm).
2. Shrink-mixed concrete is partially mixed in a stationary mixer and completed in a mixer truck (4 rpm to 16 rpm).
3. Truck-mixed concrete is mixed completely in a mixer truck (4 rpm to 16 rpm).

Truck manufacturers usually specify the speed of rotation for their equipment. Also, specifications limit the number of revolutions in a truck mixer in order to avoid segregation. Furthermore, the concrete should be discharged at the job site within 90 minutes from the start of mixing, even if retarders are used (ASTM C94).

**FIGURE 7.4** Concrete mixing truck.

## Mobile Batcher Mixed Concrete

Concrete can be mixed in a mobile batcher mixer at the job site. Aggregate, cement, water, and admixtures are fed continuously by volume, and the concrete is usually pumped into the forms.

## Pumped Concrete

Pumped concrete is frequently used for large construction projects. Special pumps deliver the concrete directly into the forms. Careful attention must be exercised to ensure well-mixed concrete with proper workability. The slump should be between 40 mm to 100 mm (1-1/2 in. to 4 in.) before pumping. During pumping the slump decreases by about 12 mm to 25 mm (1/2 in. to 1 in.) due to partial compaction. Blockage could happen during pumping due to either the escape of water through the voids in the mix or due to friction if fines content is too high (Neville 1981).

## Vibration of Concrete

Quality concrete requires thorough consolidation to reduce the entrapped air in the mix. On small jobs, consolidation can be accomplished manually by ramming and tamping the concrete. For large jobs, vibrators are used to consolidate the concrete. Several types of vibrators are available depending on the application. *Internal vibrators* are the most common type used on construction projects. These consist of an eccentric weight housed in a spud: The weight is rotated at high speed to produce vibration. The spud is slowly lowered into and through the entire layer of concrete, penetrating into the underlying layer if it is still plastic. The spud is left in place for 5 seconds to 2 minutes, depending on the type of vibrator and the consistency of the concrete. The operator judges the total vibration time required. Over-vibration causes segregation as the mortar migrates to the surface.

Several specialty types of vibrators are used in the production of precast concrete. These include *external vibrators, vibrating tables, surface vibrators, electric hammers,* and *vibratory rollers* (Neville 1981).

## Pitfalls and Precautions for Mixing Water

Since the water-cement ratio plays an important role in concrete quality, the water content must be carefully controlled in the field. Water should not be added to the concrete during transportation. Crews frequently want to increase the amount of water in order to improve workability. If water is added, the hardened concrete will suffer serious loss in quality and strength. The engineer in the field must prevent any attempt to increase the amount of mixing water in the concrete beyond that which is specified in the mix design.

## Measuring Air Content in Fresh Concrete

Mixing and handling can significantly alter the air content of fresh concrete. Thus, field tests are used to ensure that the concrete has the proper air content prior to placing. Air content can be measured with the pressure, volumetric, gravimetric, or Chace air indicator methods.

The pressure method (ASTM C231) is widely used since it takes less time than the volumetric method. The pressure method is based on Boyle's law, which relates pressure to volume. A calibrated cylinder (Figure 7.5) is filled with fresh concrete. The vessel is capped and air pressure applied. The applied pressure compresses the air in the voids of the concrete. The volume of air voids is determined by measuring the

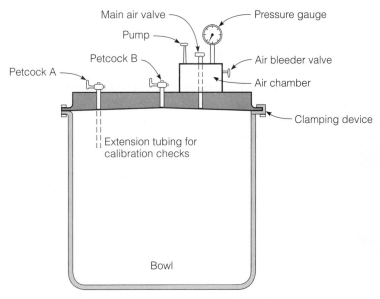

**FIGURE 7.5** Pressure method apparatus for determining air voids in fresh concrete—Type B meter (ASTM C231). Copyright ASTM. Reprinted with permission.

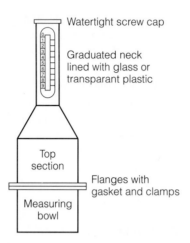

FIGURE 7.6 Volumetric method (Roll-A-Meter) apparatus for determining air voids in fresh concrete (ASTM C173). Copyright ASTM. Reprinted with permission.

amount of volume reduced by the pressure applied. This method is not valid for concrete made with lightweight aggregates since air in the aggregate voids is also compressed, confounding the measurement of the air content of the cement paste.

The volumetric method for determining air content (ASTM C173) can be used for concrete made with any type of aggregate. The basic process involves placing concrete in a fixed volume cylinder, as shown in Figure 7.6. An equal volume of water is added to the container. Agitation of the container allows the excess water to displace the air in the cement paste voids. The water level in the container falls as the air rises to the top of the container. Thus, the volume of air in the cement paste is directly measured. The accuracy of the method depends on agitating the sample enough to remove all the air from it.

The gravimetric method (ASTM C138) compares the unit weight of freshly mixed concrete to the theoretical maximum unit weight of the mix. The theoretical unit weight is computed from the mix proportions and the specific gravity of each ingredient. This method requires very accurate specific gravity measurements and thus is more suited to the laboratory rather than the field.

The Chace air indicator test (AASHTO T199) is a quick method used to determine the air content of freshly mixed concrete. The device consists of a small glass tube with a stem, a rubber stopper, and a metal cup mounted on the stopper, as shown in Figure 7.7. The metal cup is filled with cement mortar from the concrete to be tested. The indicator is filled with alcohol to a specified level and the stopper is inserted into the indicator. The indicator is then closed with a finger and gently rolled and tapped until all of the mortar is dispersed in the alcohol and all the air is displaced with alcohol. With the indicator held in a vertical position, the alcohol level in the stem is read. This reading is then adjusted using calibration tables or figures to determine the air content. The Chace air indicator test can be used to rapidly monitor air content, but it is not precise, nor does it have the repeatability required for specification control. It is

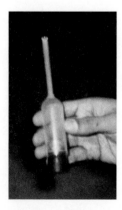

FIGURE 7.7 Chace air indicator.

especially useful for measuring the air content of small areas near the surface that may have lost air content by improper finishing.

These methods of measuring air content determine the total amount of air including entrapped air and entrained air, as well as air voids in aggregate particles. Only minute bubbles produced by air-entraining agents impart durability to the concrete. However, the current state of the art is unable to distinguish between the types of air in fresh concrete.

## Curing Concrete

Curing is the process of maintaining satisfactory moisture content and temperature in the concrete for a definite period of time. Hydration of cement is a long-term process and requires water and proper temperature. Therefore, curing allows continued hydration and consequently, continued gains in concrete strength. In fact, once curing stops the concrete dries out, and the strength gain stops, as indicated in Figure 7.8. If the concrete is not cured and is allowed to dry in air, it will gain only about 50% of the strength of continuously cured concrete. If concrete is cured for only 3 days, it will reach about 60% of the strength of continuously cured concrete; if it is cured for 7 days, it will reach 80% of the strength of continuously cured concrete. If curing stops for some time and then resumes again, the strength gain will also stop and reactivate.

Increasing temperature increases the rate of hydration and, consequently, the rate of strength development. Temperatures below 10°C (50°F) are unfavorable for hydration and should be avoided, if possible, especially at early ages.

Although concrete of high strength may not be needed for a particular structure, strength is usually emphasized and controlled since it is an indication of the concrete quality. Thus proper curing not only increases strength, but also provides other desirable properties such as durability, water tightness, abrasion resistance, volume stability, resistance to freeze and thaw, and resistance to de-icing chemicals.

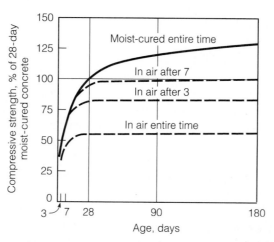

**FIGURE 7.8** Compressive strength of concrete at different ages and curing levels.

Curing should start after the final set of the cement. If concrete is not cured after setting, concrete will shrink, causing cracks. Drying shrinkage can be prevented if ample water is provided for a long period of time. An example of improper curing would be a concrete floor built directly over the subgrade, not cured at the surface, with the moisture in the soil curing it from the bottom. In this case the concrete slab may curl due to the relative difference in shrinkage.

Curing can be performed by

1. maintaining the presence of water in the concrete during early ages. Methods to maintain the water pressure include pounding or immersion, spraying or fogging, and wet coverings.

2. preventing loss of mixing water from the concrete by sealing the surface. Methods to prevent water loss include impervious papers or plastic sheets, membrane-forming compounds, and leaving the forms in place.

3. accelerating the strength gain by supplying heat and additional moisture to the concrete. Accelerated curing methods include steam curing, insulating blankets or covers, and various heating techniques.

Note that preventing loss of mixing water from the concrete by sealing the surface is not as effective as maintaining the presence of water in the concrete during early ages. The choice of the specific curing method or combination of methods depends on the availability of curing materials, size and shape of the structure, in-place versus plant production, economics, and aesthetics (Kosmatka and Panarese 1988; American Concrete Institute 1986a).

## Ponding or Immersion

Ponding involves covering the exposed surface of the concrete structure with water. Ponding can be achieved by forming earth dikes around the concrete surface to retain water. This method is suitable for flat surfaces such as floors and pavements, especially for small jobs. The method requires intensive labor and supervision. Immersion is used to cure test specimens in the laboratory, as well as other concrete members, as appropriate.

## Spraying or Fogging

A system of nozzles or sprayers can be used to provide continuous spraying or fogging. This method requires a large amount of water and could be expensive. It is most suitable in high temperature and low humidity environments. Commercial test laboratories generally have a controlled temperature and humidity booth for curing specimens.

## Wet Coverings

Moisture-retaining fabric coverings saturated with water, such as burlap, cotton mats, and rugs are used in many applications. The fabric can be kept wet either by periodic watering or covering the fabric with polyethylene film to retain moisture. On small jobs, wet coverings of earth, sand, saw dust, hay, or straw can be used. Stains or discoloring of concrete could occur with some types of wet coverings.

## Impervious Papers or Plastic Sheets

Evaporation of moisture from concrete can be reduced using impervious papers, such as kraft papers, or plastic sheets, such as polyethylene film. Impervious papers are suitable for horizontal surfaces and simply shaped concrete structures, while plastic sheets are effective and easily applied to various shapes. Periodic watering is not required when impervious papers or plastic sheets are used. Discoloration, however, can occur on the concrete surface.

## Membrane-forming Compounds

Various types of liquid membrane-forming compounds can be applied to the concrete surface to reduce or retard moisture loss. These can be used to cure fresh concrete, as well as hardened concrete, after removal of forms or after moist curing. Curing compounds can be applied by hand or by using spray equipment. Either one coat or two coats (applied perpendicular to each other) are used. Normally, the concrete surface should be damp when the curing compound is applied. Curing compounds should not be used when subsequent concrete layers are to be placed since the compound hinders the bond between successive layers. Also, some compounds affect the bond between the concrete surface and paint.

## Forms Left in Place

Loss of moisture can be reduced by leaving the forms in place as long as practical, provided that the top concrete exposed surface is kept wet. If wood forms are used, the forms should also be kept wet. After removing the forms, another curing method can be used.

## Steam Curing

Steam curing is used when early strength gain in concrete is required or additional heat is needed during cold weather. Steam curing can be attained either with or without pressure. Steam at atmospheric pressure is used for enclosed cast-in-place structures and large precast members. High-pressure steam in autoclaves can be used at small manufactured plants.

## Insulating Blankets or Covers

When the temperature falls below freezing, concrete should be insulated using layers of dry, porous material such as hay or straw. Insulating blankets manufactured of fiberglass, cellulose fibers, sponge rubber, mineral wool, vinyl foam, or open-cell polyurethane foam can be used to insulate formwork. Moisture proof commercial blankets can also be used.

## Electrical, Hot Oil, and Infrared Curing

Precast concrete sections can be cured using electrical, oil, or infrared curing techniques. Electrical curing includes electrically heated steel forms, and electrically heated blankets. Reinforcing steel can be used as a heating element, and concrete can be used as the electrical conductor. Steel forms can also be heated by circulating hot oil around the outside of the structure. Infrared rays have been used for concrete curing on a limited basis.

## Curing Period

The curing period should be as long as is practical. The minimum time depends on several factors, such as type of cement, mixture proportions, required strength, ambient weather, size and shape of the structure, future exposure conditions, and method of curing. For most concrete structures the curing period at temperatures above 5°C (40°F) should be a minimum of 7 days or until 70% of specified compressive or flexure strength is attained. The curing period can be reduced to 3 days if high early strength concrete is used and the temperature is above 10°C (50°F).

# Properties of Hardened Concrete

It is important for the engineer to understand the basic properties of hardened portland cement concrete and to be able to evaluate these properties. The main properties of hardened concrete that are of interest to civil and construction engineers include the early volume change, creep, permeability, and stress-strain relation.

## Early Volume Change

When the cement paste is still plastic it undergoes a slight decrease in volume of about 1%. This shrinkage is known as *plastic shrinkage* and is due to the loss of water from the cement paste either from evaporation or from suction by dry concrete below the fresh concrete. Plastic shrinkage may cause cracking; it can be prevented or reduced by controlling water loss.

In addition to the possible decrease in volume when the concrete is still plastic, another form of volume change may occur after setting, especially at early ages. If concrete is not properly cured and allowed to dry, it will shrink. This shrinkage is referred to as *drying shrinkage,* and it also causes cracks. Shrinkage takes place over a long period of time, although the rate of shrinkage is high early, then decreases rapidly with time. In fact, about 15% to 30% of the shrinkage occurs in the first 2 weeks, while 65% to 85% occurs in the first year. Shrinkage and shrinkage-induced cracking are increased by several factors, including lack of curing, high water-cement ratio, high cement content, low coarse aggregate content, existence of steel reinforcement, and aging. On the other hand, if concrete is cured continuously in water after setting, concrete will swell very slightly due to the absorption of water. Since swelling, if it happens, is very small, it does not cause significant problems. *Swelling* is accompanied by a slight increase in weight (Neville 1981).

How much drying shrinkage occurs depends on the size and shape of the concrete structure. Also, nonuniform shrinkage could happen due to the nonuniform loss of water. This may happen in mass concrete structures, where more water is lost at the surface than at the interior. In cases like this cracks may develop at the surface. In other cases curling might develop due to the nonuniform curing throughout the structure and, consequently, nonuniform shrinkage.

## Creep Properties

Creep is defined as the gradual increase in strain, with time, under sustained load. Creep of concrete is a long-term process, and it takes place over many years. Although the amount of creep in concrete is relatively small, it could affect the performance of structures. The effect of creep varies with the type of the structure. In simply supported reinforced concrete beams, creep increases the deflection and, therefore, increases the stress in the steel. In reinforced concrete columns, creep results in a gradual transfer of load from the concrete to the steel. Creep also could result in losing some of the prestress in prestressed concrete structures, although the use of high-tensile stress steel reduces this effect. Rheological models discussed in Chapter 1 have been used to analyze the creep response of concrete (Neville 1981).

## Permeability

Permeability is an important factor that largely affects the durability of hardened concrete. Permeable concrete allows water and chemicals to penetrate, which, in turn, reduces the resistance of the concrete structure to frost, alkali-aggregate reactivity, and other chemical attacks. Water that permeates into reinforced concrete causes corrosion of steel rebars. Furthermore, impervious concrete is a prerequisite in watertight structures, such as tanks and dams.

Typically, the air voids in the cement paste and aggregates are small and do not affect permeability. However, the air voids that do affect permeability of hardened concrete are obtained from two main sources: incomplete consolidation of fresh concrete and voids resulting from evaporation of mixing water that is not used for hydration of cement.

Therefore, increasing the water-cement ratio in fresh concrete has a severe effect on permeability. Figure 7.9 shows the typical relation between the water-cement ratio and the coefficient of permeability of mature cement paste (Powers 1954). It can be

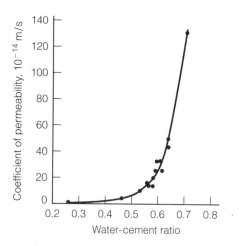

**FIGURE 7.9** Relation between water-cement ratio and permeability of mature cement paste.

seen from the figure that increasing the water-cement ratio from 0.3 to 0.7 increases the coefficient of permeability by a factor of 1000. For a concrete to be watertight, the water-cement ratio should not exceed 0.48 for exposure to fresh water and not more than 0.44 for exposure to seawater (American Concrete Institute 1975).

Other factors that affect the permeability include age of concrete, fineness of cement particles, and air-entraining agents. Age reduces the permeability since hydration products fill the spaces between cement grains. The finer the cement particles, the faster the rate of hydration and the faster the development of impermeable concrete. Air-entraining agents indirectly reduce the permeability, since they allow the use of a lower water-cement ratio.

### Stress-Strain Relation

Typical stress-strain relations of 28-day-old concrete with different water-cement ratios are shown in Figure 7.10 (Hognestad et al. 1955). It can be seen that increasing the water-cement ratio decreases both strength and stiffness of the concrete. The figure also shows that the stress-strain relation is close to linear at low stress levels, then becomes nonlinear. With a water-cement ratio of 0.50 or less and a strain up to 0.0015, the stress-strain relation is almost linear. With higher water-cement ratios the stress-strain relation becomes nonlinear at smaller strains. The curves also show that high-strength concrete has sharp peaks and sudden failure characteristics when compared to low-strength concrete.

As discussed in Chapter 1, the elastic limit can be defined as the largest stress that does not cause a measurable permanent strain. When the concrete is loaded slightly beyond the elastic range and then unloaded, a small amount of strain might remain initially but it may recover eventually due to creep. Also, since concrete is not perfectly elastic, the rate of loading affects the stress-strain relation to some extent. Therefore, a specific rate of loading is required for testing concrete. It is interesting to

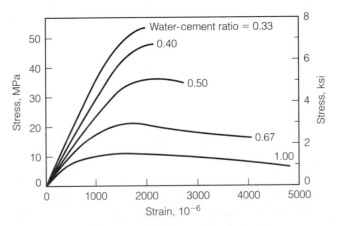

**FIGURE 7.10** Typical stress-strain relations for compressive tests on 0.15 m by 0.30 m concrete cylinders with different water-cement values at an age of 28 days.

note that the shape of the stress-strain relation of concrete is almost the same for both compression and tension, although the tensile strength is much smaller than the compressive strength. In fact, the tensile strength of concrete typically is ignored in the design of concrete structures.

The modulus of elasticity of concrete is commonly used in designing concrete structures. Since the stress-strain relationship is not exactly linear, the classic definition of modulus of elasticity (Young's modulus) is not applicable. The initial tangent modulus of concrete has little practical importance. The tangent modulus is valid only for a low stress level where the tangent is determined. Both secant and chord moduli represent "average" modulus values for certain stress ranges. The chord modulus in compression is more commonly used for concrete and is determined according to ASTM C469 (referred to as the modulus of elasticity). The method requires three or four loading and unloading cycles, after which the chord modulus is determined between a point corresponding to a very small strain value and a point corresponding to either 40% of the ultimate stress or a specific strain value. Normal-weight concrete has a modulus of elasticity of 14 GPa to 41 GPa (2000 ksi to 6000 ksi).

Poisson's ratio can also be determined using ASTM C469. Poisson's ratio is used in advanced structural analysis of shell roofs, flat-plate roofs, and mat foundations. Poisson's ratio of concrete varies between 0.11 and 0.21, depending on aggregate type, moisture content, concrete age, and compressive strength. A value of 0.15 to 0.20 is commonly used.

It is interesting to note that both aggregate and cement paste, when tested individually, exhibit linear stress-strain behavior. However, the stress-strain relation of concrete is nonlinear. The reason for this behavior is attributed to the microcracking in concrete at the interface between aggregate particles and the cement paste (Shah and Winter 1968).

The modulus of elasticity of concrete increases when the compressive strength increases, as demonstrated in Figure 7.10. There are several empirical relations between the modulus of elasticity of concrete and the compressive strength. For normal-weight concrete, the relationship used in the United States for designing concrete structures is defined by the ACI Building Code as

$$E_c = 4,731 \sqrt{f'_c} \tag{7.3a}$$

or

$$E_c = 57,000 \sqrt{f'_c} \tag{7.3b}$$

where
$E_c$ = the modulus of elasticity,
$f'_c$ = the compressive strength.

Equation 7.3a is used for SI units, where both $E_c$ and $f'_c$ are in MPa, whereas Equation 7.3b is used for the U.S. customary units, where both $E_c$ and $f'_c$ are in psi. This relation is useful since it relates the modulus of elasticity (needed for designing concrete structures) with the compressive strength, which can be measured easily in the laboratory.

**SAMPLE
PROBLEM 7.5**

A normal weight concrete has an average compressive strength of 30 MPa. What is the estimated modulus of elasticity?

*Solution:*

$$E_c = 4{,}731 \sqrt{f'_c} = 4{,}731(30)^{1/2} = 25{,}913 \text{ MPa} = 25.9 \text{ GPa}$$

# Testing of Hardened Concrete

Many tests are used to evaluate the hardened concrete properties either in the laboratory or in the field. Some of these tests are destructive, while others are nondestructive. Tests can be performed for different purposes; however, they are mostly conducted to control the quality of the concrete and to check specification compliance. Probably the most common test performed on hardened concrete is the compressive strength test, since it is relatively easy to perform and since there is a strong correlation between the compressive strength and many desirable properties (Neville 1981; Mehta and Monteiro 1993). Other tests include split-tension, flexure strength, rebound hammer, penetration resistance, ultrasonic pulse velocity, and maturity tests.

## Compressive Strength Test

The compressive strength test is the test most commonly performed on hardened concrete. The compressive strength $f'_c$ of normal-weight concrete is between 21 MPa to 34 MPa (3000 psi to 5000 psi). In the United States the test is performed on cylindrical specimens and standardized by ASTM C39. The specimen is prepared either in the lab or in the field according to ASTM C192 or C31, respectively. Cores could also be drilled from the structure following ASTM C42. The standard specimen size is 0.15 m (6 in.) in diameter and 0.30 m (12 in.) high, although other sizes with a height-diameter ratio of two can also be used. The diameter of the specimen must be at least three times the nominal maximum size of the coarse aggregate in the concrete.

In the lab specimens are prepared in three equal layers and rodded 25 times per layer. After the surface is finished, specimens are kept in the mold for the first 24 ± 8 hours. Specimens are then removed from the mold and cured at 23 ± 1.7°C (73.4 ± 3°F) either in saturated-lime water or in a moist cabinet having a relative humidity of 95% or higher until the time of testing. Before testing, specimens are capped at the two bases to ensure parallel surfaces. High-strength gypsum plaster, sulfur mortar, or a special capping compound can be used for capping and is applied with a special alignment device (ASTM C617). Using a testing machine, specimens are tested by applying axial compressive load with a specified rate of loading until failure. The compressive strength of the specimen is determined by dividing the maximum load carried by the specimen during the test by the average cross-sectional area. The number of specimens and the number of test batches depend on established practice and the nature of the test program. Usually three or more specimens are tested for each test age and test condition. Test ages often used are 7 days and 28 days.

Note that the test specimen must have a height-diameter ratio of two. The main reason for this requirement is to eliminate the end effect due to the friction between the loading heads and the specimen. Thus we can guarantee a zone of uniaxial compression within the specimen. Another reason is that previous studies have shown that a slight departure from this ratio does not seriously affect the measured value of strength. If the height-diameter ratio is less than two, a correction factor can be applied to the results as indicated in ASTM C39.

The compressive strength of the specimen is affected by the specimen size. Increasing the specimen size reduces the strength because there is a greater probability of weak elements where failure starts in large specimens than in small specimens. In general, large specimens have less variability and better representation of the actual strength of the concrete than small specimens. Therefore, the 0.15 m by 0.30 m (6 in. by 12 in.) size is the most suitable specimen size for determining the compressive strength. However, some agencies use 0.10 m (4 in.) diameter by 0.20 m (8 in.) high specimens. The advantages of using smaller specimens are the ease of handling, less possibility of accidental damage, less concrete needed, the ability to use a low-capacity testing machine, and less space needed for curing and storage. Because of the strength variability of small specimens, more specimens should be tested for smaller specimens than are tested for standard-sized specimens. In some cases, five 0.10 m by 0.20 m replicate specimens are used instead of the three replicates commonly used for the standard-sized specimens. Also, when small-sized specimens are used, the engineer should understand the limitations of the test and consider these limitations in interpreting the results.

The interface between the hardened cement paste and aggregate particles is typically the weakest location within the concrete material. When concrete is stressed beyond the elastic range, microcracks develop at the cement paste–aggregate interface and continuously grow until failure. Figure 7.11 shows a scanning electron microscope micrograph of the fractured surface of a hardened cement mortar cylinder at 500×. The figure shows that the cleavage fracture surfaces where sand particles were dislodged during loading. The figure also shows the microcracks around some sand particles developed during loading.

## Split-Tension Test

The split-tension test (ASTM C496) measures the tensile strength of concrete. In this test a 0.15-m by 0.30-m (6 in. by 12 in.) concrete cylinder is subjected to a compressive load at a constant rate along the vertical diameter until failure, as shown in Figure 7.12. Failure of the specimen occurs along its vertical diameter due to tension developed in the transverse direction. The split tensile (indirect tensile) strength is computed as

$$T = \frac{2P}{\pi L d} \tag{7.4}$$

where

$T$ = tensile strength, MPa (psi),
$P$ = load at failure, N (psi),
$L$ = length of specimen, mm (in.), and
$d$ = diameter of specimen, mm (in.).

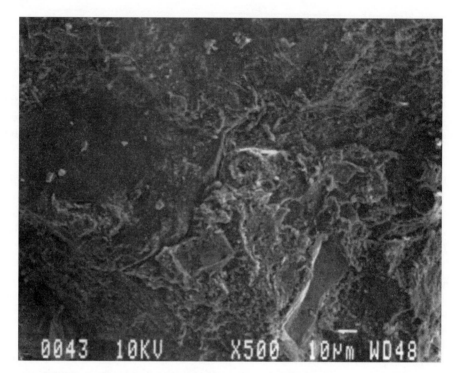

**FIGURE 7.11** Scanning electron microscope micrograph of hardened cement mortar at 500×.

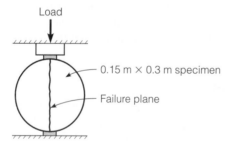

**FIGURE 7.12** Split-tension test apparatus.

Typical indirect tensile strength of concrete varies from 2.5 MPa to 3.1 MPa (360 psi to 450 psi) (Neville 1981). The tensile strength of concrete is about 10% of its compressive strength.

## Flexure Strength Test

The flexure strength test (ASTM C78) is important for design and construction of road and airport concrete pavements. The specimen is prepared either in the lab or in the field in accordance with ASTM C192 or C31, respectively. Several specimen sizes can be used. However, the sample must have a square cross section and a span of three times the specimen depth. Typical dimensions are 0.15 m by 0.15 m (6 in. by 6 in.) cross section and 0.30-m (18 in.) span. After molding, specimens are kept in the mold for the first $24 \pm 8$ hours, then removed from the mold and cured at $23 \pm 1.7°C$ ($73.4 \pm 3°F$) either in saturated-lime water or in a moist cabinet with a relative humidity of 95% or higher until testing. The specimen is then turned on its side and centered in the third-point loading apparatus, as illustrated in Figure 7.13. The load is continuously applied at a specified rate until rupture. If fracture initiates in the tension surface within the middle third of the span length, the flexure strength (modulus of rupture) is calculated as

$$R = \frac{PL}{(bd^2)} \tag{7.5}$$

where
$\quad R$ = flexure strength, MPa (psi),
$\quad P$ = maximum applied load, N (lb),
$\quad L$ = span length, mm (in.),
$\quad b$ = average width of specimen, mm (in.), and
$\quad d$ = average depth of specimen, mm (in.).

If fracture occurs slightly outside the middle third, the results can still be used with some corrections, otherwise the results are discarded.

For normal-weight concrete, the flexure strength can be approximated as

$$R = (0.62 \text{ to } 0.83)\sqrt{f'_c} \tag{7.6a}$$

$$R = (7.5 \text{ to } 10)\sqrt{f'_c} \tag{7.6b}$$

Equation 7.6a is used for SI units, where both $R$ and $f'_c$ are in MPa, whereas Equation 7.6b is used for U.S. customary units, where both $R$ and $f'_c$ are in psi.

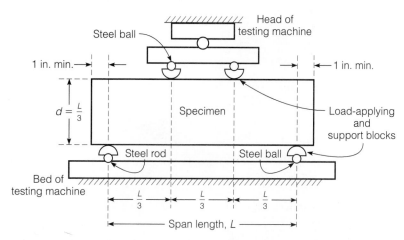

**FIGURE 7.13** Apparatus for flexure test of concrete by third-point loading method. Copyright ASTM. Reprinted with permission.

## Rebound Hammer Test

The rebound hammer test, also known as Schmidt hammer test, is a nondestructive test performed on hardened concrete to determine the hardness of the surface (Figure 7.14). The hardness of the surface can be correlated to some extent to the concrete strength. The rebound hammer is commonly used to get an indication of the concrete strength. The device is about 0.3 m (1 ft) long and encloses a mass and a spring. The spring-loaded mass is released to hit the surface of the concrete. The mass rebounds and the amount of rebound is read on a scale attached to the device. The larger the rebound, the harder the concrete surface and, therefore, the greater the strength. The device usually comes with graphs prepared by the manufacturer to relate rebound to strength. The test can also be used to check uniformity of the concrete surface.

The test is very simple to run and is standardized by ASTM C805. To perform the test, the hammer must be perpendicular to a clean, smooth concrete surface. In some cases, it would be hard to satisfy this condition. Therefore, correlations, usually provided by the manufacturer, can be used to relate the strength to the amount of rebound at different angles. Rebound hammer results are also affected by several other factors such as local vibrations, the existence of coarse aggregate particles at the surface, and the existence of voids near the surface. To reduce the effect of these factors, it is desirable to average 10 to 12 readings from different points in the test area.

## Penetration Resistance Test

The penetration resistance test, also known as the Windsor Probe test, is standardized by ASTM C803. The instrument, Figure 7.15, is a gunlike device that shoots probes into the concrete surface in order to determine its strength. The amount of penetration of

**FIGURE 7.15** Windsor Probe test device.

the probe in the concrete is inversely related to the strength of concrete. The test is almost nondestructive since it creates small holes in the concrete surface.

The device is equipped with a special template with three holes, which is placed on the concrete surface. The test is performed in each of the holes. The average of the penetrations of the three probes through these holes is determined using a scale and a special plate. Care should be exercised in handling the device to avoid injury. As a way of improving safety, the device cannot be operated without pushing hard on the concrete surface to prevent accidental shooting. The penetration resistance test is expected to provide better strength estimation than the rebound hammer, since the penetration resistance measurement is made not just at the surface but also in the depth of the sample.

## Ultrasonic Pulse Velocity Test

The ultrasonic pulse velocity test (ASTM C597) measures the velocity of an ultrasonic wave passing through the concrete (Figure 7.16). In this test the path length between transducers is divided by the travel time to determine the average velocity of wave propagation. Attempts have been made to correlate pulse velocity data with concrete strength parameters. No good correlations were found since the relationship between pulse velocity and strength data is affected by a number of variables such as age of concrete, aggregate-cement ratio, aggregate type, moisture condition, and location of reinforcement (Mehta and Moneiro 1993). This test is used to detect cracks, discontinuities, or internal deterioration in the structure of concrete.

## Maturity Test

Maturity of a concrete mixture is defined as the degree of cement hydration, which varies as a function of both time and temperature. Therefore, it is assumed that for a particular concrete mixture, strength is a function of maturity. Maturity meters

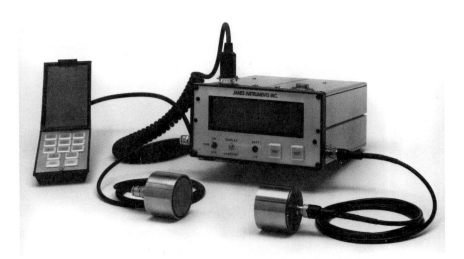

**FIGURE 7.16** Ultrasonic pulse velocity apparatus. (Courtesy of James Instrument Inc.)

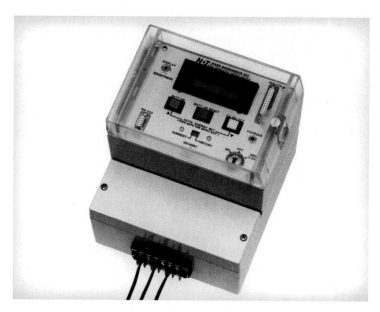

**FIGURE 7.17**  Maturity meter apparatus. (Courtesy of James Instrument Inc.)

(Figure 7.17) have been developed to provide an estimate of concrete strength by monitoring the temperature of concrete with time. This test (ASTM C1074) is performed on fresh concrete and continued for several days. The maturity meter must be calibrated for each concrete mix.

## Alternatives to Conventional Concrete

There are several alternatives that increase the flexibility and applications of concrete (Mehta and Monteiro 1993). While a technical presentation of materials for each of these technologies is beyond the scope of this book, materials the engineer should be aware of the additional capabilities of concrete. Some of these alternatives include:

lightweight concrete
high-strength concrete
shrinkage compensating concrete
fiber-reinforced concrete
heavy-weight concrete
high-workability or flowable concrete
polymers and concrete
high-performance concrete roller-compacted concrete

### Lightweight Concrete

Students competing in the annual ASCE concrete canoe competition frequently produce concrete with a unit weight less than water. The *ACI Guide for the Structural Lightweight Aggregate Concrete* requires a 28-day compressive strength of 17 MPa

**FIGURE 7.18** Aggregate used for lightweight concrete.

(2500 psi) and an air-dried unit weight of less than 1850 kg/m$^3$ (115 lb/ft$^3$) for structural lightweight concrete. Use of lightweight concrete in a structure is usually predicated on a the overall cost of the structure; the concrete may cost more, but the reduced dead weight can reduce structural and foundation costs.

The mix proportions for lightweight concrete must compensate for the absorptive nature of the aggregates. Generally, lightweight aggregates are highly absorptive and can continue to absorb water for an extended period of time (Figure 7.18). This makes determination of a water-cement ratio problematical. In addition, the lightweight aggregates tend to segregate by floating to the surface. A minimum slump mix, with air entraining, is used to mitigate this effect.

Nonstructural applications of very lightweight concrete have also been developed. Concrete made with styrofoam "aggregates" has been used for insulation in some building construction.

## Heavy-weight Concrete

Biological shielding used for nuclear power plants, medical units, and atomic research and test facilities requires massive walls to contain radiation. Concrete is an excellent shielding material. For biological shields, the mass of the concrete can be increased with the use of heavy-weight aggregates. The aggregates can be either natural or manufactured. Natural heavy-weight aggregates include barite, magnetite, hematite, geothite, illmenite, and ferrophosphorus. The specific gravity of these aggregates are from 3.4 to 6.5. Steel, with a specific gravity of 7.8, can be used as aggregate for heavy-weight concrete (Figure 7.19). However, the specific gravity of the aggregates makes workability and segregation of heavy-weight concrete problematical. Using a higher proportion of sand improves the workability. The workability problem can be avoided by preplacing aggregate, then filling the voids between aggregate particles with grout of cement, sand, water, and admixtures. ASTM C637, *Specifications for Aggregates for Radiation Shielding Concrete* and ASTM C637, *Nomenclature of Constituents of Aggregates for Radiation Shielding Concrete* provide further information on heavy-weight concrete practices.

**FIGURE 7.19** Steel used as aggregate for heavy-weight concrete.

## High-Strength Concrete

Concrete made with normal-weight aggregate and compressive strengths greater than 40 MPa (6000 psi) is considered to be high-strength concrete. Producing a concrete with more than 40 MPa compressive strength requires care in the proportioning of the components and in quality control during construction. The microstructure of concrete with a compressive strength greater than 40 MPa is considerably different from that of conventional concrete. In particular, the porosity of the cement paste and the transition zone between the cement paste and the aggregate are the controlling factors for developing high strength. This porosity is controlled by the water-cement ratio. The development of superplasticizers has permitted the development of high-strength concrete that is workable and flowable at low water-cement ratios. In addition, high-strength concrete has excellent durability due to its tight pore structure. In the United States, high-strength concrete is primarily used for skyscrapers. The high-strength, and corresponding high elastic modulus, allows for reduced structural element size.

## High-Workability Concrete

High-workability or flowable concrete has a slump in the range of 180 mm to 230 mm (7 in. to 9 in.). The workability is achieved by using superplasticizers. The benefits of using high-workability concrete include

1. reduced vibration so that concrete may be placed in areas of closely bunched reinforcement and areas with poor access,
2. rapid concrete placement in flat work, floor slabs, roof decks, etc.,
3. rapid pumping of concrete,
4. underwater concrete placement with a tremie pipe, and
5. production of uniform and compact concrete surfaces.

The ability of high-workability concrete to flow into complex shapes has increased the use of concrete for structures with complex designs and highly demanding applications.

Superplasticizers significantly increase the material cost; however, this may be offset by lower placement costs.

## Shrinkage Compensating Concrete

Normal concrete shrinks at early ages especially if it is not properly cured as discussed earlier in this chapter. The addition of alumina powders to the cement can cause the concrete to expand at early ages. Shrinkage compensating cement is marketed as Type K cement. Expansive properties can be used to advantage by restraining the concrete, either by reinforcing or by other means, at early ages. As the restrained concrete tries to expand, compressive stresses are developed. These compressive stresses reduce the tensile stresses developed by drying shrinkage, and the chance of the concrete cracking due to drying shrinkage is reduced. Details on the design and use of shrinkage compensating concrete are available in ACI Committee 223 report *Recommendations for the Use of Shrinkage-Compensating Cements.*

## Polymers and Concrete

Polymers can be used in several ways in the production of concrete. The polymer can be used as the sole binding agent to produce *polymer concrete*. Polymers can be mixed with the plastic concrete to produce *polymer–portland cement concrete*. Polymers can be applied to hardened concrete to produce *polymer-impregnated concrete*.

Polymer concrete is a mixture of aggregates and a polymer binder. There are a wide variety of polymers that can be mixed with aggregates to make polymer concrete. Some of these can be used to make rapid-setting concrete that can be put in service under an hour after placing. Others are formulated for high strength; 140 MPa (20,000 psi) is possible. Some have good resistance to chemical attack. A common characteristic is that most polymer concretes are expensive, limiting their application to situations where their unique characteristics make polymer concrete a cost-effective alternative to conventional concrete.

Polymer–portland cement concrete incorporates a polymer into the production of the portland cement concrete. The polymer is generally an elastomeric emulsion, such as latex.

## Fiber-reinforced Concrete

The brittle nature of concrete is due to the rapid propagation of microcracking under applied stress. However, with fiber-reinforced concrete failure takes place due to fiber pull-out or debonding. Unlike plain concrete, fiber-reinforced concrete can sustain load after initial cracking. This effectively improves the toughness of the material. In addition, the flexural strength of the concrete is improved by up to 30%. For further information on the design and applications of fiber-reinforced concrete, consult the *ACI Guide for Specifying, Mixing, Placing and Finishing Steel Fiber Reinforced Concrete.*

Fibers are available in a variety of sizes, shapes, and materials (Figure 7.20). The fibers can be made of steel, plastic, glass, and natural materials. Steel fibers are the most common. The shape of fibers is generally described by the aspect ratio, length/diameter. Steel fibers generally have diameters from 0.25 mm to 0.9 mm (0.01 in. to 0.035 in.) with aspect ratios of 30 to 150. Glass fiber elements' diameters range from 0.013 mm to 1.3 mm (0.005 in. to 0.05 in.).

**FIGURE 7.20** Fibers used for fiber-reinforced concrete.

The addition of fibers to concrete reduces the workability. The extent of reduction depends on the aspect ratio of the fibers and the volume concentration. Generally, due to construction problems, fibers are limited to a maximum of 2% by volume of the mix. Admixtures can be used to restore some of the workability to the mix.

Since the addition of fibers does not greatly increase the strength of concrete, its use in structural members is limited. In beams, columns, suspended floors, etc., conventional reinforcing must be used to carry the total tensile load. Fiber-reinforced concrete has been successfully used for floor slabs, pavements, slope stabilization, and tunnel linings.

### Roller-compacted Concrete

Based on the unique requirements for mass concrete used for dam construction, roller-compacted concrete (RCC) was developed. This material uses a relatively low cement factor, relaxed gradation requirements, and a water content selected for construction considerations rather than strength. RCC is a no-slump concrete that is transported, placed, and compacted with equipment used for earth and rockfill dam construction. The RCC is hauled by dump trucks, spread with bulldozers, and compacted with vibration compactors. Japanese experience using RCC in construction found several advantages.

1. The mix is economical because of the low cement content.
2. Formwork is minimal because of the layer construction method.
3. The low cement factor limits the heat of hydration, reducing the need for external cooling of the structure.
4. The placement costs are lower than conventional concrete methods due to the use of high-capacity equipment and rapid placement rates.
5. The construction period is shorter than conventional concrete.

In addition, experience in the United States has demonstrated that RCC in-place material costs are about one-third that of conventional concrete. The two primary applications of RCC have been for the construction of dams and large paved areas such as military tank parking aprons.

### High-Performance Concrete

While the current specifications for concrete has provided a material that performs reasonably well, there is concern that the emphasis on strength in the mix design process has lead to concrete that is inadequate in other performance characteristics. This has lead to an interest in developing specifications and design methods for what has been termed *high-performance concrete* (HPC). This material is defined as concrete that is enhanced in some or all of the following properties:

> ease of placement and compaction,
> long-term mechanical properties,
> early-age strength,
> toughness,
> volume stability, and
> extended life in severe environments.

These enhanced characteristics may be accomplished by altering the aggregate gradation, including special admixtures, and improving mixing and placement practices. As the need for HPC is better understood and embraced by the engineering community, there will probably be a transition in concrete specification from the current prescriptive method to the performance-based or performance-related specifications.

**SUMMARY**

The design of durable portland cement concrete materials is the direct responsibility of civil engineers. Selection of the proper proportions of portland cement, water, aggregates, and admixtures, along with good construction practices, dictates the quality of concrete used in structural applications. Using the volumetric mix design method presented in this chapter will lead to concrete with the required strength and durability. However, the proper design of portland cement concrete is irrelevant unless proper construction procedures are followed, including the appropriate mixing, transporting, placing, and curing of the concrete. To ensure these processes produce concrete with the desired properties, a variety of quality control tests are performed by civil engineers, including slump tests, air content tests , and strength gain with time tests.

While the vast majority of concrete projects are constructed with conventional materials, there are a variety of important alternative concrete formulations available for specialty applications. These alternatives are introduced in this chapter; however, the technology associated with these alternatives are relatively complex and further study is required in order to fully understand the behavior of these materials.

**QUESTIONS AND PROBLEMS**

**7.1.** The design engineer specifies a concrete strength of 5500 psi. Determine the required average compressive strength for:

  **a.** a new plant where $s$ is unknown

  **b.** a plant where $s = 500$ psi for 22 test results

  **c.** a plant with extensive history of producing concrete with $s = 400$ psi

  **d.** a plant with extensive history of producing concrete with $s = 600$ psi

**7.2.** Design the concrete mix according to the following conditions:

*Design Environment*
Building frame
Required design strength = 27.6 MPa
Minimum dimension = 150 mm
Minimum space between rebar = 40 mm
Minimum cover over rebar = 40 mm

Statistical data indicates a standard deviation of compressive strength of 2.1 MPa is expected (more than 30 samples).

Only air entrainer is allowed.

*Available Materials*
Air entrainer: Manufacture specification 6.3 ml/1% air/100 kg cement.
Coarse aggregate: 9 mm maximum size, river gravel (rounded)
   Bulk oven-dry specific gravity = 2.55, Absorption = 0.3%
   Oven-dry rodded density = 1761 kg/m$^3$
   Moisture content = 2.5%
Fine aggregate: Natural sand
   Bulk oven-dry specific gravity = 2.659, Absorption = 0.5%
   Moisture content = 2%
   Fineness modulus = 2.47

**7.3.** Design the concrete mix according to the following conditions.

*Design Environment*
Pavement slab, Bozeman, Montana (cold climate)
Required design strength = 20.7 MPa (3000 psi)
Slab thickness = 0.3 m (12 in.)

Statistical data indicates a standard deviation of compressive strength of 1.7 MPa (250 psi) is expected (more than 30 samples).

Only air entrainer is allowed.

*Available Materials*
Air entrainer: Manufacture specification is 9.4 ml/1% air/100 kg cement (0.15 fl oz/1% air/100 lb cement).
Coarse aggregate: 50 mm (2 in.) maximum size, crushed gravel (angular)
   Bulk oven-dry specific gravity = 2.573, Absorption = 0.1%
   Oven-dry rodded density = 1921 kg/m$^3$ (120 pcf)
   Moisture content = 3.5%
Fine aggregate: Natural sand
   Bulk oven-dry specific gravity = 2.54, Absorption = 0.2%
   Moisture content = 3.67%
   Fineness modulus = 2.68

**7.4.** The design of a concrete mix requires 1173 kg/m$^3$ of gravel in dry condition, 582 kg/m$^3$ of sand in dry condition, and 157 kg/m$^3$ of free water. The gravel available at the job site has a moisture content of 0.8% and absorption of 1.5%, and the available sand has a moisture content of 1.1% and absorption of 1.3%. What are the masses of gravel, sand, and water per cubic meter that should be used at the job site?

**7.5.** Design a non–air-entrained concrete mix for a small job with a maximum gravel size of 25 mm (1 in.). Show the results as follows:

    **a.** masses of components to produce 2000 kg (4400 lb) of concrete.

    **b.** volumes of components to produce 1 m$^3$ (36 ft$^3$) of concrete.

**7.6.** Why is it necessary to measure the air content of concrete at the job site rather than at the batch plant? Name one of the methods used to measure the air content of concrete.

**7.7.** What do we mean by curing concrete? What would happen if concrete is not cured?

**7.8.** Discuss five different methods of concrete curing.

**7.9.** Discuss the change in volume of concrete at early ages.

**7.10.** Discuss the creep response of concrete structures. Provide examples of the effect of creep on concrete structures.

**7.11.** On one graph, draw a sketch showing the typical relationship between the stress and strain of concrete specimens with high and low water-cement ratios. Label all axes and curves. Comment on the effect of increasing the water-cement ratio on the stress-strain response.

**7.12.** Using Figure 7.10:

    **a.** Determine the ultimate stress at each water-cement ratio.

    **b.** Determine the secant modulus at 40% of the ultimate stress at each water-cement ratio.

    **c.** Plot the relationship between the secant moduli and the ultimate stresses.

    **d.** Plot the relationship between the moduli and the ultimate stresses on the same graph of part (c), using the relation of the ACI Building Code (Equation 7.3).

    **e.** Compare the two relations and comment on any discrepancies.

**7.13.** A normal-weight concrete has an average compressive strength of 4500 psi. What is the estimated modulus of elasticity?

**7.14.** Discuss the significance of the compressive strength test on concrete.

**7.15.** What is the purpose of performing the flexure test on concrete? How are the results of this test related to the compressive strength test results?

**7.16.** The flexure strength test was performed on a concrete beam having a cross section of 0.15 m by 0.15 m and a span of 0.45 m. If the load at failure was 35.7 kN, calculate the flexure strength of the concrete.

**7.17.** A normal-weight concrete has an average compressive strength of 20 MPa. What is the estimated flexure strength?

**7.18.** Discuss two nondestructive tests to be performed on hardened concrete. Show the basic principles behind the tests and how they are performed.

**7.19.** Discuss the concept of concrete maturity meters.

**7.20.** Discuss four alternatives that increase the flexibility and application of conventional concrete.

**REFERENCES**

American Concrete Institute. 1975. *Specifications for structural concrete for buildings.* ACI Committee 301 Report, ACI 301–72. *Journal of American Concrete Institute* 72 (7). Farmington Hills, MI: American Concrete Institute.

American Concrete Institute. 1980. *Guide for concrete floor and slab construction.* ACI Committee 302 Report, ACI 302.1R-80. Farmington Hills, MI: American Concrete Institute.

American Concrete Institute. 1982. *Hot-weather concreting.* ACI Committee 305 Report, ACI 305R-77. Farmington Hills, MI: American Concrete Institute.

American Concrete Institute. 1983. *Cold-weather concreting.* ACI Committee 306 Report, ACI 306R-78. Farmington Hills, MI: American Concrete Institute.

American Concrete Institute. 1985. *Standard practice for selecting proportions for normal, heavyweight and mass concrete.* ACI Committee 211 Report, ACI 211.1–81. Farmington Hills, MI: American Concrete Institute.

American Concrete Institute. 1986. *Standard practice for curing concrete.* ACI Committee 308 Report, ACI 308–81. Farmington Hills, MI: American Concrete Institute.

American Concrete Institute. 1986. *Building code requirements for reinforced concrete.* ACI Committee 318 Report, ACI 318–83. Farmington Hills, MI: American Concrete Institute.

Hognestad, E., N. W. Hanson, and D. McHenry. 1955. *Concrete stress distribution in ultimate strength design.* Development Department bulletin DX006. Skokie, IL: Portland Cement Association.

Kosmatka, S. H. and W. C. Panarese. 1988. *Design and control of concrete mixtures.* 13th ed. Skokie, IL: Portland Cement Association.

Mehta, P. K. and P. J. M. Monteiro. 1993. *Concrete structure, properties, and materials.* 2nd ed. Englewood Cliffs, NJ: Prentice-Hall.

Neville, A. M. 1981. *Properties of concrete.* 3rd ed. London: Pitman Books Ltd.

Portland Cement Association. 1980. *Concrete for small jobs.* IS174T. Skokie, IL: Portland Cement Association.

Powers, T. C. et al. Nov 1954. Permeability of portland cement paste. *Journal of American Concrete Institute* 51 (11) 285–298.

Shah, S. P. and G. Winter. 1968. Inelastic behavior and fracture of concrete. In *Symposium on causes, mechanism, and control of cracking in concrete.* American Concrete Institute Special Publication no. 20. Farmington Hills, MI: American Concrete Institute.

# 8 Masonry

A masonry structure is formed by combining masonry units, such as stone or brick, with mortar. Masonry is one of the oldest construction materials. Examples of ancient masonry structures include the pyramids of Egypt, the Great Wall of China, and Greek and Roman ruins. Bricks of nearly uniform size became commonly used in Europe during the beginning of the thirteenth century. The first extensive use of bricks in the United States was around 1600. In the last two centuries bricks have been used in constructing sewers, bridge piers, tunnel linings, and multistory buildings. Masonry units are still being used in construction in the United States and are competing with other materials such as wood, steel, and concrete (Adams 1979).

## Masonry Units

A masonry unit can be classified as

- concrete masonry units,
- clay bricks,
- structural clay tiles,
- glass blocks, and
- stone.

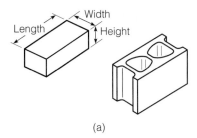
(a)

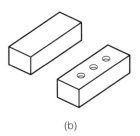
(b)

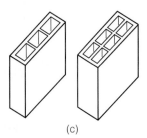
(c)

**FIGURE 8.1** Examples of masonry units: (a) concrete masonry units, (b) clay bricks, and (c) structural clay tiles.

Concrete masonry units can be either solid or hollow, but clay bricks, glass blocks, and stone are typically solid. Structural clay tiles are hollow units that are larger than clay bricks and are used for lightweight masonry such as partition walls and filler panels. They can be used with their webs in either a horizontal or a vertical direction. Figure 8.1 shows examples of concrete masonry units, clay bricks, and structural clay tiles. Concrete masonry units and clay bricks are commonly used in the United States.

## Concrete Masonry Units

Solid concrete units are commonly called concrete bricks, whereas hollow units are known as concrete blocks, hollow blocks, or cinder blocks. Hollow units have net cross-sectional area in every plane parallel to the bearing surface with less than 75% of the cross-sectional area in the same plane. If this ratio is 75% or more, the unit is categorized as solid (Portland Cement Association 1991).

Concrete masonry units are manufactured in three classes, based on their density: *lightweight units, medium-weight units, and normal-weight units,* with dry unit weights as shown in Table 8.1. Well-graded sand, gravel, and crushed stone are used to manufacture normal-weight units. Lightweight aggregates such as pumice, scoria, cinders, expanded clay, and expanded shale are used to manufacture lightweight units. Lightweight units are the most common concrete units used in masonry construction because they are easy to handle and transport and the weight of the structure is reduced. Lightweight units have higher thermal and fire resistance properties and lower sound resistance than normal weight units.

Concrete masonry units are manufactured using a relatively dry (zero-slump) concrete mixture consisting of portland cement, aggregates, water, and admixtures. Type I cement is usually used to manufacture concrete masonry units; however, Type III is sometimes used to reduce the curing time. Air-entrained concrete is sometimes used to increase the resistance of the masonry structure to freeze and thaw effects and to improve workability, compaction, and molding characteristics of the units during manufacturing. The units are molded under pressure, then cured, usually using low-pressure steam curing. After manufacturing, the units are stored under controlled conditions so that the concrete continues curing.

Concrete masonry units can be classified as load-bearing (ASTM C90) and non–load-bearing (ASTM C129). Load-bearing units must satisfy a higher minimum compressive strength requirement than non–load-bearing units, as shown in Table 8.2. The compressive strength of individual concrete masonry units is determined by capping the unit and applying load in the direction of the height of the unit until failure (ASTM

**TABLE 8.1**  Weight Classifications of Concrete Masonry Units (ASTM C129)*

| Weight Classification | Density, $Mg/m^3$ (pcf) |
| --- | --- |
| Lightweight | < 1.68 (105) |
| Medium-Weight | 1.68 to < 2.00 (105 to 125) |
| Normal-Weight | ≥ 2.00 (125) |

*Copyright ASTM. Reprinted with permission.

**TABLE 8.2**  Minimum Strength Requirements of
Concrete Masonry Units, Net Area, MPa (psi)
(ASTM C90 and C129)*

| Type | Average of Three Units | Individual Units |
|------|------------------------|------------------|
| Load-bearing | 13.1 (1900) | 11.7 (1700) |
| Non–load-bearing | 4.1 (600) | 3.5 (500) |

*Copyright ASTM. Reprinted with permission.

C140). A full-size unit is recommended for testing, although a portion of a unit can be used if the capacity of the testing machine is not large enough. The *gross area compressive strength* is calculated by dividing the load at failure by the gross cross-sectional area of the unit. The *net area compressive strength* is calculated by dividing the load at failure by the net cross-sectional area. The net cross-sectional area is calculated by dividing the net volume of the unit by its average height. The net volume is determined using the water displacement method according to ASTM C140.

Load-bearing concrete masonry units are manufactured in two types: Type I, moisture-controlled units, and Type II, non–moisture-controlled units. Type I units are required to comply with certain moisture content provisions specified by ASTM C90, whereas Type II does not have to comply with these requirements.

The moisture content is controlled in Type I units to limit the amount of shrinking due to moisture loss after construction. Type I units are used in arid areas and must have low moisture content when delivered to the job site. In addition, they must be protected from rain, snow, and other moisture before being used. If moisture content is not reduced in the units before using them, drying shrinkage will occur, which might cause cracking when climatic balance is achieved.

In humid areas, moisture control of the concrete blocks is not required. Type II units are permitted in such cases, but they should not be very moist during construction in order to avoid excessive drying shrinkage, which might cause cracking. They should be stored long enough to achieve climatic balance depending on the material used, moisture content in the units, and humidity conditions. Type II units are more commonly used in construction than Type I.

The absorption of concrete masonry units is determined by immersing the unit in water for 24 h (ASTM C140). The absorption and moisture content are calculated as follows.

$$\text{Absorption, \%} = \frac{W_s - W_d}{W_d} \times 100 \qquad (8.1)$$

$$\text{Moisture content as a percent of total absorption} = \frac{W_r - W_d}{W_s - W_d} \times 100 \qquad (8.2)$$

where

$W_s$ = saturated weight of unit,
$W_d$ = oven-dry weight of unit, and
$W_r$ = weight of unit as received

**SAMPLE PROBLEM 8.1**

A concrete masonry unit was tested according to ASTM C140 procedure and produced the following results:

mass of unit as received    = 10,354 g
saturated mass of unit     = 11,089 g
oven-dry mass of unit      = 9,893 g

Calculate the absorption and moisture content of the unit as a percent of total absorption.

*Solution:*

$$\text{Absorption, \%} = \frac{11,089 - 9,893}{9,893} \times 100 = 12.1\%$$

Moisture content as a percent of total absorption

$$= \frac{(10,354 - 9,893)}{(11,089 - 9,893)} \times 100 = 38.5\%$$

Concrete masonry units are available in different sizes, colors, shapes, and textures. Concrete masonry units are specified by their nominal dimensions. The *nominal dimension* is greater than its *specified* (or *modular*) dimension by the thickness of the mortar joint that is usually 10 mm (3/8 in.). For example, a 200 mm × 200 mm × 400 mm (8 in. × 8 in. × 16 in.) block has an actual width of 190 mm (7-5/8 in.), height of 190 mm (7-5/8 in.), and length of 390 mm (15-5/8 in.), as illustrated in Figure 8.2. Load-bearing concrete masonry units are available in nominal widths of 100 mm, 150 mm, 200 mm, 250 mm, and 300 mm (4 in., 6 in., 8 in., 10 in., and 12 in.), heights of 100 mm and 200 mm (4 in. and 8 in.), and lengths of 300 mm, 400 mm, and 600 mm (12 in., 16 in., and 24 in.). Common load-bearing blocks are 200 mm × 200 mm × 400 mm, whereas non–load-bearing blocks are 100 mm × 200 mm × 400 mm, as shown in Figure 8.3. Also, depending on the position within the masonry wall, they are manufactured as stretcher, single-corner, and double-corner units, as depicted in Figure 8.4.

Solid concrete masonry units (concrete bricks) are manufactured in two grades (N and S) and two types (Types I and II), based on strength and absorption requirements. Grade N units have higher compressive strength, resistance to moisture penetration, and resistance to frost action than grade S. According to ASTM C55, the minimum compressive strength of individual units is 20.7 MPa (3000 psi) for grade N,

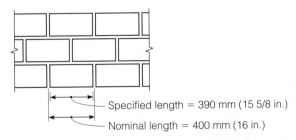

Specified length = 390 mm (15 5/8 in.)
Nominal length = 400 mm (16 in.)

**FIGURE 8.2** Nominal demensions and specified (modular) dimensions.

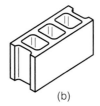

(a)                (b)

**FIGURE 8.3** Concrete masonry units:
(a) non–load-bearing and (b) load-bearing.

(a)                (b)                (c)

**FIGURE 8.4** Concrete masonry units: (a) stretcher, (b) single-corner, and
(c) double-corner.

and 13.8 MPa (2000 psi) for grade S. Grade N bricks are typically used as architectural veneers and facing units in exterior walls. Grade S bricks are for general use where moderate strength and resistance to frost action is required. As discussed for hollow concrete units, Type I bricks are moisture-controlled, whereas Type II bricks are non–moisture-controlled.

## Clay Bricks

Clay bricks are small, rectangular blocks made of fired clay. Clays for brick making vary widely in composition from one place to another. Clays are composed mainly of silica (grains of sand), alumina, lime, iron, manganese, sulfur, and phosphates, with different proportions. Bricks are manufactured by grinding or crushing the clay in mills and mixing it with water to make it plastic. The plastic clay is then molded, textured, dried, and finally fired. Bricks are manufactured with different colors, such as dark red, purple, brown, gray, pink, or dull brown, depending on the firing temperature of the clay during manufacturing. The firing temperature for brick manufacturing varies from 900°C to 1200°C (1650°F to 2200°F). Clay bricks have an average density of 2 Mg/m$^3$ (125 pcf).

Bricks are used for different purposes, including building, facing and aesthetics, floor making, and paving. *Building bricks* (*common bricks*) are used as a structural material and typically are strong and durable. *Facing bricks* are used for facing and aesthetic purposes and are available in different sizes, colors, and textures. *Floor bricks* are used on finished floor surfaces and are generally smooth and dense and have high resistance to abrasion. Finally, *paving bricks* are used as a paving material for roads, sidewalks, patios, driveways, and interior floors. Paving bricks are available in different colors such as red, gray, or brown, and typically they are abrasion resistant and could be vitrified.

Absorption is one of the important properties that determine the durability of bricks. Highly absorptive bricks can cause efflorescence and other problems in the masonry. According to ASTM C67 absorption by 24-h submersion, absorption by 5-h boiling, and saturation coefficient are calculated as follows.

$$\text{Absorption by 24-h submersion, \%} = \frac{(W_{s24} - W_d)}{W_d} \times 100 \qquad (8.3)$$

$$\text{Absorption by 5-h boiling, \%} = \frac{(W_{b5} - W_d)}{W_d} \times 100 \qquad (8.4)$$

$$\text{Saturation coefficient} = \frac{(W_{s24} - W_d)}{(W_{b5} - W_d)} \times 100 \qquad (8.5)$$

where

$W_d$   = dry weight of specimen,
$W_{s24}$ = saturated weight after 24-h submersion in cold water, and
$W_{b5}$  = saturated weight after 5-h submersion in boiling water.

Clay bricks are very durable, fire-resistant, and require very little maintenance. They have moderate insulating properties, which make brick houses cooler in summer and warmer in winter, as compared to houses built with other construction materials. Clay bricks are also noncombustible and poor conductors.

The compressive strength of clay bricks is an important mechanical property that controls their load-carrying capacity and durability. The compressive strength of clay bricks is dependent on the composition of the clay, method of brick manufacturing, and the degree of firing. The compressive strength is determined by capping and testing a half unit "flatwise" (load applied in the direction of the height of the unit) and calculated by dividing the load at failure by the cross-sectional area (ASTM C67). In determining the compressive strength, either the net or gross cross-sectional area is used. Net cross-sectional area is used only if the net cross-section is less than 75% of the gross cross section. A quarter of a brick can be tested if the capacity of the testing machine is not large enough to test a half brick. Other mechanical properties of bricks include modulus of rupture, tensile strength, and modulus of elasticity. Most clay bricks have modulus of rupture between 3.5 MPa and 26.2 MPa (500 psi and 3800 psi). The tensile strength is typically between 30% to 49% of the modulus of rupture. The modulus of elasticity ranges between 10.3 GPa and 34.5 GPa ($1.5 \times 10^6$ psi and $5 \times 10^6$ psi).

Building bricks are graded according to properties related to durability and resistance to weathering, such as compressive strength, water absorption, and saturation coefficient (ASTM C62). Table 8.3 shows the three available grades and their requirements: SW, MW, and NW, standing for severe weathering, moderate weathering, and negligible weathering, respectively. Grade SW bricks are intended for use in areas subjected to frost action, especially at or below ground level. Grade NW bricks are recommended for use in areas with no frost action and in dry locations, even where subfreezing temperatures are expected. Grade NW bricks can be used in interior construction, where no freezing occurs.

**TABLE 8.3**   Physical Requirements for Building Bricks (ASTM C62)*

| Grade | Min. Compressive Strength, Gross Area, MPa (psi) | | Max. Water Absorption by 5-h Boiling, % | | Max. Saturation Coefficient | |
|---|---|---|---|---|---|---|
| | Average of Five Bricks | Individual | Average of Five Bricks | Individual | Average of Five Bricks | Individual |
| SW† | 20.7 (3000) | 17.2 (2500) | 17.0 | 20.0 | 0.78 | 0.80 |
| MW‡ | 17.2 (2500) | 15.2 (2200) | 22.0 | 25.0 | 0.88 | 0.90 |
| NW** | 10.3 (1500) | 8.6 (1250) | No limit | No limit | No limit | No limit |

*Copyright ASTM. Reprinted with permission.
†Severe weathering
‡Moderate weathering
**Negligible weathering

**SAMPLE PROBLEM 8.2**

The 5-h boiling test was performed on a medium weathering clay brick according to ASTM C67 and produced the following masses:

> dry mass of specimen = 1.788 kg
> saturated mass after 5-h submersion in boiling water = 2.262 kg

Calculate percent absorption by 5-h boiling and check whether the brick satisfies the ASTM requirements.

*Solution:*

$$\text{Absorption by 5-h boiling} = \frac{2.262 - 1.788}{1.788} \times 100 = 26.5\%$$

From Table 8.3: the maximum allowable absorption by 5-h boiling = 25.0%
Therefore, the brick does not satisfy the ASTM requirements.

Facing bricks (ASTM C216) are manufactured in two durability grades for severe weathering (SW) and moderate weathering (MW). Each durability grade is manufactured in three appearance types: FBS, FBX, and FBA. These three types stand for *face brick standard, face brick extra,* and *face brick architecture.* Type FBS bricks are used for general exposed masonry construction. Type FBX bricks are used for general exterior and interior masonry construction where a high degree of precision and a low permissible variation in size are required. The FBA type bricks are manufactured to produce characteristic architectural effects resulting from nonuniformity in size and texture of the individual units.

Similar to concrete masonry units, bricks are designated by their nominal dimensions. The *nominal dimension* of the brick is greater than its *specified* (or *modular*) dimension by the thickness of the mortar joint, which is about 10 mm (3/8 in.) and could go up to 12.5 mm (1/2 in.). The *actual size* of the brick depends on the nominal size and the amount of shrinking that occurs during the firing process, which ranges from 4% to 15%.

Clay bricks are specified by their nominal width times nominal height times nominal length. For example, a 4 × 2-2/3 × 8 brick has nominal width of 100 mm (4 in.), height of 70 mm (2-2/3 in.), and length of 200 mm (8 in.). Clay bricks are available in nominal widths ranging from 75 mm to 300 mm (3 in. to 12 in.), heights from

50 mm to 200 mm (2 in. to 8 in.), and lengths up to 400 mm (16 in.). Bricks can be classified as either modular or nonmodular where modular bricks have widths and lengths of multiples of 100 mm (4 in.).

## Mortar

Mortar is a mixture of portland cement, lime, sand, and water. Adding a small percentage of lime to the cement mortar makes the mortar "fat" or "rich," which increases its workability. Mortar can be classified as lime mortar or cement mortar. *Lime mortar* is made of lime, sand, and water, whereas *cement* (or *cement-lime*) *mortar* is made of lime mortar mixed with portland cement (Portland Cement Association 1987).

Mortar is used for the following functions:

- bonding masonry units together
- serving as a seating material for the units
- leveling and seating the units
- providing aesthetic quality of the structure

Lime mortar gains strength slowly with a typical compressive strength of 0.7 MPa to 2.8 MPa (100 psi to 400 psi). Cement mortar is manufactured in four types: M, S, N, and O. Type M has the lowest amount of hydrated lime, whereas type O has the highest amount. The compressive strength of mortar is tested using 50-mm cubes according to ASTM C109. The minimum average compressive strengths of types M, S, N, and O at 28 days are 17.2 MPa, 12.4 MPa, 5.2 MPa, and 2.4 MPa (2500 psi, 1800 psi, 750 psi, and 350 psi) (ASTM C270).

Mortar starts to bind masonry units when it sets. During construction, bricks and blocks should be rubbed and pressed down in order to force the mortar into the pores of the masonry units to produce maximum adhesion. It should be noted, however, that mortar is the weakest part of the masonry wall. Therefore, thin mortar layers generally produce stronger walls than do thick layers.

Unlike concrete, the compressive strength is not the most important property of mortar. Since mortar is used as an adhesive and sealant, it is very important that it forms a complete, strong, and durable bond with the masonry units and with the rebars that might be used to reinforce masonry walls. The ability to bond individual units is measured by the *tensile bond strength* of mortar (ASTM C952), which is related to the force required to separate the units. The tensile bond strength affects the shear and flexural strength of masonry. The tensile bond strength is usually between 0.14 MPa and 0.55 MPa (20 psi to 80 psi) and is affected by the amount of lime in the mix. The optimum lime content that results in the highest bond strength is typically between 1 part and 1/4 part of portland cement by volume (Somayaji 1995).

Other properties that affect the performance of mortar are workability, tensile strength, compressive strength, resistance to freeze and thaw, and water retentivity. The water retentivity is a measure, according to ASTM C91, of the rate at which water is lost to the masonry units.

## Grout

Grout is a high-slump concrete consisting of portland cement, lime, sand, fine gravel, and water. Grout is used to fill the cores or voids in hollow masonry units for the purpose of: 1) bonding the masonry units, 2) bonding the reinforcing steel to the masonry, 3) increasing the bearing area, 4) increasing fire resistance, and 5) improving the overturning resistance by increasing the weight.

## Plaster

Plaster is a fluid mixture of portland cement, lime, sand, and water, which is used for finishing either masonry walls or framed (wood) walls. Plaster is used for either exterior or interior walls. Stucco is plaster used to cover exterior walls. The average compressive strength of plaster is about 13.8 MPa (2000 psi) at 28 days.

**SUMMARY**

Masonry is one of the oldest building technologies, dating back to use of sun-dried adobe blocks in ancient times. Modern masonry units are produced to high standards in the manufacturing process. While the strength of the masonry units is important for quality control, the strength of masonry construction is generally limited by the ability to bond the units together with mortar. The ability of masonry units to resist environmental degradation is an important quality consideration. This ability is closely related to the absorption of the masonry units.

**QUESTIONS AND PROBLEMS**

**8.1.** Define solid and hollow masonry units according to ASTM C90.

**8.2.** What are the advantages of masonry walls over framed (wood) walls?

**8.3.** A concrete masonry unit is tested for compressive strength and produces the following results.

> Failure load = 593 kN
> Gross area = 0.074 m$^2$
> Gross volume = 0.014 m$^3$
> Net volume = 0.006 m$^3$

Is the unit categorized as solid or hollow? Why? What is the net area compressive strength? Does the compressive strength satisfy the ASTM requirements for load bearing units shown in Table 8.2?

**8.4.** Discuss why concrete masonry units should be protected from moisture before use in arid areas.

**8.5.** A portion of a concrete masonry unit was tested according to ASTM C140 procedure and produced the following masses.

> mass of unit as received = 8152 g
> saturated mass of unit = 8666 g
> oven-dry mass of unit = 7753 g

Calculate the absorption and moisture content of the unit as a percent of total absorption.

**8.6.** Define the nominal, specified (modular), and actual dimensions of clay bricks.

**8.7.** Name and define the three grades of clay bricks.

**8.8.** A severe weathering clay brick was tested according to ASTM C67 procedure and produced the following data.

> dry mass of specimen = 1.822 kg
> saturated mass after 24-h submersion in cold water = 2.044 kg
> saturated mass after 5-h submersion in boiling water = 2.060 kg

Calculate absorption by 24-h submersion, absorption by 5-h boiling, and saturation coefficient. Does the brick satisfy the ASTM requirements?

**8.9.** What are the functions of mortar?

**REFERENCES**

Adams, J. T. 1979. *The complete concrete, masonry and brick handbook.* New York: Arco.

Portland Cement Association. 1987. *Mortars for masonry walls.* Skokie, IL: Portland Cement Association.

Portland Cement Association. 1991. *Masonry information.* Skokie, IL: Portland Cement Association.

Somayaji, S. 1995. *Civil engineering materials.* Englewood Cliffs, NJ: Prentice-Hall.

# 9 Asphalt and Asphalt Mixture

Asphalt is one of the oldest materials used in construction. Asphalt binders were used in 3000 B.C., preceding the use of the wheel by 1000 years. Before the mid 1850s asphalt came from natural pools found in various locations throughout the world, such as the Trinidad Lake asphalt that is still mined. However, with the discovery and refining of petroleum in Pennsylvania use of asphalt cement became wide-spread. By 1907 more asphalt cement came from refineries than came from natural deposits. Today practically all asphalt cement is from refined petroleum.

Bituminous materials are classified as asphalts and tars, as shown in Figure 9.1. Several asphalt products are used; asphalt is used mostly in pavement construction, but is used as sealing and waterproofing agents, as well. Tars are produced by the destructive distillation of bituminous coal or by cracking petroleum vapors. In the United States tar is used primarily for waterproofing membranes, such as roofs. Tar may also be used for pavement treatments, particularly where fuel spills may desolve asphalt cement, such as on airport aprons.

The fractional distillation process of crude petroleum is illustrated in Figure 9.2. Different products are separated at different temperatures. Figure 9.2 shows the main products such as gasoline, kerosene, diesel oil, and asphalt residue (asphalt cement). Since asphalt is a lower-valued product than other components of crude oil, refineries are set up to produce the more valuable fuels at the expense of asphalt production. The quantity and quality of the asphalt depends on the crude petroleum source and the refining method. Some crude sources, such as the Nigerian oils, produce little asphalt, while others, such as many of the Middle Eastern oils, have a high asphalt content.

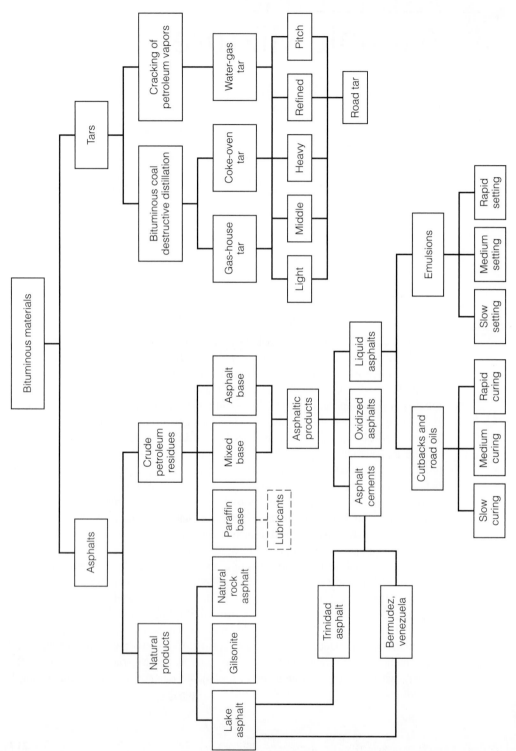

**FIGURE 9.1**  Classification of bituminous materials. (Goetz and Wood 1960)

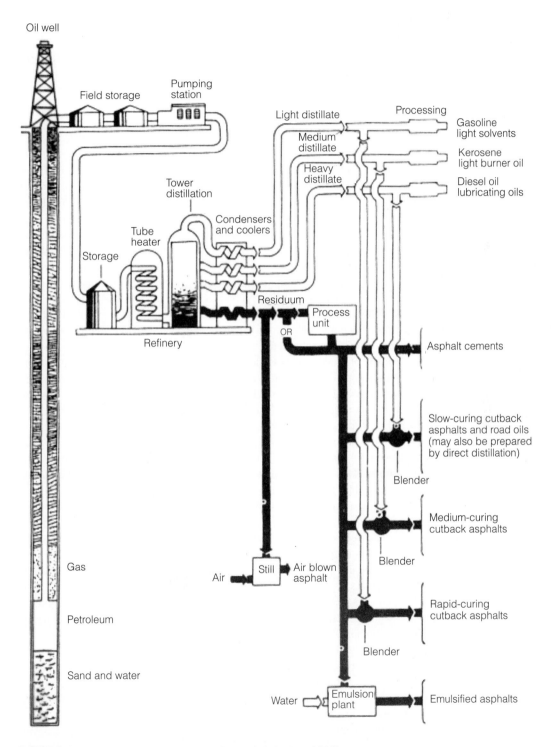

**FIGURE 9.2** Distillation of crude petroleum. (The Asphalt Institute 1989)

This chapter reviews the types, uses, and chemical and physical properties of asphalt. The asphalt concrete used in road and airport pavements, which is a mixture of asphalt and aggregates, is also presented. The chapter discusses the recently developed Superpave asphalt binder specifications and mix design. Recycling of pavement materials and additives used to modify the asphalt properties are also included.

## Types of Asphalt Products

Asphalt used in pavements is produced in three forms: *asphalt cement, asphalt cutback,* and *asphalt emulsion.* Asphalt cement is a blend of hydrocarbons of different molecular weights. The characteristics of the asphalt depend on the chemical composition and the distribution of the molecular weight hydrocarbons. As the distribution shifts toward heavier molecular weights the asphalt becomes harder and more viscous. At room temperatures asphalt cement is a semisolid material that cannot be applied readily as a binder without heating it. Liquid asphalt products, cutbacks and emulsions, have been developed and can be used without heating (The Asphalt Institute 1989).

Although the liquid asphalts are convenient, they cannot produce a quality of asphalt concrete comparable to what can be produced by heating neat asphalt cement and mixing with carefully selected aggregates. Asphalt cement has excellent adhesive characteristics, which make it a superior binder for pavement applications. In fact, it is the most common binder material used in pavements.

A cutback is produced by dissolving asphalt cement in a lighter molecular weight hydrocarbon solvent. When the cutback is sprayed on a pavement or mixed with aggregates, the solvent evaporates, leaving the asphalt residue as the binder. In the past cutbacks were widely used for highway construction. They were effective and could be applied easily in the field. However, three disadvantages have severely limited the use of cutbacks. First, as petroleum costs have escalated, the use of these expensive solvents as a carrying agent for the asphalt cement is no longer cost effective. Second, cutbacks are hazardous materials due to the volatility of the solvents. Finally, application of the cutback releases environmentally unacceptable hydrocarbons into the atmosphere. In fact, many regions with air pollution problems have outlawed the use of any cutback material.

An alternative to dissolving the asphalt in a solvent is dispersing the asphalt in water as emulsion. In this process the asphalt cement is physically broken down into micron-sized globules that are mixed into water containing an emulsifying agent. Emulsified asphalts typically consist of about 60% to 70% asphalt residue, 30% to 40% water, and a fraction of a percent of emulsifying agent. There are many types of emulsifying agents; basically they are a soap material. The emulsifying molecule has two distinct components, the head portion, which has an electrostatic charge, and the tail portion, which has a high affinity for asphalt. The charge can be either positive to produce a *cationic* emulsion or negative to produce an *anionic* emulsion. When asphalt is introduced into the water with the emulsifying agent, the tail portion of the emulsifier attaches itself to the asphalt, leaving the head exposed. The electric charge of the emulsifier causes a repulsive force between the asphalt globules, which maintains their separation in the water. Since the specific gravity of asphalt is very near that of water, the globules have a neutral buoyancy and, therefore, do not tend to float or sink. When the emulsion is mixed with aggregates or used on a pavement, the water evaporates, allowing the asphalt globs to come together, forming the binder. The phenomenon of separation between the asphalt residue and water is referred to as *breaking* or *setting*. The rate of emulsion setting can be controlled by varying the type and amount of the emulsifying agent.

Since most aggregates bear either positive surface charges (such as limestone) or negative surface charges (such as siliceous aggregates), they tend to be compatible with anionic or cationic emulsions, respectively. However, some emulsion manufacturers can produce emulsions that bond well to aggregate-specific types, regardless of the surface charges.

Although emulsions and cutbacks can be used for the same applications, the use of emulsions is increasing because they do not include hazardous and costly solvents.

## Uses of Asphalt

The main use of asphalt is in pavement construction and maintenance. In addition, asphalt is used in sealing and waterproofing various structural components, such as roofs and underground foundations.

The selection of the type and grade of asphalt depends on the type of construction and the climate of the area. Asphalt cements are used typically to make hot-mix asphalt concrete for the surface layer of asphalt pavements. Asphalt concrete is also used in patching and repairing both asphalt and portland cement concrete pavements. Liquid asphalts (emulsions and cutbacks) are used for pavement maintenance applications, such as chip seals, slurry seals, fog seals, prime coats, and tack coats (The Asphalt Institute 1989). Liquid asphalts may also be used to seal the cracks in pavements. Liquid asphalts are mixed with aggregates to produce cold mixes, as well. Cold mixtures are normally used for patching (when hot-mix asphalt concrete is not available), base and subbase stabilization, and surfacing of low-volume roads. Table 9.1 shows asphalt's common paving applications.

**TABLE 9.1**  Paving Applications of Asphalt

| Term | Description | Application |
|---|---|---|
| Hot-mix asphalt concrete | Carefully designed mixture of asphalt and aggregates | Surfacing pavement, patching |
| Cold-mix | Mixture of aggregates and liquid asphalt | Patching, surfacing low-volume road, asphalt-stabilized base |
| Fog seal | Spray of diluted asphalt emulsion on existing pavement surface | Sealing existing pavement surface |
| Prime coat | Spray coat to bond aggregate base and asphalt-concrete surface | Construction of flexible pavement |
| Tack coat | Spray coat between lifts of asphalt concrete | Construction of new pavements or between an existing pavement and an overlay |
| Chip seal | Spray coat of asphalt cement, emulsion, or cut back followed with a uniform aggregate layer | Maintenance of existing pavement or low-volume road surfaces |
| Slurry seal | Mixture of emulsion, well-graded fine aggregate, and water | Resurfacing low-volume roads |
| Microsurfacing | Mixture of polymer-modified emulsion, well-graded crushed fine aggregate, mineral filler, water, and additives | Texturing, sealing, crack filling, rut filling, and minor leveling |

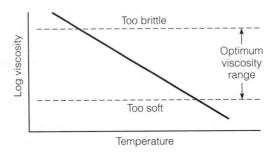

**FIGURE 9.3** Typical relation between asphalt viscosity and temperature.

## Temperature Susceptibility of Asphalt

The consistency of asphalt is greatly affected by temperature. Asphalt gets hard and brittle at low temperatures and soft at high temperatures. Figure 9.3 shows a conceptual relation between temperature and logarithm of viscosity. The viscosity of the asphalt decreases when the temperature increases. Asphalt's temperature susceptibility can be represented by the slope of the line shown in Figure 9.3. The steeper the slope the higher the temperature susceptibility of the asphalt. However, additives can be used to reduce this susceptibility.

When asphalt is mixed with aggregates, the mixture will perform properly only if the asphalt viscosity is within an optimum range. If the viscosity of asphalt is higher than the optimum range, the mixture will be too brittle and susceptible to low-temperature cracking. On the other hand, if the viscosity is below the optimum range, the mixture will flow readily, resulting in permanent deformation (rutting) and bleeding.

Due to temperature susceptibility, the grade of the asphalt cement should be selected according to the climate of the area. The viscosity of the asphalt should be mostly within the optimum range for the area's annual temperature range; soft-grade asphalts are used for cold climates and hard-grade asphalts for hot climates.

## Chemical Properties of Asphalt

Asphalt is a mixture of a wide variety of hydrocarbons primarily consisting of hydrogen and carbon atoms, with minor components such as sulfur, nitrogen, and oxygen (heteroatoms), and trace metals. The percentages of the chemical components, as well as the molecular structure of asphalt, vary depending on the crude oil source (Peterson 1984).

The molecular structure of asphalt affects the physical and aging properties of asphalt, as well as how the asphalt molecules interact with each other and with aggregate. Asphalt molecules have three arrangements, depending on the carbon atom links: 1) *aliphatic* or parraffinic, which form straight or branched chains; 2) *saturated rings,* which have the highest hydrogen to carbon ratio; and 3) *unsaturated rings* or aromatic. Heteroatoms attached to carbon alter the molecular configuration. Since the number of molecular structures of asphalt is extremely large, research on asphalt chemistry has focused on separating asphalt to major fractions that are less complex or more homogeneous. Each of these fractions is a complex chemical structure.

Asphalt cement consists of asphaltenes and maltenes (petrolenes). The maltenes consist of resins and oils. The asphaltenes are dark-brown friable solids that are chemically complex, with the highest polarity among the components. The asphaltenes are responsible for the viscosity and the adhesive property of the asphalt. If the asphaltene content is less than 10%, the asphalt concrete will be difficult to compact to the proper construction density. Resins are dark semisolid or solid, with a viscosity that is largely affected by temperature. The resins act as agents to disperse asphaltenes in the oils; the oils are clear or white liquids. When the resins are oxidized, they yield asphaltene-type molecules. Various components of asphalt interact with each other to form a balanced or compatible system. This balance of components makes the asphalt suitable as a binder.

Three fractionation schemes are used to separate asphalt components, as illustrated in Figure 9.4. The first scheme [Figure 9.4(a)] is partitioning with partial

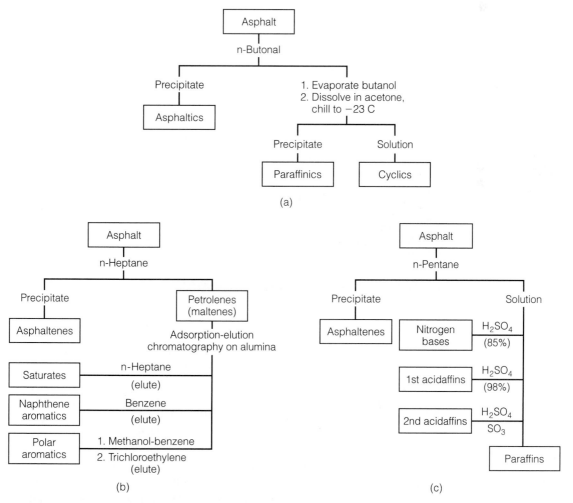

**FIGURE 9.4** Schematic diagrams of three asphalt fractionation schemes: (a) partitioning with partial solvents, (b) selective adsorbtion-description, (c) chemical precipitation. (Peterson 1984)

solvents in which *n*-butranol is added to separate (precipitate) the asphaltics. The butranol is then evaporated and the remaining component is dissolved in acetone and chilled to −23°C to precipitate the paraffinics and leave the cyclics in solution. The second scheme [Figure 9.4(b)] is selective adsorption-desorption, in which *n*-heptane is added to separate asphaltene. The remaining maltine fraction is introduced to a chromatographic column and desorbed using solvents with increasing polarity to separate other fractions. The third scheme [Figure 9.4(c)] is chemical precipitation in which *n*-pentane is added to separate the asphaltenes. A sulfuric acid ($H_2SO_4$) is added in increasing strengths to precipitate other fractions.

In addition, asphalt can be separated based on the molecular size using high pressure liquid chromatography (gel-permeation chromatography).

## Superpave

In 1987 the Strategic Highway Research Program (SHRP) began developing a new system for specifying asphalt materials and designing asphalt mixes. The final product of the SHRP research program is a new system referred to as Superpave (*Su*perior *Per*forming Asphalt *Pave*ments) (McGennis 1994; 1995). The objectives of SHRP's asphalt research were to extend the life or reduce the life-cycle costs of asphalt pavements, to reduce maintenance costs, and to minimize premature failures. An important result of this research effort was the development of performance-based specifications for asphalt binders and mixtures to control three distress modes: rutting, fatigue cracking, and thermal cracking. Note that the Superpave specifications use the term *asphalt binder,* which refers to asphalt cement with or without the addition of modifiers.

## Characterization of Asphalt

Many tests are available to characterize asphalt cement. Some tests are commonly used by highway agencies, while others are used for research. Since the properties of the asphalt are highly sensitive to temperature, all asphalt tests must be conducted at a specified temperature within very tight tolerances (The Asphalt Institute 1989).

Before Superpave the asphalt cement specifications typically were based on measurements of viscosity, penetration, ductility, and softening point temperature. These measurements are not sufficient to properly describe the viscoelastic and failure properties of asphalt cement that are needed to relate asphalt binder properties to mixture properties and to pavement performance. The new Superpave binder specifications were designed to provide performance-related properties that can be related in a rational manner to pavement performance (McGennis 1994).

### Superpave Binder Characterization Approach

The Superpave tests used to characterize the asphalt binder are performed at pavement temperatures to represent the upper, middle, and lower range of service temperatures. The measurements are obtained at temperatures in keeping with the distress mechanisms. Therefore, unlike previous specifications that require performing the test at a fixed temperature and varying the requirements for different grades of asphalt, the Superpave specifications require performing the test at the critical pavement temperature and fixing the criteria for all asphalt grades. Thus, the Superpave philosophy ensures that the asphalt properties meet the specification criteria at the critical pavement temperature.

Three pavement design temperatures are required for the binder specifications: a maximum, an intermediate, and a minimum temperature. The maximum and minimum pavement temperatures for a given geographical location in the United States can be generated using algorithms contained within the Superpave software based on weather information from 7500 weather stations. The maximum pavement design temperature is selected as the highest successive seven-day average maximum pavement temperature. The minimum pavement design temperature is the minimum pavement temperature expected over the life of the pavement. The intermediate pavement design temperature is the average of the maximum and minimum pavement design temperatures plus 4°C.

Laboratory tests that evaluate rutting potential use the maximum pavement design temperature, whereas tests that evaluate fatigue potential use the intermediate pavement design temperature. Thermal-cracking tests use the minimum pavement design temperature plus 10°C (18°F). The minimum pavement design temperature is increased by 10°C to reduce the testing time. These results are corrected to the minimum temperature using the time-temperature shift factor (McGennis 1994).

## Superpave Binder Characterization Tests

Several tests are used in the Superpave method to characterize the asphalt binder. Some of these tests have been used before for asphalt testing, while others are new. The following discussion summarizes the main steps and the significance of SHRP tests. With the exception of the rotational (Brookfield) viscometer test, the test temperatures are selected based on the temperature at the design location. The binder specification indicates the specific test temperatures used for various binders for each test (McGennis 1994).

**Flash Point**  At high temperatures, asphalt can flash or ignite in the presence of open flame or spark. The flash point test is a safety test that measures the temperature at which the asphalt flashes; asphalt cement may be heated to a temperature below this without becoming a fire hazard. The Cleveland open cup method (ASTM D92) requires partially filling a standard brass cup with asphalt cement. The asphalt is then heated at a specified rate and a small flame is periodically passed over the surface of the cup, as shown in Figure 9.5. The flash point is the temperature of the asphalt when the volatile fumes coming off the sample will sustain a flame for a short period of time. The minimum temperature where there are sufficient volatile fumes to sustain a flame for an extended period of time is the fire point.

**Rolling Thin-Film Oven Conditioning Procedures**  The engineer must know how the asphalt cement properties change when it is heated to mix with aggregates. The rolling thin-film oven (RTFO) procedure is used to simulate the short-term aging that occurs in the asphalt during production of asphalt concrete. In the RTFO method (ASTM D2872) the asphalt binder is poured in special bottles, as shown in Figure 9.6. The

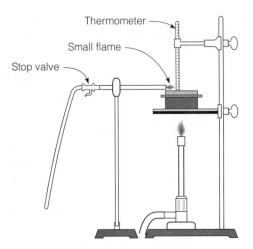

**FIGURE 9.5** Cleveland open cup flash point test apparatus.

bottles are placed in a rack in a forced draft oven at a temperature of 163°C (325°F) for 75 min. The rack rotates vertically, continuously exposing fresh asphalt. The binder in the rotating bottles is also subjected to an air jet to speed up the aging process. The aged binder is usually tested for penetration or viscosity and the results compared with those of new asphalt. The Superpave specifications limit the amount of mass loss during RTFO conditioning.

**Pressure Aging Vessel Procedure** Another test system consists of a pressure aging vessel (PAV), temperature-controlled chamber, and pressure- and temperature-controlling and measuring devices, as illustrated in Figure 9.7. In this procedure, the asphalt binder is first aged using the rolling thin-film oven (RTFO) (ASTM D2872). A specified thickness of residue from the RTFO is placed in the PAV pans. The asphalt is

**FIGURE 9.6** Rolling thin-film oven test apparatus.

**FIGURE 9.7** Pressure aging vessel apparatus.

then aged at the specified aging temperature for 20 hours in a vessel under 2.10 MPa (305 psi) of air pressure. Aging temperature, which ranges between 90°C and 110°C, is selected according to the grade of the asphalt binder.

The PAV test is designed to simulate the oxidative aging that occurs in asphalt binders during pavement service. Residue from this process may be used to estimate the physical or chemical properties of an asphalt binder after 5 to 10 years in the field.

**Rotational Viscometer Test** The rotational (Brookfield) viscometer test is standardized by ASTM D4402. The apparatus consists of a rotational coaxial cylinder viscometer and a unit to control the temperature, as shown in Figure 9.8. The test is performed on unaged binders. In this test, the asphalt binder is placed in the sample chamber at 135°C (275°F); then both are placed in the thermocontainer. A spindle is placed in the asphalt sample and rotated at a specified speed. The sample size depends on the spindle size used; the rotational speed of the spindle depends on the viscosity of the binder. Typically, soft binders require high speed (and vice versa) in order to maintain an acceptable viscosity torque range of 2% to 98%. The viscosity is read from the viscometer in units of centipoises (cP) and then converted to Pascal · seconds (Pa · s) by dividing cP by 1000. The viscosity is recorded as the average of three readings at one-minute intervals to the nearest 0.1 Pa · s.

**FIGURE 9.8** Rotational viscometer.

**Dynamic Shear Rheometer Test** The dynamic shear rheometer test system consists of two parallel metal plates, an environmental chamber, a loading device, and a control and data acquisition system (ASTM P246) (Figure 9.9). The test temperature is selected according to the grade of the asphalt binder. Two upper-plate sizes are used.

**FIGURE 9.9** Dynamic shear rheometer apparatus.

A small plate, 8 mm (0.31 in.) in diameter, is used to perform tests at moderate temperatures of about 34°C (93°F) or below. A large plate, 25 mm (1 in.) in diameter, is used to perform tests at higher temperatures, those greater than 52°C (126°F). Test specimens 2 mm thick by 8 mm in diameter, or 1 mm thick by 25 mm in diameter, are formed between the parallel plates. During testing, one of the parallel plates is oscillated with respect to the other at preselected frequencies and rotational deformation amplitudes (or torque amplitudes). The required amplitude depends upon the value of the complex shear modulus of the asphalt binder being tested. Specification testing is performed at a test frequency of 10 rads/s. The complex shear modulus ($G^*$) and phase angle ($\delta$) are calculated automatically by the rheometer's computer software.

The complex shear modulus and the phase angle define the asphalt binder's resistance to shear deformation in the linear viscoelastic region. The complex shear modulus and the phase angle are related to rutting and fatigue of the asphalt mixture.

**Bending Beam Rheometer Test** The bending beam rheometer measures the midpoint deflection of a simply supported prismatic beam of asphalt binder subjected to a constant load applied to its midpoint (ASTM P245). The bending beam rheometer test system consists of a loading frame, a controlled temperature bath, and a computer-controlled automated data acquisition unit, as shown in Figure 9.10. The controlled temperature liquid bath maintains the temperature between –40°C (–40°F) and 25°C (77°F). In this test the asphalt binder beam is placed in the bath and loaded with a constant load of 980 ± 50 mN for 240 s. The test temperature is selected according to the grade of the asphalt binder. As the beam creeps, the midpoint deflection is monitored after 8, 15, 30, 60, 120, and 240 s. The constant maximum stress in the beam is

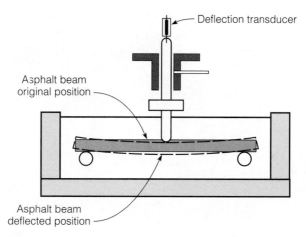

**FIGURE 9.10** Schematic of the bending beam rheometer.

calculated from the load magnitude and the dimensions of the beam. The maximum strain is calculated from the deflection and the dimensions of the beam. The flexural creep stiffness of the beam is then calculated by dividing the maximum stress by the maximum strains for the loading times specified above.

The low-temperature thermal-cracking performance of paving mixtures is related to the creep stiffness and the slope of the logarithm of the creep stiffness versus the logarithm of the time curve of the asphalt binder contained in the mix.

**Direct Tension Test** The direct tension test system consists of a displacement-controlled tensile loading machine with gripping system, a temperature controlled chamber, measuring devices, and a data acquisition system, as shown in Figure 9.11 (ASTM P252). In this test, an asphalt binder specimen is pulled at a constant rate of deformation of 1 mm/min. The test temperature is selected according to the grade of the asphalt binder. A noncontact extensometer measures the elongation of the specimen. The maximum load developed during the test is monitored. The tensile strain and stress in the specimen when the load reaches a maximum is reported as the failure strain and failure stress, respectively.

The strain at failure is a measure of the amount of elongation that the asphalt binder can sustain without cracking. Strain at failure is used as a criterion for specifying the low-temperature properties of the binder.

## Superpave Testing Procedure

The Superpave tests are performed on both unaged and aged asphalt binders. The unaged (tank) asphalt binder is tested for viscosity using the rotational viscometer to ensure pumpability during storage, transportation, and at the mixing plant. The Cleveland open cup test is also used to check the flash point. Moreover, the dynamic shear rheometer test results are used to estimate rutting potential. The asphalt is then subjected to short-term aging using the rolling thin-film oven procedure and tested using the dynamic shear rheometer to further evaluate the rutting potential. The asphalt is also subjected to long-term aging using the pressure-aging vessel procedure, then

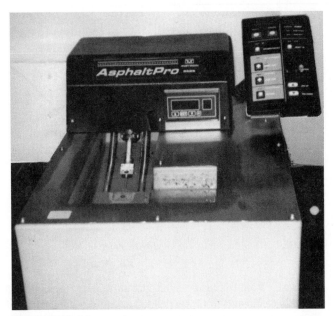

**FIGURE 9.11** Direct tension test apparatus.

tested using the dynamic shear rheometer test to evaluate the fatigue-cracking potential. The bending beam rheometer and the direct tension test are used on long-term aged asphalt to evaluate the thermal-cracking potential.

## Traditional Asphalt Characterization Tests

Traditional tests that have been used to characterize asphalt before the development of Superpave methods include the penetration, absolute and kinematic viscosities, ductility, and solubility tests.

**Penetration** The penetration test (ASTM D5) measures asphalt cement consistency. An asphalt sample is prepared and brought to 25°C (77°F). A standard needle with a total mass of 100 g is placed on the asphalt surface. The needle is released and allowed to penetrate the asphalt for 5 s, as shown in Figure 9.12. The depth of penetration, in units of 0.1 mm, is recorded and reported as the penetration value. A large penetration value indicates soft asphalt.

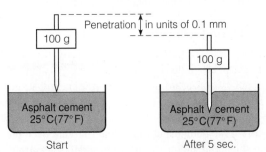

**FIGURE 9.12** Penetration test.

**FIGURE 9.13**
Absolute viscosity test apparatus.

**Absolute and Kinematic Viscosities** Similar to the penetration test, the viscosity test is used to measure asphalt consistency. Two types of viscosity are commonly measured: absolute and kinematic. The absolute viscosity procedure (ASTM D2171) requires heating the asphalt cement and pouring it in a viscometer placed in a water or oil bath at a temperature of 60°C (140°F) (Figure 9.13). The viscometer is a U-tube, with a reservoir where the asphalt is introduced and a section with a calibrated diameter and timing marks. For absolute viscosity tests vacuum is applied at one end. The time during which the asphalt flows between two timing marks on the viscometer is measured using a stop watch. The flow time, measured in seconds, is multiplied by the viscometer calibration factor to obtain the absolute viscosity in units of poises. Different-sized viscometers are used for different asphalt grades to meet minimum and maximum flow time requirements of the test procedure.

The kinematic viscosity test procedure (ASTM D2170) is similar to that of the absolute viscosity, except the test temperature is 135°C (275°F). Since the viscosity of the asphalt at 135°C is fairly low, vacuum is not used. The time it takes the asphalt to flow between the two timing marks is multiplied by the calibration factor to obtain the kinematic viscosity in units of cenistokes (cSt).

**Ductility** The ductility test (ASTM D113) measures asphalt cement's ability to deform without breaking. It provides a measure of tensile properties of the asphalt cement. In this test a standard briquette of asphalt cement is prepared and placed in a bath at a temperature of 25°C (77°F), as illustrated in Figure 9.14. The test specimen is stretched at a rate of 5 cm/min until it breaks. The distance it stretches before breaking is the ductility. The bath must be at the same specific gravity as the asphalt so that the stretched specimen will not float or sink. Salt or alcohol is added to water to increase or decrease the specific gravity of the bath. Sometimes the ductility test is run at other temperatures, such as 16°C (60°F) or 4°C (39.2°F).

**FIGURE 9.14** Ductility test apparatus.

**Solubility**  The solubility test (ASTM D2042) measures the purity of the asphalt cement. In essence, a sample of the material is washed through a filter using trichloroethylene (TCE). Inert materials in the sample are captured on the filter and weighed. The result of the test is expressed as the percent weight of the sample that is soluble. Generally, asphalt cement is required to be 99% pure. Also the asphalt must not contain moisture; otherwise, it will foam when heated to 100°C (212°F).

### Characterization of Emulsion and Cutback

Common methods used to characterize emulsion include distillation and Saybolt Furol viscosity tests. Cutback is characterized by distillation.

**FIGURE 9.15** Saybolt Furol viscometer.

**Distillation of Cutback and Emulsion**  The distillation test of cutback asphalt (ASTM D402) measures the amount and character of volatile constituents it contains. The procedure requires that the percentages, by volume, of the distillate fractions at specified temperatures be determined. The distillation test of emulsified asphalt (ASTM D244) determines the percent of residue and oil distillates by weight.

**Saybolt Furol Viscosity of Emulsion**  Emulsion viscosity is an important factor in field applications. When applied in a spray, the emulsion must be thin enough to be uniformly applied through the spray bar of the distributor truck, yet viscous enough that it will not flow from the crown or grade of the road. Emulsion viscosity is measured using the Saybolt Furol viscometer, as shown in Figure 9.15 (ASTM D244). In this test the emulsion is bought to a temperature of either 25°C (77°F) or 50°C (122°F) and allowed to flow through a specific orifice. The Saybolt Furol viscosity is the time (in seconds) required to fill a special flask.

## Classification of Asphalt

Several methods are used to characterize asphalt binders, asphalt cutbacks, and asphalt emulsions.

### Asphalt Binders

Asphalt binder is produced in several grades or classes. There are four methods for classifying asphalt binders:

1. performance grading (Superpave)
2. penetration grading
3. viscosity grading
4. viscosity of aged residue grading

**Superpave Binder Specifications and Selection**  Several grades of binder are available based on their performance in the field. Names of grades start with PG (Performance Graded) followed by two numbers representing the maximum and minimum pavement design temperatures in Celsius. For example, an asphalt binder PG 52-28 would meet the specification for a design high pavement temperature up to 52°C (126°F) and a design low temperature warmer than –28°C (–18°F). The high temperature is calculated 20 mm (0.75 in.) below the pavement surface, and the lowest

**TABLE 9.2** Superpave Performance-Graded Asphalt Binder Grades

| High Temperature Grades, °C | Low Temperature Grades, °C |
| --- | --- |
| PG 46 | −34, −40, −46 |
| PG 52 | −10, −16, −22, −28, −34, −40, −46 |
| PG 58 | −16, −22, −28, −34, −40 |
| PG 64 | −10, −16, −22, −28, −34, −40 |
| PG 70 | −10, −16, −22, −28, −34, −40 |
| PG 76 | −10, −16, −22, −28, −34 |
| PG 82 | −10, −16, −22, −28, −34 |

temperature is calculated at the pavement surface. The high and low pavement temperatures are related to the air temperature as well as other factors. Table 9.2 shows the binder grades in the Superpave specifications. PG 76 and 82 are intended to accommodate only slow transient or standing loads, such as those that occur near intersections or in truck climbing lanes.

The performance-graded asphalt binder specifications are shown in Table 9.3 (ASTM P248). The table shows the design criteria of various test parameters at the specified test temperatures. One important difference between the Superpave specifications and the old specifications is in the way the specifications work. As shown in Table 9.3 (on pages 234–235), the physical properties (criteria) remain constant for all grades, but the temperatures at which these properties must be achieved vary depending on the climate at which the binder is expected to be used. The temperature ranges shown in Table 9.3 encompass all pavement temperature regimes that exist in the United States and Canada.

The binder is selected to satisfy the maximum and minimum design pavement temperature requirements. The average seven-day maximum pavement temperature is used to determine the design maximum, whereas the design minimum pavement temperature is the lowest pavement temperature. Since the maximum and minimum pavement temperatures vary from one year to another, a reliability level is considered. As used in Superpave, reliability is the percent probability in a single year that the actual pavement temperature will not exceed the design high pavement temperature or be lower than the design low pavement temperature.

It is assumed the design high and design low pavement temperatures throughout the years follow normal distributions as illustrated in Figure 9.16(a). In this example, the average seven-day maximum pavement temperature is 56°C and the standard deviation is 2°C. Similarly, the average one-day minimum pavement temperature is −23°C and the standard deviation is 4°C. Since the area under the normal distribution curve represents the probability as illustrated in Figure 1.18, the range of temperature that satisfies the assumed probability can be calculated. For example, the range between −23°C and 56°C results in a 50% reliability for both high and low temperatures. By subtracting 2 standard deviations from the minimum pavement temperature and adding 2 standard deviations to the maximum pavement temperature, the range between −31°C and 60°C results in 98% reliability. In selecting the appropriate grade, the designer should select the standard PG grade that most

**TABLE 9.3**  Performance-Graded Asphalt Binder Specifications

| Performance Grade | PG 46– | | | PG 52– | | | | | | | PG 58– | | | | | PG 64– | | | | | |
|---|---|---|---|---|---|---|---|---|---|---|---|---|---|---|---|---|---|---|---|---|---|
| | 34 | 40 | 46 | 10 | 16 | 22 | 28 | 34 | 40 | 46 | 16 | 22 | 28 | 34 | 40 | 10 | 16 | 22 | 28 | 34 | 40 |
| Average seven-day max. pavement design temperature, °C | <46 | | | <52 | | | | | | | <58 | | | | | <64 | | | | | |
| Min. pavement design temperature, °C | >−34 | >−40 | >−46 | >−10 | >−16 | >−22 | >−28 | >−34 | >−40 | >−46 | >−16 | >−22 | >−28 | >−34 | >−40 | >−10 | >−16 | >−22 | >−28 | >−34 | >−40 |
| **Original Binder** | | | | | | | | | | | | | | | | | | | | | |
| Flash point temperature, min. °C | 230 | | | | | | | | | | | | | | | | | | | | |
| Viscosity, ASTM D4402: max., 3 Pa · s, test temperature, °C | 135 | | | | | | | | | | | | | | | | | | | | |
| Dynamic shear; G*/sinδ, min., 1.00 kPa test temperature at 10 rad/s, °C | 46 | | | 52 | | | | | | | 58 | | | | | 64 | | | | | |
| **Rolling Thin-Film Oven Residue** | | | | | | | | | | | | | | | | | | | | | |
| Mass loss, max., % | 1.00 | | | | | | | | | | | | | | | | | | | | |
| Dynamic shear; G*/sinδ, min., 2.20 kPa test temperature at 10 rad/s, °C | 46 | | | 52 | | | | | | | 58 | | | | | 64 | | | | | |
| **Pressure Aging Vessel (PAV) Residue** | | | | | | | | | | | | | | | | | | | | | |
| PAV aging temperature, °C | 90 | | | 90 | | | | | | | 100 | | | | | 100 | | | | | |
| Dynamic shear; G*/sinδ, max., 5000 kPa test temperature at 10 rad/s, °C | 10 | 7 | 4 | 25 | 22 | 19 | 16 | 13 | 10 | 7 | 25 | 22 | 19 | 16 | 13 | 31 | 28 | 25 | 22 | 19 | 16 |
| **Physical Hardening Report** | | | | | | | | | | | | | | | | | | | | | |
| Creep stiffness S, max., 300 MPa, m-value: min., 0.300 test temperature at 60 s, °C | −24 | −30 | −36 | 0 | −6 | −12 | −18 | −24 | −30 | −36 | −6 | −12 | −18 | −24 | −30 | 0 | −6 | −12 | −18 | −24 | −30 |
| Direct tension: failure strain, min., 1.0% test temperature at 1.0 mm/min, °C | −24 | −30 | −36 | 0 | −6 | −12 | −18 | −24 | −30 | −36 | −6 | −12 | −18 | −24 | −30 | 0 | −6 | −12 | −18 | −24 | −30 |

**TABLE 9.3** Performance-Graded Asphalt Binder Specifications *(continued)*

| | \multicolumn{16}{c}{Performance Grade} |
| Property | PG 70- | | | | | | PG 76- | | | | | PG 82- | | | | |
| --- | --- | --- | --- | --- | --- | --- | --- | --- | --- | --- | --- | --- | --- | --- | --- | --- |
| | 10 | 16 | 22 | 28 | 34 | 40 | 10 | 16 | 22 | 28 | 34 | 10 | 16 | 22 | 28 | 34 |
| Average seven-day max. pavement design temperature, °C | \multicolumn{6}{c}{<70} | | | | | | \multicolumn{5}{c}{<76} | | | | | \multicolumn{5}{c}{<82} | | | | |
| Min. pavement design temperature, °C | >−10 | >−16 | >−22 | >−28 | >−34 | >−40 | >−10 | >−16 | >−22 | >−28 | >−34 | >−10 | >−16 | >−22 | >−28 | >−34 |
| *Original Binder* | | | | | | | | | | | | | | | | |
| Flash point temperature, min. °C | \multicolumn{16}{c}{230} | | | | | | | | | | | | | | | |
| Viscosity, ASTM D4402: max., 3 Pa·s, test temperature, °C | \multicolumn{16}{c}{135} | | | | | | | | | | | | | | | |
| Dynamic shear, $G^*/\sin\delta$, min., 1.00 kPa test temperature at 10 rad/s, °C | \multicolumn{6}{c}{70} | | | | | | \multicolumn{5}{c}{76} | | | | | \multicolumn{5}{c}{82} | | | | |
| *Rolling Thin-Film Oven Residue* | | | | | | | | | | | | | | | | |
| Mass loss, max., % | \multicolumn{16}{c}{1.00} | | | | | | | | | | | | | | | |
| Dynamic shear, $G^*/\sin\delta$, min., 2.20 kPa test temperature at 10 rad/s, °C | \multicolumn{6}{c}{70} | | | | | | \multicolumn{5}{c}{76} | | | | | \multicolumn{5}{c}{82} | | | | |
| *Pressure Aging Vessel (PAV) Residue* | | | | | | | | | | | | | | | | |
| PAV aging temperature, °C | \multicolumn{6}{c}{100 (110)} | | | | | | \multicolumn{5}{c}{100 (110)} | | | | | \multicolumn{5}{c}{100 (110)} | | | | |
| Dynamic shear, $G^*/\sin\delta$, max., 5000 kPa test temperature at 10 rad/s, °C | 34 | 31 | 28 | 25 | 22 | 19 | 37 | 34 | 31 | 28 | 25 | 40 | 37 | 34 | 31 | 28 |
| *Physical Hardening Report* | | | | | | | | | | | | | | | | |
| Creep stiffness S, max., 300 MPa, m-value; min., 0.300 test temperature at 60 s, °C | 0 | −6 | −12 | −18 | −24 | −30 | 0 | −6 | −12 | −18 | −24 | 0 | −6 | −12 | −18 | −24 |
| Direct tension, failure strain, min., 1.0% test temperature at 1.0 mm/min, °C | 0 | −6 | −12 | −18 | −24 | −30 | 0 | −6 | −12 | −18 | −24 | 0 | −6 | −12 | −18 | −24 |

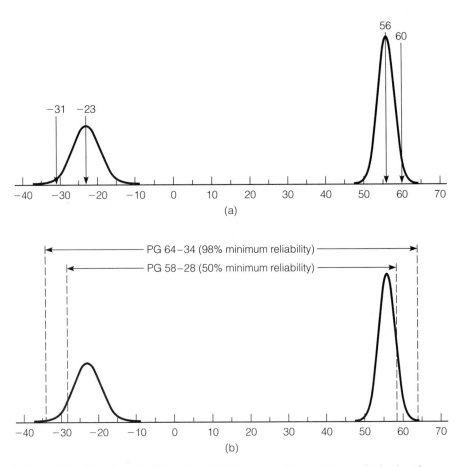

**FIGURE 9.16** Example of the distribution of design pavement temperatures and selection of binder grades: (a) distribution of high and low design pavement temperatures, and (b) binder grade selection.

closely satisfies the required reliability level. This "rounding" typically results in a higher reliability level than is intended, as shown in Figure 9.16(b). Note that the reliability levels at the high- and low-temperature grades do not need to be the same, depending on the specific pavement conditions.

**SAMPLE PROBLEM 9.1**

What standard PG asphalt binder grade should be selected under the following conditions.

The seven-day maximum pavement temperature has a mean of 57°C and a standard deviation of 2°C.

The minimum pavement temperature has a mean of –6°C and a standard deviation of 3°C.

Reliability is 98%.

***Solution:***

$$\text{High-temperature grade} \geq 57 + (2 \times 2) \geq 61°C$$

$$\text{Low-temperature grade} \leq -6 - (2 \times 3) \leq -12°C$$

The closest standard PG asphalt binder grade that satisfies the two temperature grades is PG 64–16. ◆

**Other Asphalt Binder Grading Methods** Table 9.4 shows various asphalt cement grades based on penetration and on their properties (ASTM D946). The grades correspond to the allowable penetration range; that is, the penetration of a 40–50 grade must be in the range of 40 to 50. Various grades based on viscosity and their properties are shown in Table 9.5 (ASTM D3381). The AC grade numbers are 1/100 of the middle of the allowable viscosity range; that is, an AC-5 has an absolute viscosity of 500 ± 100 poises. Therefore, high viscosity asphalt cements have a high designation number. Aged-residue grades are based on the absolute viscosity of the asphalt after it has been conditioned to simulate the effects of the aging that occur when the asphalt cement is heated to make asphalt concrete. The aged-residue grade numbers are at the middle of the allowable viscosity range after conditioning, as shown in Table 9.6 (ASTM D3381). For example, an AR-1000 has an absolute viscosity of 1000 ± 250 poises.

**TABLE 9.4** Penetration Grading System of Asphalt Cement

| Grade | Penetration Min. | Penetration Max. | Flash Point, °C (°F) | Ductility, cm |
|-------|------|------|------------------|-----------|
| 40–50 | 40 | 50 | 232 (450) | 100 |
| 60–70 | 60 | 70 | 232 (450) | 100 |
| 85–100 | 85 | 100 | 232 (450) | 100 |
| 120–150 | 120 | 150 | 219 (425) | 100 |
| 200–300 | 200 | 300 | 177 (350) | 100 |

**TABLE 9.5** Viscosity Grading System of Asphalt Cement

| Grade | Viscosity Absolute, poises | Viscosity Kinematic,* cSt | Penetration* | Flash Point,* °C (°F) |
|-------|-----------------|-----------------|--------------|-------------------|
| AC-2.5 | 250 ± 50 | 125 | 220 | 163 (325) |
| AC-5 | 500 ± 100 | 175 | 140 | 177 (350) |
| AC-10 | 1000 ± 200 | 250 | 80 | 219 (425) |
| AC-20 | 2000 ± 400 | 300 | 60 | 232 (450) |
| AC-30 | 3000 ± 600 | 350 | 50 | 232 (450) |
| AC-40 | 4000 ± 800 | 400 | 40 | 232 (450) |

*Specification is for the minimum acceptable values.

**TABLE 9.6**   Aged Residue Grading System of Asphalt Cement

| Grade | Viscosity | | Penetration* | Flash Point,*† °C (°F) |
|---|---|---|---|---|
| | Absolute, poises | Kinematic,* cSt | | |
| AR-1000 | 1000 ± 250 | 140 | 65 | 205 (400) |
| AR-2000 | 2000 ± 500 | 200 | 40 | 219 (425) |
| AR-4000 | 4000 ± 1000 | 275 | 25 | 227 (440) |
| AR-8000 | 8000 ± 2000 | 400 | 20 | 232 (450) |
| AR-16000 | 16000 ± 4000 | 550 | 20 | 238 (460) |

*Specification is for the minimum acceptable values.
†Flash point specification is for the asphalt cement before rolling thin-film oven conditioning. All the other specifications are for samples that have been conditioned.

## Asphalt Cutbacks

Three types of cutbacks are produced, depending on the hardness of the residue and the type of solvent used. *Rapid-curing cutbacks* are produced by dissolving hard residue in a highly volatile solvent, such as gasoline. *Medium-curing cutbacks* use medium hardness residue and a less volatile solvent, such as kerosene. *Slow-curing cutbacks* are produced by either diluting soft residue in nonvolatile or low-volatile fuel oil or by simply stopping the refining process before all of the fuel oil is removed from the stock.

Curing the cutback refers to the evaporation of the solvent from the asphalt residue. Rapid-curing (RC) cutbacks cure in about 5 to 10 minutes, while medium-curing (MC) cutbacks cure in a few days. Slow-curing (SC) cutbacks cure in a few months. In addition to the three types, cutbacks have several grades defined by the kinematic viscosity at 60°C (140°F). Grades of 30, 70, 250, 800, and 3000 are manufactured, with higher grades indicating higher viscosities. Thus cutback asphalts are designated by letters (RC, MC, or SC), representing the type, followed by a number that represents the grade. For example, MC-800 is a medium-curing cutback with a grade of 800. The different grades of cutback are produced by varying the amounts and types of solvent and base asphalt. The specifications of cutbacks are standardized by ASTM D2026, D2027, and D2028.

## Asphalt Emulsions

As indicated earlier, asphalt emulsions can be either anionic or cationic, depending on the electric charge. Also, emulsions set (break) at different rates. Three types of emulsion are produced: rapid-setting (RS), medium-setting (MS), and slow-setting (SS). Rapid-setting emulsion sets in about 5 to 10 minutes, medium-setting in several hours, and slow-setting in a few months.

In addition to the three types, emulsions are graded based on the Saybolt Furol viscosity at 60°C (140°F) (Figure 9.15). Asphalt emulsions are designated by letters (RS, MS, SS, CRS, CMS, or CSS), representing the type, followed by a number (1 or 2) that represents the grade. An emulsion type with grade 2 is more viscous than an emulsion type with grade 1. For example, SS-2 is an anionic slow-setting emulsion with high viscosity, while CRS-1 is a cationic rapid-setting emulsion with low viscosity.

Other emulsion types are also produced, such as the high float residue emulsion and the quick-set emulsion. Different types and grades of emulsions are used for different pavement applications. The specifications of various asphalt emulsions are standardized by ASTM D977.

# Asphalt Concrete

Asphalt concrete, also known as hot-mix asphalt (HMA), consists of asphalt cement and aggregates mixed together at a high temperature and placed and compacted on the road while still hot. Asphalt (flexible) pavements cover approximately 93% of the 2 million miles of paved roads in the United States, while the remaining 7% of the roads are portland cement concrete (rigid) pavements. The performance of asphalt pavements is largely a function of the asphalt concrete surface material.

## Desired Properties

The objective of the asphalt concrete mix design process is to provide (Roberts et al. 1991)

1. stability or resistance to permanent deformation under the action of traffic loads, especially at high temperatures;
2. fatigue resistance to prevent fatigue cracking under repeated loadings;
3. resistance to thermal cracking that might occur due to contraction at low temperatures;
4. resistance to hardening or aging during production in the mixing plant and in service;
5. resistance to moisture-induced damage that might result in stripping of asphalt from aggregate particles;
6. skid resistance by providing enough texture at the pavement surface;
7. workability to reduce the effort needed during mixing, placing and compaction.

Regardless of the set of criteria used to state the objectives of the mix design process, the design of asphalt concrete mixes requires compromises. For example, extremely high stability often is obtained at the expense of lower durability, and vice versa. Thus, in evaluating and adjusting a mix design for a particular use, the aggregate gradation and asphalt content must strike a favorable balance between the stability and durability requirements. Moreover, the produced mix must be practical and economical.

## Asphalt Concrete Production

Asphalt concrete is produced in either a batch plant or a continuous (drum) plant (The Asphalt Institute 1989). In the United States batch plants were used extensively in the past; however, more energy efficient continuous plants are now preferred.

In continuous plants (Figure 9.17) aggregates of different gradations are placed in cold bins. The gradation proportions needed are taken from the cold bins by a cold feed elevator. Aggregates are transferred to the the first part of the drum where they are dried and heated. Hot asphalt cement is introduced in the last one-third of the

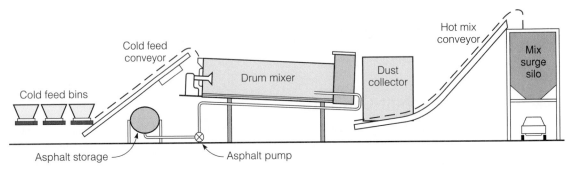

**FIGURE 9.17** Layout of a continuous (drum) mix asphalt concrete plant.

drum; then aggregates and asphalt are mixed. Since asphalt concrete is produced continuously in this type of plant, it is transferred to a storage silo until placed in a truck and transported to the job site.

## Asphalt Concrete Mix Design

The purpose of asphalt concrete mix design is to determine the design asphalt content using the available asphalt and aggregates. The design asphalt content varies for different material types, material properties, loading levels, and environmental conditions. To produce good-quality asphalt concrete it is necessary to accurately control the design asphalt content. If the appropriate design asphalt content is not used, the pavement will lack durability or stability, resulting in premature pavement failure. Typical design asphalt contents range from 4% to 7% by weight of total mix.

Before Superpave, there were two common methods of designing the asphalt concrete mixture: the Marshall (ASTM D1559) and the Hveem (ASTM D1560) methods. The Marshall method was more commonly used than the Hveem method due to its relative simplicity and its ability to be used for field control. Both methods are empirical in nature; that is, they are based on previous observations. Both methods have been used satisfactorily for several decades and have produced long-lasting pavement sections. However, due to their empirical nature they are not readily adaptable to new conditions, such as modified binders, large-sized aggregates, and heavier traffic loads.

The Superpave design system is performance-based and more rational than the Marshall and Hveem methods. Many highway agencies are implementing the Superpave system.

### Specimen Preparation in the Laboratory

Asphalt concrete specimens are prepared in the laboratory for mix-design and quality-control tests. To prepare specimens in the lab, aggregates are batched and heated, according to a specified gradation. Asphalt cement is also heated separately and added to the aggregate at a specified rate. Aggregates and asphalt are mixed with a mechanical mixer until the aggregate particles are completely coated with asphalt. Three compaction machines are commonly used.

1. gyratory compactor
2. Marshall hammer
3. California kneading compactor

Regardless of the compaction method, the procedure for preparing specimens basically follows the same four steps.

1. Heat and mix the aggregate and asphalt cement.
2. Place the material into a mold.
3. Apply compactive force.
4. Allow specimen to cool and extrude from the mold.

The specific techniques for placing the material into the mold vary among the three compaction methods, and the standards for the test must be followed.

The greatest difference between the compaction procedures is the manner in which the compactive force is applied. For the gyratory compaction, the mixture in the mold is placed in the compaction machine at an angle to the applied force. As the force is applied the mold is gyrated, creating a shearing action in the mixture. Gyratory compaction devices have been available for a long time but their use has been limited due to the lack of a mix-design procedure based on this type of compaction. However, the recently developed Superpave method of mix design (FHWA, 1995) uses a gyratory compactor; thus the use of this compaction method is expected to increase. Figure 9.18 shows the Superpave gyratory compactor.

In the Marshall procedure (Figure 9.19) a slide hammer weighing 4.45 kg (10 lb) is dropped from a height of 0.46 m (18 in.) to create an impact compaction force (ASTM D1559). The head of the Marshall hammer has a diameter equal to the specimen size, and the hammer is held flush with the specimen at all times.

In the California kneading compactor method, Figure 9.20 on page 242, the area of the compactor foot is smaller than the area of the mold. When the compaction force is applied, the mold rotates and the asphalt mixture is subjected to a kneading action (ASTM D1561). After the kneading compaction is complete, the specimen is reheated while still in the mold; then a compression machine is used to apply a static force to level the face of the specimen.

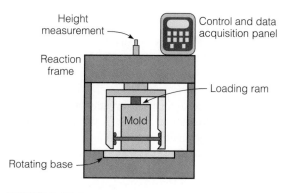

**FIGURE 9.18** Superpave gyratory compactor.

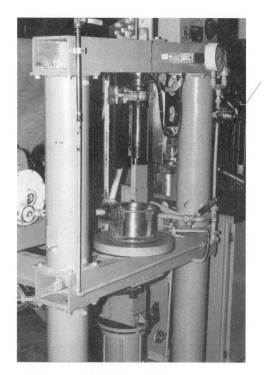

**FIGURE 9.19** Marshall
compactor.

**FIGURE 9.20** California kneading compactor.

The Superpave gyratory compactor is used for the Superpave mix design, whereas the Marshall hammer and the California kneading compactor are used for the Marshall and Hveem methods of mix design, respectively. The Superpave gyratory compactor produces specimens 150 mm (6 in.) in diameter and 95 mm to 115 mm (3.75 in. to 4.5 in.) high, allowing the use of aggregates with a maximum size of more than 25 mm (1 in.). Specimens prepared with both Marshall and California kneading compactors, as well as some gyratory compactors, are typically 101.6 mm (4 in.) in diameter and 63.5 mm (2.5 in.) high.

### Density and Voids Analysis

It is important to understand the density and voids analysis of compacted asphalt mixtures for both mix design and construction control. Regardless of the method used, mix design is a process to determine the volume of asphalt binder and aggregates required to produce a mixture with the desired properties. However, since volumes are difficult and not practical to measure, weights are used instead; the specific gravity is used to convert from weight to volume. Figure 9.21 shows that the asphalt mixture consists of aggregates, asphalt binder, and air voids. Note that a portion of the asphalt is absorbed by aggregate particles. Three important parameters commonly used are percent of air voids (voids in total mix) (VTM), voids in the mineral aggregate (VMA), and voids filled with asphalt (VFA). These are defined as

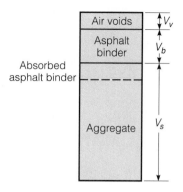

**FIGURE 9.21** Components of compacted asphalt mixture.

$$\text{VTM} = \frac{V_v}{V_t} \times 100 \qquad (9.1)$$

$$\text{VMA} = \frac{V_v + V_b}{V_t} \times 100 \qquad (9.2)$$

$$\text{VFA} = \frac{V_b}{V_b + V_v} \times 100 \qquad (9.3)$$

where
  $V_v$ = volume of air voids
  $V_b$ = volume of effective asphalt binder
  $V_t$ = total volume of the mixture

The effective asphalt is the total asphalt minus the absorbed asphalt.

**SAMPLE PROBLEM 9.2**

A compacted asphalt concrete specimen contains 5% asphalt binder (Sp. Gr. 1.023) by weight of total mix, and aggretage with a specific gravity of 2.755. The bulk density of the specimen is 2.441 Mg/m³. Ignoring absorption, compute VTM, VMA, and VFA.

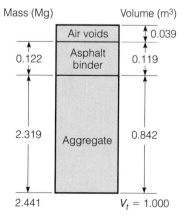

### Solution:

Assume $V_t = 1 \text{ m}^3$

Total mass $= 1 \times 2.441 = 2.441 \text{ Mg}$

Mass of asphalt binder $= 0.05 \times 2.441 = 0.122 \text{ Mg}$

Mass of aggregate $= 0.95 \times 2.441 = 2.319 \text{ Mg}$

$$V_b = \frac{0.122}{1.023} = 0.119 \text{ m}^3$$

$$V_s = \frac{2.319}{2.755} = 0.842 \text{ m}^3$$

$$V_v = V_t - V_b - V_s = 1 - 0.119 - 0.842 = 0.039 \text{ m}^3$$

$$\text{VTM} = \frac{V_v}{V_t} \times 100 = \frac{0.039}{1} \times 100 = 3.9\%$$

$$\text{VMA} = \frac{V_v + V_b}{V_t} \times 100 = \frac{(0.039 + 0.119)}{1} \times 100 = 15.8\%$$

$$\text{VFA} = \frac{V_b}{V_b + V_v} \times 100 = \frac{0.119}{(0.119 + 0.039)} \times 100 = 75\%$$

To determine these parameters, laboratory tests are performed to estimate specific gravities of the mixture components to use to convert weights to volumes. The density and void analysis requires using the effective specific gravity of the asphalt-coated aggregate, determined from the theoretical maximum specific gravity of the mix. The theoretical maximum specific gravity of the mix is performed according to ASTM D2041 procedure. The weight of the loose mixture specimen in air $A$ is measured along with the weight of the measurement bowl filled with water $D$ and the weight of the bowl containing the asphalt mix and filled with water $E$. When the loose mixture specimen is submerged in water, a vacuum is used to remove all air from the sample. The theoretical maximum specific gravity is

$$G_{mm} = \frac{A}{A + D - E} \tag{9.4}$$

It is necessary to determine only the theoretical maximum specific gravity of the sample at one asphalt content. However, the result should be based on the average of three samples (with a minimum of two). By definition, the theoretical maximum specific gravity of asphalt concrete is

$$G_{mm} = \frac{100}{\left( \dfrac{P_s}{G_{se}} + \dfrac{P_b}{G_b} \right)} \tag{9.5}$$

Solving this equation for $G_{se}$ produces

$$G_{se} = \frac{P_s}{\left( \dfrac{100}{G_{mm}} - \dfrac{P_b}{G_b} \right)} \tag{9.6}$$

where

$G_{mm}$ = theoretical maximum specific gravity of the asphalt concrete
$P_s$   = percent weight of the aggregate
$P_b$   = percent weight of the asphalt cement
$G_{se}$ = effective specific gravity of aggregate coated with asphalt
$G_b$   = specific gravity of the asphalt binder

Although $G_{se}$ is determined for only one asphalt content, we assume that it remains constant for all asphalt contents. Thus, once $G_{se}$ is determined based on the results of the theoretical maximum specific gravity test, it can be used in Equation 9.5 to calculate $G_{mm}$ for the different asphalt contents.

The next step in the process is to determine the bulk specific gravity $G_{mb}$ (ASTM D2726) of each of the compacted specimens. This requires weighing the dry specimen when it is saturated-surface dry and submerged. The bulk specific gravity is computed as

$$G_{mb} = \frac{\text{Weight in air}}{(\text{Weight SSD} - \text{Weight in water})} \qquad (9.7)$$

The unit weight of each specimen is computed by multiplying the bulk specific gravity by the density of water, 1 Mg/m³ (62.4 lb/ft³). The average bulk specific gravity and unit weight for each asphalt content are computed and used to calculate VTM as follows.

$$\text{VTM} = 100\left(1 - \frac{G_{mb}}{G_{mm}}\right) \qquad (9.8)$$

The percent voids in the mineral aggregate (VMA), is a measure of the space available in the aggregates for the addition of the asphalt cement. The percent VMA is the volume of the mix minus the volume of the aggregates divided by the volume of the mix and converted to a percent. VMA is commonly computed from the bulk specific gravity of the aggregate $G_{sb}$, the bulk specific gravity of the mix $G_{mb}$, and the percent weight of aggregate as

$$\text{VMA} = \left(100 - G_{mb}\frac{P_s}{G_{sb}}\right) \qquad (9.9)$$

The percent of the voids filled with asphalt, %VFA, is determined as

$$\text{VFA} = 100\frac{(\text{VMA} - \text{VTM})}{\text{VMA}} \qquad (9.10)$$

**SAMPLE PROBLEM 9.3**

An asphalt concrete specimen has the following properties:

asphalt content = 5.9% by total weight of mix
bulk specific gravity of the mix = 2.457
theoretical maximum specific gravity = 2.598
bulk specific gravity of aggregate = 2.692

Calculate the percents VTM, VMA, and VFA.

**Solution:**

$$\text{VTM} = 100\left(1 - \frac{G_{mb}}{G_{mm}}\right) = 100 \times \left(1 - \frac{2.457}{2.598}\right) = 5.4\%$$

$$\text{VMA} = \left(100 - G_{mb}\frac{P_s}{G_{sb}}\right) = 100 - 2.457 \times \frac{100 - 5.9}{2.692} = 14.1\%$$

$$\text{VFA} = 100\frac{(\text{VMA} - \text{VTM})}{\text{VMA}} = 100 \times \frac{14.1 - 5.4}{14.1} = 61.7\%$$

## Superpave Mix Design

The Superpave asphalt concrete mix design system guides the selection of aggregates. The following four main aggregate properties must be achieved:

- coarse aggregate angularity measured by the percentage of fractured faces,
- fine aggregate angularity (AASHTO TP 33),
- flat and elongated particles (ASTM D4791), and
- clay content (ASTM D2419).

Specification limits for these properties depend on the traffic level and how deep under pavement surface the materials will be used. In addition to these properties, highway agencies may consider other factors that are critical to the specific local conditions.

Aggregate used in asphalt concrete must be well graded. Superpave recommends using the 0.45 power chart discussed in Chapter 5 on page 126. The gradation curve should go through control points specified by Superpave. In addition, the gradation curve should not go through a specified restricted zone in the gradation chart in order to limit the amount of rounded sand in the aggregate blend. Figure 9.22 shows the gradation requirements for the 12.5 mm (1/2 in.) nominal-sized mix.

The binder is selected based on the maximum and minimum pavement temperatures, as discussed earlier. In addition to the specification tests, the specific gravity and the rotational viscosity versus temperature relationship for the selected asphalt binder must be measured. The specific gravity is needed for the void analysis. The viscosity temperature relationship is needed to determine the required mixing and compaction temperatures. The Superpave method requires mixing the asphalt and aggregates at a temperature where the rotational viscosity of the asphalt binder is $0.170 \pm 20$ Pa $\cdot$ s and where the compacting temperature corresponds to a viscosity of $0.280 \pm 30$ Pa $\cdot$ s.

The Superpave asphalt concrete mixture design system varies depending on the design traffic level. Three levels of mix design are available: *volumetric, intermediate,* and *complete.* At the lowest traffic volumes, the volumetric mixture design is used, based on the results of the gyratory compactor and the associated void analysis. At intermediate and high traffic volumes, the intermediate and complete designs are used, in which shear and indirect tension tests are performed on specimens prepared with the gyratory compactor. The three levels of the mix design system are integrated in such a way that the complete design contains all properties measured in the

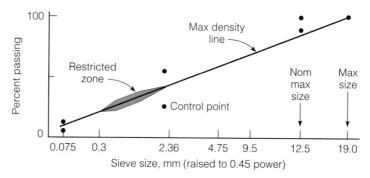

**FIGURE 9.22** Superpave gradation limits for 12.5 mm nominal maximum size. (McGennis 1995)

volumetric and intermediate designs, and the intermediate design contains all properties measured in the volumetric design. The intermediate and complete design analyses are currently undergoing revisions and, therefore, have not been implemented.

**Volumetric Mix Design** The volumetric mix design plays a central role in Superpave mix design. After selecting the appropriate aggregate, binder, and modifiers (if any), trial specimens are prepared with different aggregate gradations and asphalt contents. Specimens are compacted using the Superpave gyratory compactor (Figure 9.18) with a gyration angle of 1.25 degrees and a constant vertical pressure of 600 kPa (87 psi). The number of gyrations used for compaction is determined based on the design high air temperature of the paving location and the traffic level as shown in Table 9.7. Three critical numbers of gyrations are determined, namely, initial, design, and maximum numbers of gyrations ($N_{ini}$, $N_{des}$, and $N_{max}$, respectively).

During compaction, the height of the specimen is continuously monitored by the compactor, as shown in the example in Table 9.8. Knowing the mass of the mix, the diameter of the mold, and the measured height at any gyration, the bulk specific gravity of the specimen $G_{mb}$ can be estimated throughout the compaction process. This is accomplished by dividing the mass of the specimen by its volume, which is represented by the volume of a smooth-sided cylinder of known diameter and measured

**TABLE 9.7**  Superpave Gyratory Compaction Effort Based on Average Design High Air Temperature

| Design ESALs, $10^6$ | < 39 °C | | | 39 °C–40 °C | | | 41 °C–42 °C | | | 43 °C–44 °C | | |
|---|---|---|---|---|---|---|---|---|---|---|---|---|
| | $N_{ini}$ | $N_{des}$ | $N_{max}$ | $N_{ini}$ | $N_{des}$ | $N_{max}$ | $N_{ini}$ | $N_{des}$ | $N_{max}$ | $N_{ini}$ | $N_{des}$ | $N_{max}$ |
| < 0.3 | 7 | 68 | 104 | 7 | 74 | 114 | 7 | 78 | 121 | 7 | 82 | 127 |
| 0.3–1 | 7 | 76 | 117 | 7 | 83 | 129 | 7 | 88 | 138 | 8 | 93 | 146 |
| 1–3 | 7 | 86 | 134 | 8 | 95 | 150 | 8 | 100 | 158 | 8 | 105 | 167 |
| 3–10 | 8 | 96 | 152 | 8 | 106 | 169 | 8 | 113 | 181 | 9 | 119 | 192 |
| 10–30 | 8 | 109 | 174 | 9 | 121 | 195 | 9 | 128 | 208 | 9 | 135 | 220 |
| 30–100 | 9 | 126 | 204 | 9 | 139 | 228 | 9 | 146 | 240 | 10 | 153 | 253 |
| > 100 | 9 | 142 | 233 | 10 | 158 | 262 | 10 | 165 | 275 | 10 | 172 | 288 |

**TABLE 9.8**   Example of Densification Data of a Trial Blend[*]

| Number of Gyrations | Height, mm | Estimated $G_{mb}$[†] | Corrected $G_{mb}$[‡] | Percent $G_{mm}$ |
|---|---|---|---|---|
| 8 ($N_{ini}$) | 127.0 | 2.170 | 2.217 | 86.5 |
| 50 | 118.0 | 2.334 | 2.385 | 93.0 |
| 100 | 115.2 | 2.392 | 2.444 | 95.4 |
| 109 ($N_{des}$) | 114.9 | 2.398 | 2.450 | 95.6 |
| 150 | 113.6 | 2.425 | 2.478 | 96.7 |
| 174 ($N_{max}$) | 113.1 | 2.436 | 2.489 | 97.1 |

[*]McGennis (1995)
[†]$G_{mm}$ (measured) = 2.563; total mass = 4869 g
[‡]$G_{mb}$ (measured) = 2.489.

height. After compaction is complete, the specimen is extruded from the mold and the bulk specific gravity $G_{mb}$ is determined according to ASTM D2726 procedure. The theoretical maximum specific gravity of the mix $G_{mm}$ is performed according to ASTM D2041 procedure.

The estimated bulk specific gravity of the specimen at any given gyration is then corrected by a factor that is the ratio of the measured bulk specific gravity to the estimated bulk specific gravity at the maximum number of gyrations, as shown in the example in Table 9.8. In this example, the correction factor is 2.489/2.436, or 1.0218. The corrected bulk specific gravity at any given gyration $G_{mb}$ (corrected) is then calculated as a percent of $G_{mb}$. Two specimens are prepared for each trial blend and their data are averaged, as shown in Table 9.8. Densification curves are plotted as illustrated in Figure 9.23.

The following three main steps are used in the testing and analysis process:

- selection of design aggregate structure
- selection of design asphalt content
- evaluation of moisture sensitivity of design mixture

A total of 20 specimens, 150 mm (6 in.) in diameter and 95 mm to 115 mm (3.75 in. to 4.5 in.) high, are required in these three steps as summarized in Figure 9.24. To select the proper aggregate structure [Figure 9.24(a)], three trial

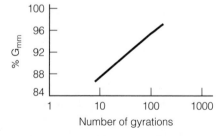

**FIGURE 9.23** Example of Superpave densification curves (see Table 9.8).

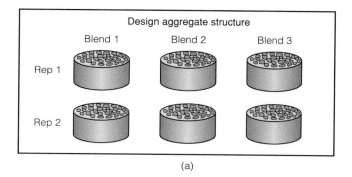

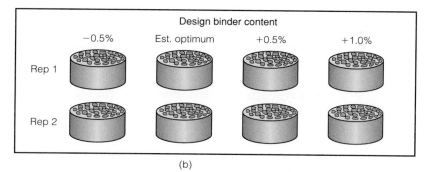

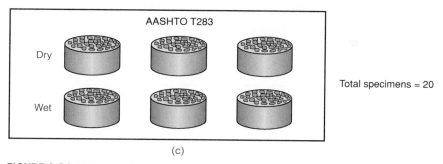

**FIGURE 9.24** Testing requirements for Superpave volumetric mix design: (a) selecting design aggregate structure, (b) selecting design asphalt content, (c) evaluating moisture sensitivity.

blends are made using an estimated optimum asphalt content and three different aggregate gradations; 2 replicate specimens are prepared for each trial blend. The trial blend that proves to be the closest to the required design criteria is selected. The design criteria include 4% air voids, as well as other requirements for voids in mineral aggregate and voids filled with asphalt. In addition, the density at the initial number of gyrations must be less than 89% of the maximum theoretical density. The density at the maximum number of gyrations must be less than 98% of the maximum theoretical density. These density requirements are specified to enhance resistance to permanent deformation.

The design binder content is obtained by preparing 8 specimens; 2 replicates at each of four binder contents: estimated optimum binder content, 0.5% less than the optimum, 0.5% more than the optimum, and 1% more than the optimum, as shown in Figure 9.24(b). The voids and density data are plotted against binder content and compared with the design criteria. The binder content that satisfies all design criteria is selected as the design binder.

The moisture sensitivity of the design mixture is determined using the AASHTO T283 procedure on 6 specimens prepared at the design binder content and 7% air voids. Three specimens are conditioned by vacuum-saturation, then freezing and thawing, whereas 3 other specimens are not conditioned [as illustrated in Figure 9.24(c)]. The tensile strength ratio is determined as the ratio of the average tensile strength of conditioned specimens to that of unconditioned specimens. The minimum Superpave criterion for tensile strength ratio is 80%.

**Intermediate Mix Design**  In the intermediate mix design, some performance-based properties are used. After satisfying the volumetric requirements, more specimens are compacted using the Superpave gyratory compactor. Typically a total of 23 specimens are required for the intermediate design. The laboratory tests to be performed and the types of potential pavement distress all:

| Test | Distress |
| --- | --- |
| Simple shear | Fatigue and permanent deformation |
| Repeated load shear | Fatigue and permanent deformation |
| Frequency sweep | Fatigue and permanent deformation |
| Indirect tensile strength | Permanent deformation |
| Low-temperature creep | Low-temperature cracking |
| Low-temperature fracture | Low-temperature cracking |
| Bending beam rheometer on asphalt binder | Low-temperature cracking |

**Complete Mix Design**  The complete mix design uses fundamental material models to predict the amount of distress and in what time frame it occurs. This level is similar to the intermediate design except that more tests are required to obtain better prediction for pavement performance. Typically a total of 59 specimens are required for the complete design. The laboratory tests to be performed and the types of potential pavement distress are:

| Test | Distress |
| --- | --- |
| Simple shear | Fatigue and permanent deformation |
| Repeated load shear | Fatigue and permanent deformation |
| Frequency sweep | Fatigue and permanent deformation |
| Uniaxial | Fatigue and permanent deformation |
| Hydrostatic | Fatigue and permanent deformation |
| Indirect tensile strength | Fatigue and permanent deformation |
| Low-temperature creep | Low-temperature cracking |
| Low-temperature fracture | Low-temperature cracking |

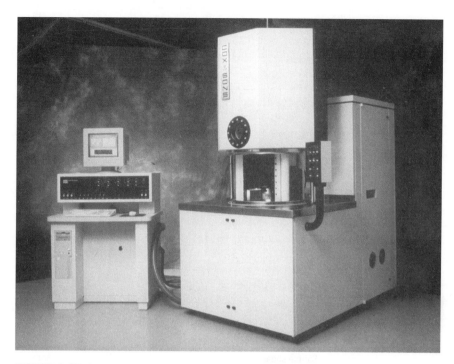

**FIGURE 9.25** Closed-loop electrohydrualic shear test system.

Tests required for both intermediate and complete mix designs use specimens prepared using the Superpave gyratory compactor. Typical specimens for all tests are 152 mm (6 in.) in diameter by 51 mm (2 in.) high. Three replicate specimens are used for each test; some specimens are used for more than one test. A closed-loop electrohydraulic shear test machine with two actuators is used to perform the simple shear, repeated load shear, frequency sweep, uniaxial, and hydrostatic tests. The machine is equipped with a temperature-controlled chamber and a computerized data acquisition system, as shown in Figure 9.25. Figure 9.26 illustrates loading configurations of the Superpave intermediate and complete mix designs (McGennis 1995).

**Superpave Software Support System**   The Superpave software is used to perform all calculations needed in the mix design and analysis system. The complete software package includes weather database files and other associated subroutines. The software contains results of fundamental material research in asphalt binder and asphalt mixture. It ties together new performance-based tests and analysis methods with the elements of current mixture design that can be used in the future. The software is capable of analyzing the asphalt mixture according to the design procedure discussed earlier.

## Marshall Method of Mix Design

The basic steps required for performing Marshall mix design are (The Asphalt Institute 1995):

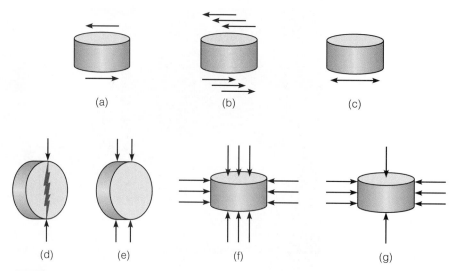

**FIGURE 9.26** Superpave tests for intermediate and complete levels: (a) simple shear, (b) repeated load shear, (c) frequency sweep, (d) indirect tensile strength, (e) low-temperature creep, (f) hydrostatic, (g) uniaxial.

1. aggregate evaluation
2. asphalt cement evaluation
3. prepare specimens
4. measure Marshall stability and flow
5. density and voids analysis
6. determine design asphalt content.

**1. Aggregate Evaluation** The aggregates' characteristics that must be evaluated before it can be used for an asphalt concrete mix include the durability, soundness, presence of deleterious substances, polishing, shape, and texture. Agency specifications define the allowable ranges for aggregate gradation. The Marshall method is applicable to densely graded aggregates with a maximum size of not more than 25 mm (1 in.).

**2. Asphalt Cement Evaluation** The grade of asphalt cement is selected based on the expected temperature range and traffic conditions. Most highway agencies have specifications that prescribe the grade of asphalt for the design conditions.

**3. Prepare Specimens** The full Marshall mix-design procedure requires 18 specimens 101.6 mm (4 in.) in diameter and 63.5 mm (2.5 in.) high. The stability and flow is measured for 15 specimens. In addition, 3 specimens are used to determine the theoretical maximum specific gravity $G_{mm}$. This value is needed for the void and density analysis. The specimens for the theoretical maximum specific gravity determination are prepared at the estimated design asphalt content. Samples are also required for each of five different asphalt contents; the expected design asphalt content, $\pm 0.5\%$ and $\pm 1.0\%$. Engineers use experience and judgment to estimate the design asphalt content.

**FIGURE 9.27** Marshall stability machine.

Specimen preparation for the Marshall method uses the Marshall compactor discussed earlier (Figure 9.19). The Marshall method requires mixing the asphalt and aggregates at a temperature where the kinematic viscosity of the asphalt cement is $170 \pm 20$ cSt and compacting temperature corresponds to a viscosity of $280 \pm 30$ cSt.

The Asphalt Institute permits three different levels of energy to be used for the preparation of the specimens: 35, 50, and 75 blows on each side of the sample. Most mix designs for heavy-duty pavements use 75 blows, since this better simulates the required density for pavement construction.

**4. Measure Marshall Stability and Flow** The Marshall stability of the asphalt concrete is the maximum load the material can carry when tested in the Marshall apparatus, Figure 9.27. The test is performed at a deformation rate of 51 mm/min. (2 in./min.) and a temperature of 60°C (140° F). The Marshall flow is the deformation of the specimen when the load starts to decrease. Stability is reported in newtons (pounds) and flow is reported in units of 0.25 mm (0.01 in.) of deformation. The stability of specimens that are not 63.5 mm thick are adjusted by multiplying by the factors in Table 9.9. All specimens are tested and the average stability and flow determined for each asphalt content.

**5. Density and Voids Analysis** The values of VTM, VMA, and VFA are determined as discussed in Equations 9.8, 9.9, and 9.10.

**6. Determine Design Asphalt Content** Traditionally test results and calculations are tabulated and graphed to help determine the factors that must be used to choose the optimum asphalt content. Table 9.10 presents example mix design measurements and calculations. Figure 9.28 shows plots of results obtained from Table 9.10 which includes asphalt content versus air voids, VMA, VFA, unit weight, Marshall stability, and Marshall flow.

**TABLE 9.9**   Marshall Stability Adjustment Factors

| Approximate Thickness of Specimen, mm (in.) | Adjustment Factor | Approximate Thickness of Specimen, mm (in.) | Adjustment Factor |
|---|---|---|---|
| 50.8 (2) | 1.47 | 65.1 (2-9/16) | 0.96 |
| 52.4 (2-1/16) | 1.39 | 66.7 (2-5/8) | 0.93 |
| 54.0 (2-1/8) | 1.32 | 68.3 (2-11/16) | 0.89 |
| 55.6 (2-3/16) | 1.25 | 69.8 (2-3/4) | 0.86 |
| 57.2 (2-1/4) | 1.19 | 71.4 (2-13/16) | 0.83 |
| 58.7 (2-5/16) | 1.14 | 73.0 (2-7/8) | 0.81 |
| 60.3 (2-3/8) | 1.09 | 74.6 (2-15/16) | 0.78 |
| 61.9 (2-7/16) | 1.04 | 76.2 (3) | 0.76 |
| 63.5 (2-1/2) | 1.00 | | |

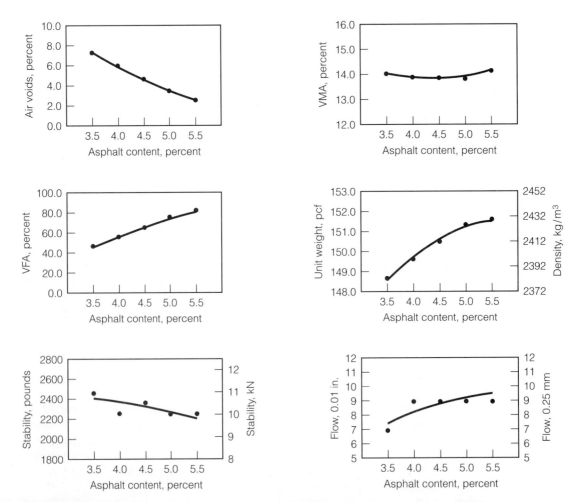

**FIGURE 9.28**   Graphs used for Marshall mix design analysis. (see Table 9-10). (The Asphalt Institute 1995)

**TABLE 9.10** Example of Mix Design Measurements and Calculations by the Marshall Method*†

| AC by Wt. of Mix, % —Spec. No. | Specimen Height, mm | Mass in Air, g | Mass in Water, g | SSD Mass, g | Bulk Vol., cm³ | Bulk Sp. Gr. | Max. Theo. Sp. Gr. (loose mix) | Air Voids, % | VMA, % | VFA, % | Measured Stability, kN | Adjusted Stability, kN | Flow, 0.25 mm |
|---|---|---|---|---|---|---|---|---|---|---|---|---|---|
| 3.5 — A | | 1240.6 | 726.4 | 1246.3 | 519.9 | 2.386 | | | | | 10.9 | 10.9 | 8 |
| 3.5 — B | | 1238.7 | 723.3 | 1242.6 | 519.3 | 2.385 | | | | | 10.8 | 10.8 | 7 |
| 3.5 — C | | 1240.1 | 724.1 | 1245.9 | 521.8 | 2.377 | | | | | 11.2 | 11.2 | 7 |
| Average | | | | | | 2.383 | 2.570 | 7.3 | 14.0 | 48.0 | | 10.9 | 7 |
| 4.0 — A | | 1244.3 | 727.2 | 1246.6 | 519.4 | 2.396 | | | | | 9.7 | 9.7 | 9 |
| 4.0 — B | | 1244.6 | 727.0 | 1247.6 | 520.6 | 2.391 | | | | | 10.1 | 10.1 | 9 |
| 4.0 — C | | 1242.6 | 727.9 | 1244.0 | 516.1 | 2.408 | | | | | 10.3 | 10.3 | 8 |
| Average | | | | | | 2.398 | 2.550 | 6.0 | 13.9 | 57.1 | | 10.0 | 9 |
| 4.5 — A | | 1249.3 | 735.8 | 1250.2 | 414.4 | 2.429 | | | | | 10.8 | 10.8 | 9 |
| 4.5 — B | | 1250.8 | 728.1 | 1251.6 | 523.5 | 2.389 | | | | | 10.7 | 10.3 | 9 |
| 4.5 — C | | 1251.6 | 735.3 | 1253.1 | 517.8 | 2.417 | | | | | 10.4 | 10.4 | 9 |
| Average | | | | | | 2.412 | 2.531 | 4.7 | 13.9 | 66.1 | | 10.5 | 9 |
| 5.0 — A | | 1256.7 | 739.8 | 1257.6 | 517.8 | 2.427 | | | | | 10.2 | 10.2 | 9 |
| 5.0 — B | | 1258.7 | 742.7 | 1259.3 | 516.6 | 2.437 | | | | | 9.7 | 9.7 | 8 |
| 5.0 — C | | 1258.4 | 737.5 | 1259.1 | 521.6 | 2.413 | | | | | 10.0 | 10.0 | 9 |
| Average | | | | | | 2.425 | 2.511 | 3.4 | 13.8 | 75.2 | | 10.0 | 9 |
| 5.5 — A | | 1263.8 | 742.6 | 1264.3 | 521.7 | 2.422 | | | | | 9.8 | 9.8 | 9 |
| 5.5 — B | | 1258.8 | 741.4 | 1259.4 | 518.0 | 2.430 | | | | | 10.2 | 10.2 | 10 |
| 5.5 — C | | 1259.0 | 742.5 | 1259.5 | 517.0 | 2.435 | | | | | 9.8 | 10.0 | 9 |
| Average | | | | | | 2.429 | 2.493 | 2.5 | 14.1 | 82.1 | | 10.0 | 9 |

*The Asphalt Institute (1995).
†AC-20 binder; $G_b$ = 1.030; $G_{sb}$ = 2.674; absorbed AC of aggregate: 0.6%; $G_{se}$ = 2.717; compaction: 75 blows.

**TABLE 9.11**  Asphalt Institute Criteria for Marshall Mix Design

| | Traffic Level | | | | | |
|---|---|---|---|---|---|---|
| | Light | | Medium | | Heavy | |
| Compaction, blows | 35 | | 50 | | 75 | |
| | Minimum | Maximum | Minimum | Maximum | Minimum | Maximum |
| Stability, kN | 3.34 | — | 5.34 | — | 8.01 | — |
| Flow, 0.25 mm | 8 | 18 | 8 | 16 | 8 | 14 |
| Air Voids, % | 3 | 5 | 3 | 5 | 3 | 5 |
| VMA, % | | | Use the criteria in Table 9.12 | | | |
| VFA, % | 70 | 80 | 65 | 78 | 65 | 75 |

**TABLE 9.12**  Minimum Percent Voids in Mineral Aggregate (VMA), %[*]

| | Design Air Voids[‡] | | |
|---|---|---|---|
| Nominal Maximum Particle Size[†] | 3.0 | 4.0 | 5.0 |
| 2.36 mm (No. 8) | 19.0 | 20.0 | 21.0 |
| 4.75 mm (No. 4) | 16.0 | 17.0 | 18.0 |
| 9.5 mm (3/8 in.) | 14.0 | 15.0 | 16.0 |
| 12.5 mm (1/2 in.) | 13.0 | 14.0 | 15.0 |
| 19.0 mm (3/4 in.) | 12.0 | 13.0 | 14.0 |
| 25.0 mm (1.0 in.) | 11.0 | 12.0 | 13.0 |

[*]The Asphalt Institute (1995).
[†]The nominal maximum particle size is one size larger than the first sieve to retain more than 10%.
[‡]Interpolate minimum VMA to find design air void values between those listed.

The design asphalt content is usually the most economical one that will also satisfactorily meet all of the established criteria. Different criteria are used by different agencies. Table 9.11 and 9.12 depict the mix design criteria recommended by The Asphalt Institute. Figure 9.29 shows an example of the narrow range of acceptable asphalt contents. The asphalt content selection can be adjusted within this narrow range to achieve a mix that satisfies the requirements of a specific project. Other agencies, such as the National Asphalt Paving Association, use the asphalt cement content at 4% air voids as the design value, and then check that the other factors meet the criteria. If the Marshall stability, flow, VMA, or VFA fall outside the allowable range, the mix must be redesigned using an adjusted aggregate gradation or new material sources.

The laboratory-developed mixture design forms the basis for the initial job mix formula (JMF). The initial JMF should be adjusted to consider the slight differences between the laboratory-supplied aggregates and those used in the field.

## Hveem Method of Mix Design

The basic steps required for performing Hveem mix design are (The Asphalt Institute 1995)

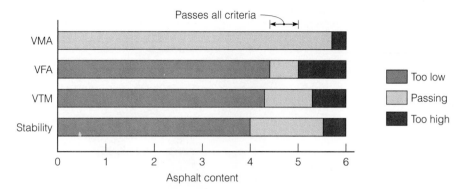

**FIGURE 9.29** An example of the narrow range of acceptable asphalt contents. (The Asphalt Institute 1995)

1. aggregate evaluation
2. asphalt cement evaluation
3. evaluation of centrifuge kerosene equivalent of fine aggregate
4. evaluation of surface capacity of coarse aggregate
5. estimation of optimum asphalt content
6. specimen preparation
7. measurement of the Hveem stability
8. density and voids analysis
9. determination of design asphalt content.

The evaluation of aggregate and asphalt cement is performed as in the Marshall method of mix design. The Hveem method requires measuring aggregate properties and using a series of charts to estimate the design asphalt content (The Asphalt Institute 1995).

Three cylindrical specimens 102 mm (4 in.) in diameter and 63.5 mm (2.5 in.) high are prepared using the California kneading compactor (Figure 9.20) according to ASTM D1561. Three asphalt contents near the estimated design value are used to fabricate the specimens. The Hveem stability of the specimens is determined using the Hveem stabilometer (Figure 9.30), according to ASTM D1560. The Hveem stabilometer is a device that allows for applying a lateral pressure on the specimen while applying vertical load using a compression machine.

As in the Marshall method, the bulk specific gravity, theoretical maximum specific gravity, percent air voids (VTM), and density of all specimens are determined. The Hveem stability, density, and air voids are tabulated and plotted versus asphalt content. The optimum asphalt content for the design mix should be the highest asphalt content the mix will accommodate without reducing the stability or void content below the minimum values required by the design criteria.

The laboratory-developed mixture design forms the basis for the initial JMF. The initial JMF should be adjusted to consider the slight differences between the laboratory supplied aggregates and those used in the field.

**FIGURE 9.30**  Hveem stabilometer.

## Evaluation of Moisture Susceptibility

Since loss of bond between asphalt and aggregates (*stripping*) has become a signi-ficant form of asphalt pavement distress, several methods have been developed for evaluating the susceptibility of a mix to water damage. Most of the popular methods require the specimens to be at the optimum asphalt content and mix gradation.

The specimens are divided into two lots, reference specimens and conditioned specimens. A strength test is used to evaluate the strength before and after condition-ing; the retained strength, the ratio of conditioned strength to reference strength, expressed in percent, is computed. Criteria are used to determine if the retained strength is adequate. The different techniques for evaluating moisture susceptibility vary depending on the specimen preparation, conditioning procedures, and strength.

The immersion-compression test (ASTM D1075) has been used to evaluate mois-ture susceptibility. The method evaluates the retained compressive strength after vac-uum saturation. Other methods use Marshall specimens, freezing and water soaking to condition the samples, and determining diametral strength and modulus values to evaluate the retained strength. Freezing the samples greatly increases the severity of the test.

There are several ways to alter asphalt concrete's susceptibility to water damage. Methods identified by the Asphalt Institute include

1. increasing asphalt content
2. using a higher viscosity asphalt cement
3. cleaning aggregate of any dust and clay
4. adding antistripping additives
5. altering aggregate gradation

In addition, portland cement and lime have been used by some agencies as anti-stripping agents. Generally, when water damage susceptibility is a problem, the additive is added to the mix at three levels, and the water damage test is performed to determine the minimum amount of additive that can be used to increase the retained strength to an acceptable level. If an acceptable mix can be developed, Marshall or Hveem specimens are prepared and the mix is tested to determine if it meets the design criteria.

# Characterization of Asphalt Concrete

Tests used to characterize asphalt concrete are somewhat different from those used to characterize other civil engineering materials such as steel, portland cement concrete, and wood. One of the main reasons for this difference is that asphalt concrete is a nonlinear viscoelastic or visco-elastoplastic material. Thus its response to loading is greatly affected by the rate of loading and temperature. Also, asphalt pavements are typically subjected to dynamic loads applied by traffic. Moreover, asphalt pavements do not normally fail due to sudden collapse under the effect of vehicular loads, but due to accumulation of permanent deformation in the wheel path (rutting), cracking due to repeated bending of the asphalt concrete layer (fatigue cracking), thermal cracking, excessive roughness of the pavement surface, migration of asphalt binder at the pavement surface (bleeding or flushing), loss of flexibility of asphalt binder due to aging and oxidation (raveling), loss of bond between the asphalt binder and aggregate particles due to moisture (stripping), or other factors. Therefore, most of the tests used to characterize asphalt concrete try to simulate actual field conditions.

Many laboratory tests have been used to evaluate asphalt concrete properties and to predict its performance in the field. These tests are performed on either laboratory-prepared specimens or cores taken from in-service pavements. These tests measure the response of the material to load, deformation, or environmental conditions such as temperature, moisture, or freeze and thaw cycles. Some of these tests are based on empirical relations, while others evaluate fundamental properties. All tests on asphalt concrete are performed at accurately controlled test temperatures and rates of loading since asphalt response is largely affected by the two parameters.

The Superpave tests used for mix design, as well as Marshall or Hveem tests discussed earlier, have been used to characterize asphalt concrete mixtures. Other tests are also being used, some of which are standardized by ASTM or AASHTO, while others have been used mostly for research. The following sections discuss some of the common tests.

## Indirect Tensile Strength

When traffic loads are applied on the pavement surface, tension is developed at the bottom of the asphalt concrete layer. Therefore, it is important to evaluate the tensile strength of asphalt concrete for the design of the layer thickness. In this test a cylindrical specimen 102 mm (4 in.) in diameter and 64 mm (2.5 in.) high is used. A compressive vertical load is applied along the vertical diameter, using a loading device similar to that shown in Figure 9.31. The load is applied using two curved loading strips moving with a rate of deformation of 51 mm/min. (2 in./min.). Tensile stresses are developed in the horizontal direction, and when these stresses reach the tensile strength the specimen fails in tension along the vertical diameter. The test is performed at a specified temperature. Using 12.5 mm (0.5 in.) loading strips the indirect tensile strength is computed as

$$\sigma_t = \frac{0.159P}{t} \tag{9.11}$$

where

$\sigma_t$ = tensile strength, MPa (psi)
$P$ = load at failure, N (lb)
$t$ = thickness of specimen, mm (in.)

## Diametral Tensile Resilient Modulus

To evaluation the structural response of the asphalt pavement system, the modulus of asphalt concrete material is needed. Since asphalt concrete is not a linear visco-elastic material, the modulus of elasticity, Young's modulus, is not applicable. The diametral tensile resilient modulus test (ASTM D4123) provides an analogous modulus, known as resilient modulus. The test uses a cylindrical specimen 102 mm (4 in.) in diameter and 63.5 mm (2.5 in.) high. A pulsating load is applied along the vertical diameter using a load guide device similar to that shown in Figure 9.31. The load is

**FIGURE 9.31** Diametral loading device for indirect tensile strength and resilient modulus tests.

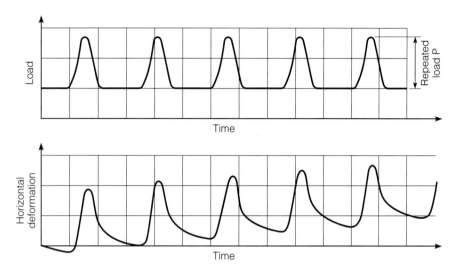

**FIGURE 9.32** Typical (a) load versus time and (b) deformation versus time during the resilient modulus test.

commonly applied with a duration of 0.1 s and a rest period of 0.9 s. After a few hundred repetitions, the recoverable horizontal deformation is measured using two linear variable differential transducers (LVDTs). Figure 9.32 shows typical load and horizontal deformation versus time relationships. Since the test is nondestructive, the test is repeated on the same specimen after rotating it 90°. The test is commonly performed at three temperatures: 5°C, 25°C, and 40°C (41°F, 77°F, and 104°F). The diametral tensile resilient modulus is computed using the following equation

$$M_R = \frac{P(0.27 + \nu)}{t \times \Delta H} \tag{9.12}$$

where

$M_R$ = indirect tensile resilient modulus, MPa (or psi)

$P$ = repeated load, N (lb)

$\nu$ = Poisson's ratio, typically 0.3, 0.35, and 0.4 at temperatures of 5°C, 25°C, and 40°C, respectively

$t$ = thickness of specimen, mm (in.)

$\Delta H$ = sum of recoverable horizontal deformations on both sides of specimen, mm (in.)

Typical resilient modulus values of asphalt concrete are 6.89 GPa, 4.13 GPa, and 1.38 GPa (1000 ksi, 600 ksi, and 200 ksi) at temperatures of 5°C, 25°C, and 40°C, respectively. The diametral tensile resilient modulus test is very sophisticated because it measures very small deformations. Therefore, extreme caution must be exercised to align the specimen between the loading trips and to reduce possible rocking. Also, the load magnitude must be small enough to reduce the possibility of permanent deformation in the specimen, yet large enough to obtain measurable deformation.

**SAMPLE PROBLEM 9.4**

The resilient modulus test was performed on an asphalt concrete specimen and the following data were obtained:

> diameter = 4.000 in.
> thickness = 2.523 in.
> repeated load = 559 lb
> sum of recoverable horizontal deformations = $254 \times 10^{-6}$ in.

Assuming a Poisson's ratio of 0.35, calculate the resilient modulus.

*Solution:* :

$$M_R = \frac{P(0.27 + \nu)}{t \times \Delta H} = 559 \times \frac{(0.27 + 0.35)}{(2.523 \times 254 \times 10^{-6})} = 541{,}000 \text{ psi} \qquad \blacklozenge$$

### Freeze and Thaw Test

The freeze and thaw test is performed to evaluate the effect of freeze and thaw cycles on the stiffness properties of asphalt concrete. Cylindrical specimens 102 mm (4 in.) in diameter and 64 mm (2.5 in.) high are used. Three specimens are tested for resilient modulus as discussed earlier, while the other three specimens are subjected to cycles of freeze and thaw, after which the resilient modulus is determined. The tensile strength ratio is computed by dividing the average resilient modulus of conditioned specimens by the average resilient modulus of unconditioned specimens, expressed in percent. A minimum tensile strength ratio is usually required to identify mixes that are not severely affected by freeze and thaw cycles.

### Creep Compliance

Each time a wheel load is applied on asphalt concrete pavements a small permanent deformation might develop; these accumulate throughout the years and cause appreciable rutting. Rutting in the asphalt pavement could be due to the compressibility of either asphalt concrete, base, subbase, or subgrade. The creep test on asphalt concrete provides good indication of the rutting potential that might occur due to the compressibility of the asphalt concrete layer. Currently, there is no ASTM or AASHTO test for asphalt concrete creep. Usually cylindrical specimens with 102 mm (4 in.) diameter are used. The specimen height varies from one study to another with typical values of 64 mm, 102 mm, or 204 mm (2.5 in., 4 in., or 8 in.). The creep test is performed by applying a constant axial compressive load, while the specimen deformation is recorded at different times; typically 1 s, 10 s, 100 s, 1000 s and 3600 s. The load is then removed and the deformation is measured at similar intervals. Figure 9.33 shows load and deformation versus time during the creep test. The figure shows that when the load is suddenly applied, an instantaneous deformation occurs after which the material creeps in a nonlinear form under constant load. When the load is suddenly removed, an instantaneous recovery occurs, after which the material creeps again under zero load. The creep compliance is computed at different times during the loading portion as follows:

$$J(t) = \frac{\epsilon(t)}{\sigma} \qquad (9.13)$$

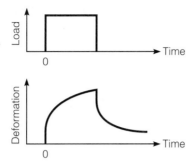

**FIGURE 9.33** Typical (a) load versus time and (b) deformation versus time during the creep test.

where

$J(t)$ = creep compliance at different times, mm²/N (in.²/lb)

$\epsilon(t)$ = strain at different times, mm/mm (in./in.)

$\sigma$  = applied constant stress, MPa (psi)

### Use of Rheological Models to Analyze Time-Dependent Response

Asphalt concrete is a viscoelastic material exhibiting a time-dependent response under load. Rheological models consisting of combinations of Hookean (spring) and Newtonean (dashpots) elements have been used to analyze the response of time-dependent materials as discussed in Chapter 1. The Burgers model illustrated in Figure 1.12 can closely approximate the response of asphaltic mixtures (Mamlouk 1984). Laboratory tests, such as the creep test, are used to obtain the parameters of the Burgers model using a curve-fitting procedure. Once these parameters are determined, the model can be used to predict the response of the material under different loading conditions. For example, Burgers model has been used to predict rutting of asphalt concrete pavement under the action of traffic loads.

## Recycling of Asphalt Concrete

Recycling pavement materials has a long history. However, recycling became more important in the mid 1970s, after the oil embargo, due to the increase in asphalt prices. In an effort to efficiently use available resources, there was a need to recycle or reuse old pavement materials (The Asphalt Institute 1989). Although the pavement could be badly deteriorated, the old asphalt concrete materials could be successfully reused in new pavements. Currently, recycling of old pavement materials is becoming a normal practice due to the following advantages:

1. economic saving of about 25% of the price of materials
2. energy saving in manufacturing and transporting raw materials
3. environmental saving by reducing the amount of required new materials and by eliminating the problem of discarding old materials

**4.** eliminating the problem of reconstruction of utility structures, curbs, and gutters associated with overlays

**5.** reducing the dead load on bridges due to overlays

**6.** maintaining the tunnel clearance when compared with overlays

Recycling can be divided to three types: *surface recycling, central plant recycling,* and *in-place recycling.*

## Surface Recycling

Surface recycling is defined as the reworking of the top 25 mm (1 in.) of the pavement surface using a heater-scarifier. The heater planing machine heats the pavement surface, which repairs minor cracks and roughness. Usually, a rejuvenating agent is added after heating, followed by slight scratching of the surface and compaction.

## Central Plant Recycling

Central plant recycling is performed by milling the old pavement and sending the reclaimed asphalt pavement (RAP) to a central asphalt concrete plant where it is mixed with some form of rejuvenating agent or soft asphalt and aggregates to produce hot-mixed asphalt concrete. If the RAP materials are mixed with the aggregates in a conventional asphalt concrete plant, they will burn and produce smoke, causing significant environmental problems. Therefore, the RAP materials are added to the pugmill (mixer) in the batch plant or added at midlength in the drum at the drum mix plant. The amount of recycled materials varies from 20% to 70%. The gradation of new aggregates is selected to correct any deficiency in the gradation of recycled aggregate. Typically, the grade of new asphalt cement is soft so that when it is mixed with the old, hard asphalt an appropriate consistency will result. The mix design of asphalt concrete, including recycled materials, is usually performed using either the Marshall or Hveem procedure.

In addition to hot-mix central plant recycling, cold central plant recycling can use new emulsified or cutback asphalt. However, the cold process will not have the quality of the hot-mixed material.

## In-Place Recycling

In-place recycling is performed by ripping and pulverizing the old pavement surface and adding new aggregate, water, and asphalt emulsion. The old and new materials are mixed together in place, graded, and compacted. The surface is left to cure and then used as a surface layer for low-volume roads. The recycled layer can also be used as a stabilized base, covered by an asphalt concrete surface.

# Additives

Many types of additives (modifiers) are used to improve the properties of asphalt or to add special properties to the asphalt concrete mixtures (Roberts et al. 1991). Laboratory tests are usually performed and field performance is observed in order to evaluate the effect of the additives and to justify their cost. The effects of using additives should be carefully evaluated, otherwise premature pavement failure might result. The recyclability of modified asphalt mixtures is still being evaluated.

## Fillers

Several types of fillers, such as crushed fines, portland cement, lime, fly ash, and carbon black can be added to asphalt concrete. Fillers are used to satisfy gradation requirements of materials passing the 0.075 mm (No. 200) sieve, to increase stability, to improve bond between aggregates and asphalt, or to fill the voids and thus reduce the required asphalt.

## Extenders

Extenders such as sulfur and lignin are used to reduce the asphalt requirements, thus reducing the cost.

## Rubber

Rubber has been used in asphalt concrete mixture in the form of natural rubber, styrene-butadiene (SBR), styrene-butadiene-styrene (SBS), or recycled tire rubber. Rubber increases elasticity and stiffness of the mix and increases the bond between asphalt and aggregates. Scrap rubber tires can be added to the asphalt cement (wet method) or added as crumb rubber to the aggregates (dry method).

## Plastics

Plastics have been used to improve certain properties of asphalt. Plastics used include polyethylene, polypropylene, ethyl-vinyl-acetate (EVA), and polyvinyl chloride (PVC). They increase the stiffness of the mix, thus reducing the rutting potential. Plastics also may reduce the temperature susceptibility of asphalt and improve its performance at low temperatures.

## Antistripping Agents

Antistripping agents are used to improve the bond between asphalt cement and aggregates, especially for water susceptible mixtures. Lime is the most commonly used antistripping agent and can be added as a filler or a lime slurry and mixed with the aggregates. Portland cement can be used as an alternative to lime.

## Others

Other additives such as fibers, oxidants, antioxidants, and hydrocarbons have been used to modify certain asphalt properties' tensile strength and stiffness.

**SUMMARY**

Asphalt produced from crude oil is a primary road-building material. The civil engineer is directly involved with the specification and requirements for both the asphalt cement binder and the asphalt concrete mixtures. There are several methods for grading asphalt cements. The current trend is toward the use of the performance-grading method used in the Superpave process developed through the Strategic Highway Research Program. This grading method directly ties the binder properties to pavement-performance parameters. Similarly, the Superpave mix design method uses performance tests to evaluate the mixture characteristics relative to expected field performance. This method will continue to be refined by highway agencies. With the support of the Federal Highway Administration, there is a concerted effort being placed on replacing the traditional Marshall and Hveem mix design methods with the Superpave procedures.

**QUESTIONS AND PROBLEMS**

**9.1.** What is the difference between tar and asphalt cement?

**9.2.** Discuss the main uses of asphalt.

**9.3.** Define what is meant by temperature susceptibility of asphalt. Discuss the effect on the performance of asphalt concrete pavements. Are soft asphalts used in hot or cold climates?

**9.4.** Briefly discuss the chemical composition of asphalt.

**9.5.** What is the significance of each one of these tests:

   **a.** flash point test

   **b.** RTFO procedure

   **c.** rotational viscometer test

   **d.** dynamic shear rheometer test

   **e.** solubility test.

**9.6.** Discuss the effect of aging that occurs in asphalt cement during mixing with aggregates and in service. How can the short-term aging of asphalt cement be simulated in the laboratory?

**9.7.** Show how various Superpave tests used to characterize the asphalt binder are related to pavement performance.

**9.8.** Define the four methods used to grade asphalt binders. Which method is used in your state?

**9.9.** As a materials engineer working for a highway department, what standard PG asphalt binder grade would you specify for each of the following conditions? (Fill in the following table.)

|  |  | 50% Reliability | 98% Reliability |
|---|---|---|---|
| Seven-day maximum pavement temperature, °C | mean = 50 SD = 4 | _____ | _____ |
| Minimum pavement temperature, °C | mean = −12 SD = 3 | _____ | _____ |

**9.10.** What are the differences between CRS-2 and SS-1 emulsions?

**9.11.** Discuss how asphalt emulsions work as a binder in asphalt mixtures.

**9.12.** What are the components of asphalt concrete? What is the function of each component in the mix?

**9.13.** What are the objectives of the asphalt concrete mix design process?

**9.14.** Explain why the strength of asphalt concrete is not necessarily the most important property of the material.

**9.15.** An asphalt concrete specimen has a mass in air of 1249.3 g, mass in water of 735.8 g, and saturated surface-dry mass of 1250.2 g. Calculate the bulk specific gravity of the specimen.

**9.16.** For asphalt concrete, define:

**a.** air voids

**b.** voids in the mineral aggregate

**c.** voids filled with asphalt

**9.17.** An aggregate blend is composed of 53% coarse aggregate by weight (Sp. Gr. 2.702), 43% fine aggregate (Sp. Gr. 2.621), and 4% filler (Sp. Gr. 2.779). The compacted specimen contains 6% asphalt binder (Sp. Gr. 1.052) by weight of total mix, and has a bulk density of 145.2 lb/ft$^3$. Ignoring absorption, compute the percent voids in total mix, percent voids in mineral aggregate, and the percent voids filled with asphalt.

**9.18.** An asphalt concrete specimen has the following properties:

Asphalt content = 5.3 % by total weight of mix
Bulk specific gravity of the mix = 2.442
Theoretical maximum specific gravity = 2.535
Bulk specific gravity of aggregate = 2.703

Calculate the percents VTM, VMA, and VFA.

**9.19.** Briefly describe the volumetric mix design procedure of Superpave.

**9.20.** An asphalt concrete specimen was compacted using the Superpave gyratory compactor and produced heights similar to those shown in Table 9.8. If the total mass of the mix was 4880g, $G_{mm}$ (measured) was 2.502, and $G_{mb}$ (measured) was 2.437, determine percent of $G_{mm}$ at different numbers of gyration, and plot the corresponding densification curve.

**9.21.** What are the six typical graphs used in the Marshall mix design process and how are they used to determine the optimum asphalt content?

**9.22.** An asphalt concrete mixture is to be designed according to the Marshall procedure. An AC-20 asphalt cement with a specific gravity of 1.00 is to be used. A dense aggregate blend is to be used with a maximum size of 19 mm and bulk specific gravity of 2.696. The theoretical maximum specific gravity of the mix is 2.470. Trial mixes were made with average results, as shown below.

| Asphalt Content, % by Wt | Bulk Specific Gravity | Stability, N | Flow, 0.25 mm |
|---|---|---|---|
| 4.0 | 2.303 | 7076 | 9 |
| 4.5 | 2.386 | 8411 | 10 |
| 5.0 | 2.412 | 7565 | 12 |
| 5.5 | 2.419 | 5963 | 15 |
| 6.0 | 2.421 | 4183 | 22 |

Plot the appropriate graphs necessary for Marshall procedure and select the optimum asphalt content using the Asphalt Institute design criteria for medium traffic.

**9.23.** Describe the process for determining the stripping potential for an asphalt concrete mix.

**9.24.** Briefly discuss how the indirect tensile resilient modulus is determined in the lab.

**9.25.** The resilient modulus test was performed on an asphalt concrete specimen and the following data were obtained:

> Diameter = 4.029 in.
> Height = 2.497 in.
> Repeated load = 559 lb
> Sum of recoverable horizontal deformations = $254 \times 10^{-6}$ in.

Assuming a Poisson's ratio of 0.35, calculate the resilient modulus.

**9.26.** State six advantages of recycling asphalt pavement materials. Why can we not mix the RAP materials with aggregates in a conventional hot-mix asphalt concrete plant? Show the proper ways of recycling the RAP materials in the two types of hot-mix asphalt plants.

**9.27.** State four different asphalt modifiers that can be added to asphalt or asphalt mixtures and indicate the effect of each.

**9.28.** When is portland cement used in asphalt concrete?

**REFERENCES**

The Asphalt Institute. 1989. *The asphalt handbook.* MS-4. Lexington, KY: The Asphalt Institute.

The Asphalt Institute. 1995. *Mix design methods for asphalt concrete and other hot-mix types.* MS-2. Lexington, KY: The Asphalt Institute.

Goetz, W. H. and L. E. Wood. 1960. Bituminous materials and mixtures. In *Highway engineering handbook,* Section 18. New York: McGraw-Hill.

Mamlouk, M. S. and L. E. Wood. 1981. "Characterization of asphalt emulsion treated bases." *ASCE, Transportation Engineering Journal* 107 (TE2): 183–196.

McGennis, R. B. et al. 1994. *Background of Superpave asphalt binder test methods.* Publication no. FHWA-SA-94-069. Washington, DC: Federal Highway Administration.

McGennis, R. B. et al. 1995. *Background of Superpave asphalt mixture design and analysis.* Publication no. FHWA-SA-95-003. Washington, DC: Federal Highway Administration.

Peterson, J. C. 1984. *Chemical composition of asphalt as related to asphalt durability—state of the art.* Transportation Research Record no. 999. Washington, DC: Transportation Research Board.

Roberts, F. L. et al. 1991. *Hot mix asphalt materials, mixture design, and construction.* Lanham, MD: NAPA Education Foundation.

# 10 Wood

Wood, because of its availability, relatively low cost, ease of use, and durability, if properly maintained, continues to be an important civil engineering material. Wood is used extensively for buildings, bridges, utility poles, floors, roofs, trusses, and piles. Civil engineers use both natural wood and engineered wood products, such as laminates, plywood, and strand board. In order to use wood efficiently, it is important to understand its basic properties and limitations (U.S. Department of Agriculture 1980).

This chapter covers the properties and characteristics of wood. In the design of a wood structure, joints and connections often limit the design elements. These are generally covered in a wood design class and, therefore, are not considered in this text.

Wood is a natural, renewable product from trees. Biologically, a tree is a woody plant that attains a height of at least 6 m (20 ft), normally has a single self-supporting trunk with no branches for about 1.5 m (4 ft) above the ground, and has a definite crown. There are over 600 species of trees in the United States.

Trees are classified as either endogenous or exogenous based on the type of growth. Endogenous trees, such as bamboo, grow with intertwined fibers. Exogenous trees grow from the center out by adding concentric layers of wood around the central core. Wood from endogenous trees is generally not used for engineering applications in the United States. Thus, this book considers only exogenous trees.

Exogenous trees are broadly classified as deciduous and conifers; they are hardwoods and softwoods, respectively. The terms *hardwood* and *softwood* are classifications within the tree family, not a description of the woods' characteristics. In general, softwoods are softer, less dense, and easier to cut than hardwoods. However, exceptions (such as balsa) exist; balsa is a very soft, lightweight wood that is technically a hardwood.

Deciduous trees generally shed their leaves at the end of each growing season. Commercial hardwood production in the United States comes from 40 different tree species. Hardwoods are generally used for furniture and decorative veneers due to

**TABLE 10.1**   Examples of Hardwood and Softwood Species

| Hardwood | Softwood |
|---|---|
| Ash | Cedar |
| Aspen | Incense |
| Basswood | Port Orford |
| Cherry | Douglas-fir |
| Cottonwood | Fir |
| Elm | Hemlock |
| Hickory | Larch |
| Maple | Pine |
| Oak | Redwood |
| Sycamore | Spruce |
| Walnut | Tamarack |

their pleasing grain pattern. Because hardwoods are expensive, they have limited use in construction.

Conifers, also known as evergreens, have needlelike leaves and normally do not shed them at the end of the growing season. Conifers grow continuously through the crown, producing a uniform stem and homogenous characteristics (Panshin and De Zeeuw 1980). Softwood production in the United States comes from about 20 individual species of conifers. Conifers grow in large stands, permitting economical harvesting. They mature rapidly, making them a renewable resource. Conifers are widely used for construction. Table 10.1 shows examples of hardwood and softwood species.

## Structure of Wood

Wood has a distinguished structure that affects its use as a construction material. Civil and construction engineers must understand the way the tree grows and the anaistropic nature of wood in order to properly design and construct wood structures.

### Growth Rings

The concentric layers in the stem of exogenous trees are called *tree rings* or *annual rings*, as shown in Figure 10.1(a). The wood produced in one growing season makes up a single growth ring. Each annual ring is composed of earlywood, produced by rapid growth during the spring, and latewood from summer growth. Latewood consists of dense, dark, and thick-walled cells, which produce a stronger structure than earlywood, as shown in Figure 10.1(b).

The predominant physical features of the tree stem, as shown in Figure 10.2, include the *bark, cambium, wood,* and *pith.* The bark is the exterior covering of the tree and has an outer and an inner layer. The outer layer is dead and corky and has great variability in thickness depending on the species and age of the tree. The inner bark layer is part of the bark not part of the wood section of the tree. The cambium is a thin layer of cells situated between the wood and the bark and is the location of all wood growth.

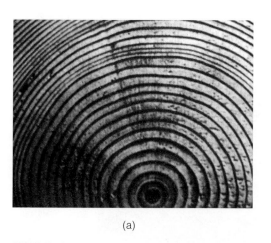

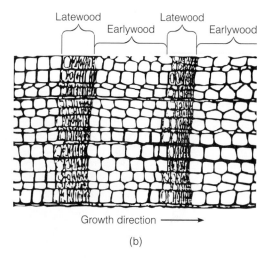

(a)                                                            (b)

**FIGURE 10.1** Cross section of a typical tree stem: (a) annual rings (photo courtesy of American Forest and Paper Association, Washington, DC) and (b) earlywood and latewood.

The wood section of the tree is made up of *sapwood* and *heartwood*. Sapwood functions as a storehouse for starches and as a pipeline to transport sap. Generally, faster growing species have thick sapwood regions. In its natural state sapwood is not durable when exposed to conditions that promote decay. Heartwood is not a living part of the tree. It is composed of cells that have been physically and chemically altered by mineral deposits. The heartwood provides structural strength for the tree. Since the heartwood does not contain sap, it is naturally resistant to decay.

The pith is the central core of the tree. Its size varies with the tree species, ranging from barely distinguishable to large and conspicuous. The color ranges from blacks to almost white, depending on the tree species and locality. The pith structure can be solid, porous, chambered, or hollow.

## Anisotropic Nature of Wood

Wood is an anisotropic material; it has different and unique properties in each direction. The three axis orientations in wood (illustrated in Figure 10.2) are longitudinal or parallel to the grain, radial or across the growth rings, and tangential or tangent to the growth rings. The anisotropic nature of wood affects physical and mechanical properties such as shrinkage, stiffness, and strength.

The anisotropic behavior of wood is the result of the tubular geometry of the wood cells. The wood cells have a rectangular cross section. The centers of the tubes are hollow, whereas the ends of the tubes are tapered. The length-to-width ratio can be as large as 100. The long dimension of most of the cells is parallel to the tree's trunk. However, a few cells, in localized bundles, grow radically from the center to the outside of the trunk. Because the preponderance of cell orientation is in one direction, wood is anisotropic. The hollow tube structure is very efficient in resisting stresses parallel to its length, but readily deforms when loaded on its side. Also, fluctuations in moisture contents flex the tube walls but have little effect on the length of the tube.

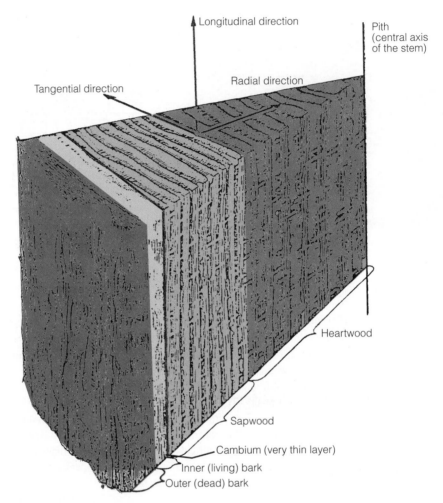

**FIGURE 10.2** Main structural features of a typical tree stem.

## Chemical Composition

Wood is composed of cellulose, lignin, hemicellulose, extractives, and ash-producing minerals. Cellulose accounts for 50% of the wood substance by weight. It is a linear polymer (aliphatic carbon compound) having a high molecular weight. The main building block of cellulose is sugar-glucose. As the tree grows, linear cellulose molecules arrange themselves into highly ordered strands, called *fibrils*. These ordered strands form the large structural elements that make up the cell walls of wood fibers.

Lignin accounts for 23% to 33% of softwood and 16% to 25% of hardwood by weight. Lignin is mostly an intercellular material. Chemically, lignin is an intractable, insoluble material that is loosely bonded to the cellulose. Basically, lignin is the glue that holds the tubular cells together. The longitudinal shear strength of wood is limited by the strength of the lignin bounds.

Hemicelluloses are polymeric units made from sugar molecules. Hemicellulose is different from cellulose in that it has several sugars tied up in its cellular structure. Hardwood contains 20% to 30% hemicellulose, and softwood averages 15% to 20%. Xylose is the main sugar unit in hardwood and mannose in softwood.

The extractives make up 5% to 30% of the wood substance. Included in this group are tannins and other poly-phenolics, coloring matters, essential oils, fats, resins, waxes, gums, starches, and simple metabolic intermediates. These materials can be removed with simple inert neutral solvents such as water, alcohol, acetone, and benzene. The amount contained in an individual tree depends on the species, growth conditions, and time of year the tree is harvested.

The ash-forming materials account for 0.1% to 3.0% of the wood material and include calcium, potassium, phosphate, and silica.

## Moisture Content

The moisture content of a wood specimen is the weight of water in the specimen expressed as a percentage of the oven-dry weight of the wood. An oven-dried wood sample is one that has been dried in an oven at 100°C to 105°C (212°F to 220°F) until the wood attains a constant weight. Physical properties such as weight, shrinkage, and strength depend on the moisture content of wood.

Moisture exists in wood as either *bound* or *free water*. Bound water is held within the cell wall by adsorption forces, whereas free water exists as either condensed water or water vapor in the cell cavities. In green wood, the cell walls are saturated. However, the cell cavities may or may not contain free water. The level of saturation at which the cell walls are completely saturated but no free water exists in the cell cavities is called the *fiber saturation point* (FSP). FSP varies from species to species but is typically in the range of 21% to 32%. The FSP is significant because adding or removing moisture below the FSP has a great effect on practically all physical and mechanical properties of wood; above the FSP the properties are independent of moisture content.

When the moisture content of wood is above the FSP, the wood is dimensionally stable. However, moisture fluctuations below that point always result in dimensional changes. Shrinkage is caused by moisture loss from the cell walls and, conversely, swelling is caused by moisture gain in the cell walls. Figure 10.3 shows that the changes in wood dimensions vary from one direction to another. Dimensional changes in the radial direction are generally one-half the change in the tangential direction. Swelling and shrinking in the longitudinal direction is minimal, typically 0.1% to 0.2% for a change in the moisture content from the FSP to oven dry. This anisotropy of wood's dimensional changes causes warping, checking, splitting, and structural performance problems (discussed in later in the chapter). It is also the reason that the sawing pattern of boards affects the amount the board distorts when subjected to moisture changes.

The moisture content in wood varies depending on air temperature and humidity. However, the natural change of moisture content is a slow process; thus, as atmo-

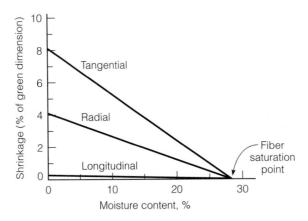

**FIGURE 10.3** Relation between shrinkage and moisture content.

spheric conditions change, the moisture content in wood tends to adjust to conditions near the average. The moisture content for the average atmospheric conditions is the *equilibrium moisture content* (EMC). The U.S. Forest Service has developed values of the EMC as a function of temperature and humidity. The EMC ranges from less than 1% at temperatures greater than 55°C (130°F) and 5% humidity to over 20% at temperatures less than 27°C (80°F) and 90% humidity.

**◆SAMPLE**
**PROBLEM 10.1**

A 250-mm-wide red spruce plank is cut in such a way that the width is in the tangential direction of annual rings. Compute the change in width as the moisture content changes from 15% to 32%. (The fiber saturation point of red spruce is 27% and it shrinks 7.8% in drying from FSP to oven dry in the tangential direction.)

*Solution:* Swell does not occur above FSP. Increasing the moisture content from 15% to 27% causes swell. From the data given, red spruce swells 7.8% for a 27% change in moisture content (from zero moisture content to FSP).

$$\text{Percent swell} = \frac{7.8}{27} \times (27 - 15) = 3.5\%$$

$$\text{Change in dimension} = 250 \times \frac{3.5}{100} = 8.7 \text{ mm}$$

$$\text{New dimension} = 258.7 \text{ mm} \qquad \blacklozenge$$

# Wood Production

Trees are harvested in the fall or winter because of their water content and because of environmental concerns connected with fire hazard and other plant growth. A vast industry has developed to harvest and process wood. Wood is harvested from forests as *logs*. They are transported to sawmills where they are cut into dimensional shapes to produce a variety of products for engineering applications.

1. *Dimensional lumber*—wood from 50 mm to 125 mm (2 in. to 5 in.) thick, sawn on all four sides. Common shapes include $2 \times 4$, $2 \times 6$, $2 \times 8$, $2 \times 10$, $2 \times 12$, and $4 \times 4$.* These sizes refer to the rough-sawn dimensions of the lumber in inches. The rough-sawn lumber is surfaced to produce smooth surfaces; this removes 5 mm to 10 mm (1/4 in. to 3/8 in.) per side. For example, the actual dimensions of a $2 \times 4$ are 40 mm by 90 mm (1-1/2 in. by 3-1/2 in.). Dimensional lumber is produced in lengths of 2.4 m to 7.2 m (8 ft to 24 ft) in 0.6 m (2 ft) increments. Dimensional lumber is typically used for studs, sill and top plates, joints, beams, rafters, trusses, and decking.

2. *Heavy timber*—wood sawn on all four sides. Common shapes include $4 \times 6$, $6 \times 6$, $8 \times 8$, and larger. As the case of dimensional lumber, these sizes specify rough-sawn dimensions in inches. Surfacing generally removes 10 mm (3/8 in.) per side. Heavy timbers are used for heavy-frame construction, landscaping, railroad ties, and marine construction.

3. *Round stock*—posts and poles used for building poles, marine piling, and utility poles.

4. *Engineering wood*—products manufactured by bonding together wood strands, veneers, lumber, and other forms of wood fiber to produce a larger and integral composite unit. These products are engineered and tested to have specific mechanical responses to loads. Structural engineered wood products include:

   structural panels including plywood, oriented strand-board, and composite panels
   glued laminated timber (glulam)
   structural composite lumber
   wood I-joists

5. Specialty items—milled and fabricated products to reduce on-site construction time. These items include lattice, hand-rails, spindles, radius-edge decking, turned posts, etc.

Sawn wood production includes the following steps.

- Sawing into desired shape
- Seasoning
- Surfacing
- Grading
- Treating with preservatives (optional)

Surfacing (planing) of a wood surface, to produce a smooth face, can be done before or after drying. Post-drying surfacing is superior because it removes small defects developed during the drying process. When surfacing is done before seasoning, the dimensions are slightly increased to compensate for shrinkage during seasoning.

---

*The standards for dimensional lumber and heavy timber standards were implemented in about 1970. When remodeling older structures, the dimension of the existing lumber must be measured.

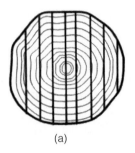 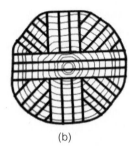 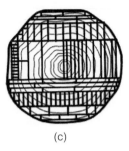

(a)                              (b)                              (c)

**FIGURE 10.4**  Common log sawing patterns: (a) plain sawing, (b) quarter sawing, and (c) combination sawing.

## Cutting Techniques

The harvested wood is cut into lumber and timber at sawmills using circular saws, band saws, or frame saws. The most common patterns for sawing a log are *plain (slash)*, *quarter,* and *combination sawing,* as shown on Figure 10.4. (Levin 1972).

The quality of the boards depends on how the angle of the growth ring is positioned in relation to the face of the board (i.e., the angle between the growth ring and the saw blade). There are three categories as illustrated, in Figure 10.5.

1. flat-sawn (45° or less)
2. rift-sawn (45° to 80°)
3. vertical- or edge-sawn (80° to 90°)

Flat-sawn boards have desirable exposure of grain for decorative applications. However, flat-sawn boards tend to distort more than vertical-sawn boards in response to moisture fluctuations. Hence, vertical-sawn boards are generally better for structural applications.

## Seasoning

The sawing pattern selected depends on the cross section of the tree, the capability of the mill, and the desired product. Plain sawing is rapid and economic, whereas quarter sawing maximizes the amount of vertical–sawn cuts. Green wood, in living trees, contains between 30% and 200% moisture by the oven-dry weight. Seasoning removes the excess moisture from wood. For structural wood, the recommended moisture content

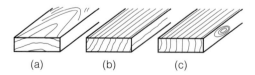

(a)          (b)          (c)

**FIGURE 10.5**  Types of board cut: (a) flat-sawn, (b) rift-sawn, and (c) vertical- or edge-sawn.

varies from 7% in the dry southwestern states to 14% in the damp coastal regions. However, as it leaves the mill, framing lumber typically has an average moisture content of 15%.

Wood is seasoned by air and kiln drying. Air drying is inexpensive but slow. The green lumber is stacked in covered piles to dry. These piles of lumber are made of successive layers of boards separated by 25-mm (1 in.) strips so that air can flow between the layers. The time required for drying varies with the climate of the area. Normally, three to four months is the maximum air-drying time used in the United States. Air drying is complete when the moisture content of the wood equals the air humidity. However, the optimum moisture may not be achieved through air drying alone.

After air drying the lumber may be kiln dried. A kiln is a large oven where all variables can be closely monitored. Drying in a kiln typically requires 4 to 10 days at temperatures in the range of 20°C to 50°C (70°F to 120°F). Care must be taken to slowly reduce the moisture content of wood; drying too rapidly can result in cracking and warping. Kiln-dried lumber will take on moisture again if exposed to water; therefore, care must be used when storing and transporting wood.

## Lumber Grades

The final step in wood production requires grading the lumber according to quality. Typically, lumber is graded according to the number of flaws that affect strength, durability, or workability. The most common grade-reducing qualities of lumber are knots, checks, pitch pockets, shakes, and stains. Due to the high degree of natural variability within lumber it is nearly impossible to develop an exact, uniform set of grading standards. As a result, grading techniques and standards can and do vary between organizations. Organizations such as the National Bureau of Standards and the United States Department of Agriculture have spent many years trying to develop a simple, uniform method of lumber sizing, common nomenclature, and grading standards.

The following agencies are certified by the American Lumber Standards Committee Board of Review in Germantown, MD, for inspecting and grading of untreated lumber.

- Northeastern Timber Manufacturer Association (NELMA), Falmouth, MN
- Northern Hardwood and Pine Manufacturer Association (NHPMA), Green Bay, WI
- Redwood Inspection Service (RIS), San Francisco, CA
- Southern Pine Inspection Bureau (SPIB), Pensacola, FL
- West Coast Lumber Inspection Bureau (WCLIB), Portland, OR
- Western Wood Products Association (WWPA), Portland, OR
- National Lumber Grader Authority (NLGA), Ganges, BC, Canada

Each of these agencies writes standards and specifications for particular species or species combinations that are produced within their operating region. For example, the SPIB provides the grading rules for all species of southern pine, the WWPA governs grading of ponderosa pine, and the NLGA provides grading rules for all grades produced in Canada.

## Hardwood Grades

The National Hardwood Lumber Association bases the grading of hardwood on the amount of usable lumber in each piece of standard-length lumber. The inspection is performed on the poorest side of the material and the grade is based on the number and size of clear "cuttings" that can be produced from a given piece of lumber. Cuttings must have one face clear of strength-reducing imperfections, and the other side must be sound. Based on cutting quantity the wood is given a classification of *Firsts, Seconds, Selects,* and *Common* (No. 1, No. 2, No. 3A, or 3B), with Firsts being the best. Frequently, Firsts and Seconds are grouped into one grade of First and Seconds, FAS.

## Softwood Grades

Softwood used for structural applications is either graded by visual inspection or is machine-stress graded. The purpose of grading is to ensure that all lumber within a specific grade has at least the minimum mechanical or load-carrying capability critical to design parameters. Under machine-stress grading, each piece of wood is subjected to a bending stress. Then, based on the mechanical response, the wood is graded as shown in Table 10.2. The grade designation identifies the minimum extreme fiber-bending stress, tensile stress parallel to grain, compressive stress parallel to grain, and modulus of elasticity of the wood.

For visual classification, the basic mechanical properties of the wood are determined by testing small, clear wood specimens. These results are then adjusted for allowable defects and characteristics for each class of wood. Unlike stress-graded lumber, visual stress-grade properties are defined for each species of softwood. Table 10.3 is an example of design values for grades of eastern white pine.

Visual grade designations include *Yard, Structural,* or *Factory and Shop,* with subgrades of Select, Select B, Select C, and No. 1, No. 2, and No. 3 commons, appearance, and studs. Other commonly used ratings are *Construction, Standard,* or *Utility*

**TABLE 10.2**   Sample of Stress Grading of Softwood for Structural Applications*

| Grade Designation | Design Values,[†‡] MPa | | | |
| --- | --- | --- | --- | --- |
| | Extreme Fiber in Bending** | Tension Parallel to Grain | Compression Parallel to Grain | Modulus of Elasticity |
| 900f-1.0E | 6.2 | 2.4 | 7.2 | 6900 |
| 1650f-1.3E | 11.4 | 7.0 | 11.7 | 9000 |
| 1950f-1.5E | 13.4 | 9.5 | 12.4 | 10,300 |
| 2250f-1.7E | 15.5 | 12.1 | 13.3 | 11,700 |
| 2400f-2.0E | 16.5 | 13.3 | 13.6 | 13,800 |
| 2850f-2.3E | 19.6 | 15.8 | 14.8 | 15,800 |
| 3000f-2.4E | 20.7 | 16.5 | 15.2 | 16,500 |

*Courtesy of American Forest and Paper Association, Washington, DC.
†Stresses apply to lumber used at 19% maximum moisture content. When lumber 38 mm thick or less is designed for use where the moisture content will exceed 19% for an extended period of time, the values shown herein shall be multiplied by certain *wet service* factors.
‡For a complete list of grade designations and more detailed design values see reference (American Forest and Paper Association 1997).
**Tabulated extreme fiber in bending values are applicable to lumber loaded on edge. When loaded flatwise, these values may be increased by multiplying by certain *flat use* factors.

**TABLE 10.3** Design Values of Eastern White Pine*

| Grade Designation | Size Classification | Design Values, MPa† | | | | | |
|---|---|---|---|---|---|---|---|
| | | Extreme fiber in Bending $F_b$ | Tension Parallel to Grain $F_t$ | Horizontal Shear $F_v$ | Compression Perpendicular to Grain $F_{c\perp}$ | Compression Parallel to Grain $F_c$ | Modulus of Elasticity $E$ |
| Select structural | | 8.6 | 4.0 | 0.5 | 2.4 | 8.3 | 8300 |
| No. 1 | 38–89 mm thick | 5.3 | 2.4 | 0.5 | 2.4 | 6.9 | 7600 |
| No. 2 | | 4.0 | 1.9 | 0.5 | 2.4 | 5.7 | 7600 |
| No. 3 | 38 mm and wider | 2.4 | 1.0 | 0.5 | 2.4 | 3.3 | 6200 |
| Stud | | 3.1 | 1.4 | 0.5 | 2.4 | 3.6 | 6200 |
| Construction | 38–89 mm thick | 4.7 | 2.1 | 0.5 | 2.4 | 7.2 | 6900 |
| Standard | | 2.6 | 1.2 | 0.5 | 2.4 | 5.9 | 6200 |
| Utility | 38–89 mm wide | 1.2 | 0.5 | 0.5 | 2.4 | 3.8 | 5500 |

*Courtesy of American Forest and Paper Association, Washington, DC.
†The design values herein are applicable to lumber that will be used under dry conditions such as in most covered structures. The design values herein may require adjustments for size and flat use. For a complete list of grade designations and more detailed design values see (American Forest and Paper Association 1997).

and combinations such as No. 2&BTR (Number 2 and better) and STD&BTR (Standard and better). Not all grades are used for all species of wood.

Yard lumber frequently refers to some specialty stress-grades of lumber such as those used for light structural framing. Structural lumber, typically pieces 50 mm to 125 mm (2 in. to 5 in.) thick, is graded according to its intended use. Grading categories include light framing, joists and planks, beams and stringers, and posts and timbers. Factory and shop lumber includes siding, flooring, casing, shingles, shakes, and finish lumber.

## Defects in Lumber

Lumber may include defects that affect either its appearance or mechanical properties, or both. These defects can have many causes, such as natural growth of the wood, wood diseases, animal parasites, too-rapid seasoning, or faulty processing. Some common defect types are shown in Figure 10.6.

### Knots

A knot is a branch base that has become incorporated into the wood of the tree trunk or another limb. Knots degrade the mechanical properties of lumber, affecting the tensile and flexural strengths. However, the presence of sound, tight knots may increase the compressive strength, hardness, and shear characteristics of the wood.

### Shakes

Shakes are lengthwise separations in the wood that occur between annual rings. They could be due to the tree bending under the force of heavy winds.

### Wane

A wane is bark or other soft material left on the edge of the board.

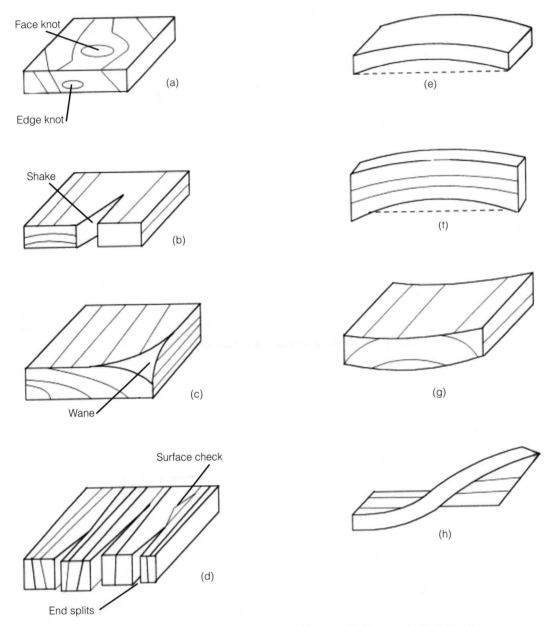

**FIGURE 10.6** Common defects in lumber: (a) knots, (b) shakes, (c) wanes, (d) checks and splits, (e) bowing, (f) crooking, (g) cupping, and (h) twisting.

## Sap Streak

A sap streak is a heavy accumulation of sap in the fibers of the wood that produces a distinctive colored streak.

## Reaction Wood

Reaction wood is the abnormally woody tissue that forms in crooked stems or limbs. Reaction wood causes the pith to be off-center from the neutral axis of the tree. It creates internal stresses that can cause warping and longitudinal cracking.

## Pitch Pockets

Pitch pockets are well-defined openings between annual rings that contain free resin. Normally, only Douglas-fir, pines, spruces, and western larches have pitch pockets.

## Bark Pockets

Bark pockets are small patches of bark embedded in the wood. These pockets form as a result of an injury to the tree, causing death to a small area of the cambium. The surrounding tree continues to grow, eventually covering the dead area with a new cambium layer.

## Checks

Checks are ruptures in wood along the grain, which develop during seasoning. They can occur on the surface or end of a board. Surface checking results from the separation of the thinner-walled earlywood cells and is confined mostly to planar surfaces. Cracks due to end checking normally follow the grain and result in end splitting.

## Splits

Splits are lengthwise separations of the wood caused by either mishandling or seasoning.

## Warping

Warping is the distortion of wood from the desired true plane. The four major types of warping are *bowing, crooking, cupping,* and *twisting.* Bowing is a longitudinal curvature from end to end; and crooking is the longitudinal curvature from side to side. Both of these defects are the result of uneven longitudinal shrinkage. In cupping, both edges roll up or down. Twisting occurs when one corner lifts out of the plane of the other three. Warping is the result of uneven drying due to the production environment or the release of internal tree stress.

## Raised, Loosened, or Fuzzy Grain

Raised, loosened, or fuzzy grain may occur during cutting and dressing of lumber.

## Chipped or Torn Grain

Chipped or torn grain occurs when pieces of wood are scooped out of the board surface or chipped away by the action of the cutting and planing tools.

## Machine Burn

A machine burn is a darkened area resulting from overheating during cutting.

# Physical Properties

Important physical properties include specific gravity and density, thermal properties, and electrical properties.

## Specific Gravity and Density

Specific gravity of wood depends on cell size, cell-wall thickness, and number and types of cells. Regardless of species, the cell material has a specific gravity of 1.5. Because of this consistency, specific gravity is an excellent index for the amount of substance a dry piece of wood actually contains, and it is nearly constant within each species. Therefore, specific gravity (or density) is a commonly cited property and is an indicator of mechanical properties within a clear, straight-grained wood.

The dry density of wood ranges from 160 kg/m$^3$ (10 pcf) for balsa to 1000 kg/m$^3$ (65 pcf) for other species. The majority of wood types have densities in the range of 300 kg/m$^3$ to 700 kg/m$^3$ (20 pcf to 45 pcf). Within common domestic species, density may vary ± 10%.

## Thermal Properties

Thermal conductivity, specific heat, thermal diffusivity, and coefficient of thermal expansion are the four significant thermal properties of wood.

**Thermal Conductivity** Thermal conductivity is a measure of the rate at which heat flows through a material. The reciprocal of thermal conductivity is the thermal resistance (insulating) value (R). Wood has a thermal conductivity that is a fraction of that of most metals and three to four times greater than common insulating materials. The thermal conductivity ranges from 0.06 W/(m °K) [0.34 Btu/(h · ft · °F)] for balsa to 0.17 W/(m °K) [1.16 Btu/(h · ft · °F)] for rock elm. Structural woods average 0.12 W/(m °K) [0.07 Btu/(h · ft · °F)] as compared to 200 W/(m °K) [115 Btu/(h · ft · °F)] for aluminum and 0.04 W/(m °K) [0.025 Btu/(h · ft · °F)] for wool. The thermal conductivity of wood depends on several items, including: 1) grain orientation, 2) moisture content, 3) specific gravity, 4) extractive content, and 5) structural irregularities such as knots.

Heat flow in wood across the radial and tangential directions (in relation to the growth rings) is nearly uniform. However, heat flow through wood in the longitudinal direction (parallel to the grain) is 2.0 to 2.8 times greater than in the radial direction.

Moisture content has a strong influence on thermal conductivity. When the wood is dry, the cells are filled with air, and the thermal conductivity is very low. As the moisture content increases, thermal conductivity increases. As the moisture content increases from 0% to 40%, the thermal conductivity increases by about 30%.

Because of the solid cell-wall material in heavy woods, they conduct heat faster than light woods. This relationship between specific gravity and thermal conductivity for wood is linear. Also affecting the heat transfer in wood are increases in extractive content and density (i.e., knots), which increase thermal conductivity.

**Specific Heat** Specific heat of a material is the ratio of the quantity of heat required to raise the temperature of the material 1° to that required to raise the temperature of an equal mass of water 1°. Temperature and moisture content largely control the specific heat of wood, with species and density having little or no effect. When wood contains water, the specific heat is increased because the specific heat of water is higher than that of dry wood. However, the value of specific heat for the wet wood is higher than the sum of the specific heats for the wood and water due to the wood-water bonds absorbing energy. An increase in temperature increases the energy absorption of wood and results in an increase in the specific heat.

**Thermal Diffusivity** Thermal diffusivity is a measure of the rate at which a material absorbs heat from its surroundings. The thermal diffusivity for wood is much smaller than that of other common building materials. Generally, wood has a thermal diffusivity value averaging 0.006 mm/s (0.00025 in./s) as compared to steel, which is 0.5 mm/s (0.02 in./s). It is because of the low thermal diffusivity that wood does not feel hot or cold to the touch compared to other materials. The small thermal conductivity, moderate density, and moderate specific heat contribute to the low value of thermal diffusivity in wood.

**Coefficient of Thermal Expansion** The coefficient of thermal expansion is a measure of dimensional changes caused by a temperature variance. Thermal expansion coefficients for completely dry wood are positive in all directions. For both hard and soft woods the longitudinal (parallel to the grain) coefficient values range from 0.009 mm/m/°C to 0.0014 mm/m/°C (0.0000017 in./in./°F to 0.0000025 in./in./°F). The expansion coefficients are proportional to density and therefore are 5 to 10 times greater across the grain than those parallel to it.

When moist wood is heated, it expands due to thermal expansion, and then shrinks because of moisture loss (below the FSP). This combined swelling and shrinking often results in a net shrinkage. Most woods, at normal moisture levels, react in this way.

## Electrical Properties

Air-dry wood is a good electrical insulator. As the moisture content of the wood increases, the resistivity decreases by a factor of three for each percentage of moisture content. However, when wood reaches the FSP it takes on the resistivity of water alone.

# Mechanical Properties

It is necessary to understand the mechanical properties of wood before a wood structure can be properly designed. Typical mechanical properties of interest to civil and construction engineers include modulus of elasticity, strength properties, creep, and damping capacity.

## Modulus of Elasticity

The typical stress-strain relation of wood is linear up to a certain limit; then a small nonlinear curve occurs after which the wood fails (as shown on Figure 10.7). The modulus of elasticity of wood is the slope of the linear portion of the representative stress-

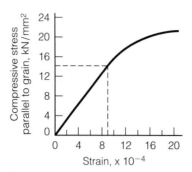

**FIGURE 10.7** Typical stress-strain relationship for wood.

strain curve. The stress-strain relation of wood varies within and between species and is affected by variation in moisture content and specific gravity. Also, since wood is anisotropic, different stress-strain relations exist for different directions. The moduli of elasticity along the longitudinal, radial, and tangential axes typically are different.

## Strength Properties

Strength properties of wood vary widely depending on the direction of grain relative to the direction of force. For example, the tensile strength in the longitudinal direction (parallel to grain) is more than 20 times the tensile strength in the radial direction (perpendicular to grain). Also, tensile strength in the longitudinal direction is larger than the compressive strength in the same direction. Common strength properties for wood include modulus of rupture in bending, compressive strength parallel and perpendicular to the grain, and shear strength parallel to the grain. Some of the less common strength properties are tensile strength parallel to the grain, torsion, toughness, fatigue strength, and rolling-shear strength.

## Creep

Under sustained loads wood continues to deform or creep. The design values of material properties consider fully stressing the member to the design values for a period of 10 years and/or the application of 90% of the full maximum load continuously throughout the life of the structure. If the maximum stress levels are exceeded, the structure can deform prematurely.

## Damping Capacity

Damping is the phenomenon in which the amplitude of vibration in a material decreases with time. Reduction in amplitude is due to internal friction within the material and resistance of the support system. Moisture content and temperature largely govern the internal friction in wood. At normal ambient temperatures an increase in moisture content produces a proportional increase in internal friction up to the FSP. Under normal conditions of temperature and moisture content the internal friction in wood (parallel to the grain) is 10 times that of structural metals. Because of these qualities, wood structures dampen vibrations more quickly than metal structures of similar design.

## Testing to Determine Mechanical Properties

Standard mechanical testing methods for wood are designed almost exclusively to obtain data to predict performance. In order to ensure that test results can be reproduced, specifications include methods of material selection and preparation, testing equipment and techniques, and computational methods for data reduction. Standards for testing wood and wood composites are published by ASTM, the U.S. Department of Commerce, the National Standard Institute (NSI), and various other trade associations, such as the Western Woods Product Association.

Due to the many variables affecting the test results, it very important to correctly select the specimen and type of test. There are two main testing techniques for establishing strength parameters: the testing of representative, small, clear specimens and the testing of timbers of structural sizes.

The primary purposes for testing small, clear specimens is to obtain the mechanical properties of various species and provide a means of control and comparison in production activities. The testing of structural timbers provides relationships between mechanical and physical properties, working stress data, correlations between environmental conditions, wood imperfections, and mechanical properties. ASTM D143 presents the complete testing standards for small, clear wood specimens. This standard gives full descriptions of sample collection, preparation, and testing techniques. Mechanical tests included in this standard are:

- static bending
- impact bending
- compression perpendicular to the grain
- shear parallel to the grain
- tension parallel to the grain
- nail withdrawal
- radial and tangential shrinkage
- compression parallel to the grain
- toughness
- hardness
- cleavage
- tension perpendicular to the grain
- specific gravity and shrinkage in volume
- moisture determination

Figure 10.8 shows a schematic of test specimens of wood tested in tension, compression, bending, and hardness. Static and impact bending, compression, and tension parallel and perpendicular to the grain, and shear parallel to the grain are commonly used.

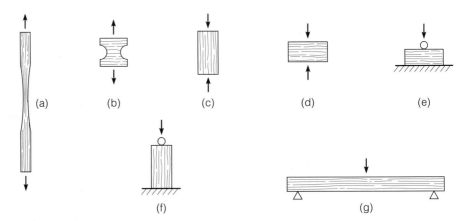

**FIGURE 10.8** Test specimens of wood: (a) tension parallel to grains, (b) tension perpendicular to grains, (c) compression parallel to grains, (d) compression perpendicular to grains, (e) hardness perpendicular to grains, (f) hardness parallel to grains, and (g) bending.

## Static Bending Test

The static-bending test is performed on either 50 mm × 50 mm × 760 mm (2 in. × 2 in. × 30 in.) or 25 mm × 25 mm × 410 mm (1 in. × 1 in. × 16 in.) specimens. For the large-sized specimens, the loading head is placed on the center of the specimen and over a span of 710 mm (28 in.), whereas the load is applied at a rate of 2.5 mm/min (0.1 in./min). For the small-sized specimens, the loading head is placed on the center of the specimen and over a span of 360 mm (14 in.), and the load is applied at a rate of 1.3 mm/min (0.05 in./min). Load-deflection data are recorded to or beyond the maximum load. Within the proportional limit, readings are taken to 0.02 mm (0.001 in.). After the proportional limit, deflection readings are usually measured with a dial gauge, to the limit of the gauge, usually 25 mm (1 in.). Load and deflection of the first failure, the maximum load, and points of sudden change are recorded. The failure appearance is described as either brash or fibrous. Brash indicates an abrupt failure and fibrous indicates a failure showing splinters.

**SAMPLE PROBLEM 10.2**

A static bending test was performed on a 50 mm × 50 mm × 760 mm wood sample according to ASTM D143 procedure. If the load at failure was 2.67 kN, calculate the modulus of rupture.

*Solution:*

$$\text{Modulus of rupture} = \frac{Mc}{I}$$

where

$M$ = bending moment at failure
$c$ = 1/2 of the specimen height
$I$ = moment of inertia of the specimen cross section

According to ASTM D143, the span is 710 mm and the load is applied at the center.

$$\text{Reaction at each support at failure} = \frac{2.67}{2} = 1.335 \text{ kN}$$

$$\text{Bending moment at failure} = 1.335 \times 355 = 473.9 \text{ N} \cdot \text{m}$$

$$I = \frac{(0.05)(0.05)^3}{12} = 6.25 \times 10^{-6} \text{ m}^4$$

$$\text{Modulus of rupture} = \frac{473.9 \times 0.025}{(6.25 \times 10^{-6})} = 1.90 \text{ MPa}$$

◆

## Compression Tests

The compression test parallel to the grain is performed on either 50 mm × 50 mm × 200 mm (2 in. × 2 in. × 8 in.) or 25 mm × 25 mm × 100 mm (1 in. × 1 in. × 4 in.) specimens, as shown in Figure 10.9. The load is applied at a rate equal to 0.003 mm/mm (in./in.) of the nominal specimen length per minute. The deformations are recorded to 0.002 mm (0.0001 in.) over a gauge length of not more than 150 mm (6 in.) for the large-sized specimens or 50 mm (2 in.) for the small-sized specimens. Load-compression readings are recorded until well past the proportional limit. The failures should occur in the center portion of the sample. If failures are occurring near the ends, the samples can be stacked such that the ends dry at the same rate as the middle. This will increase the strength of the ends of the sample. The tests are then repeated on the conditioned samples. The type of failure can be classified as crushing, wedge split, shearing, splitting, compression and shearing, and brooming and end-rolling, as shown in Figure 10.10.

**FIGURE 10.9** Compression parallel to grain test.

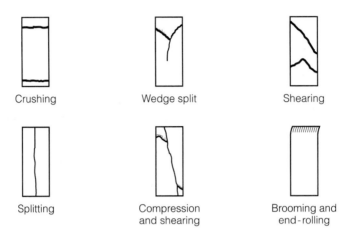

| | | |
|---|---|---|
| Crushing | Wedge split | Shearing |
| Splitting | Compression and shearing | Brooming and end-rolling |

**FIGURE 10.10** Types of failure in the compression parallel to grain test. (ASTM D143). Reprinted with permission of ASTM.

The compression test perpendicular to the grain is performed on 50 mm × 50 mm × 150 mm (2 in. × 2 in. × 6 in.) specimens. The load is applied through a metal bearing plate 50 mm (2 in.) wide, centered across the upper surface of the specimen. The load is applied at a rate of 0.305 mm/min (0.012 in./min). Deflection readings are taken to 0.002 mm (0.0001 in.). Load and deformation are measured until the deformation is 2.5 mm (0.1 in.).

## Design Considerations

Lab measurements do not reflect all the factors that affect the behavior of wood in engineering applications. For design of wood structures, the strength properties given in Tables 10.2 and 10.3 must be adjusted for (American Forest and Paper Association 1997; ASTM D2555):

- wet service
- temperature
- beam stability
- size
- volume
- flat use

- repetitive member
- curvature
- form
- column stability
- shear stress
- bearing area

In addition, sustained loads will cause wood to creep. In design applications this means that the wood can carry higher stresses of short time durations than the same wood element can carry for a long-term application. Generally, a load duration of 10 years is used for design. Typical load duration factors that would be applied to the values in Tables 10.2 and 10.3 (except for compression perpendicular to the grain) are shown in Figure 10.11. For example, if a floor joist is designed for a temporary stage used only for a one-day presentation, the allowable bending fiber stress can be increased by 33% over the allowable stress for a normal application.

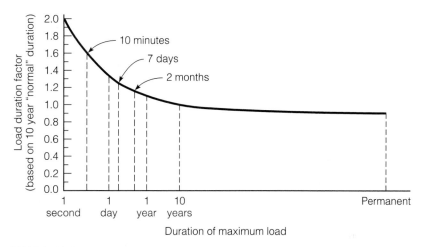

**FIGURE 10.11** Adjustment factors for load-duration working-stress for wood design based on 10-year normal duration. Reprinted with permission of American Forest and Paper Association (1997).

## Organisms That Degrade Wood

Wood can experience degradation due to attack of fungi, bacteria, insects, or marine organisms.

### Fungi

Most forms of decay and sap-stains are the result of fungal growth. Fungi need four essential conditions to exist: food, proper temperature range, moisture, and oxygen. Fungi feed on either the cell structure or the cell contents of woody plants, depending on the fungus type. The temperature range conducive to fungal growth is from 5°C to 40°C (40°F to 100°F). A moisture content above the FSP is required for fungal growth. Fungi are plants and, as such, require oxygen for respiration. Fungi attack produces *stains* and/or *decay damage*.

To protect against fungal attack one of the four essential conditions for fungus growth must be removed. The most effective protection measure is to keep the wood dry by using coatings or by correct placement during storage and in the structure. Fungi can also be contained by treating the wood fibers with chemical poisons through a pressure treatment process.

Construction procedures that limit decay in buildings include:

1. building with dry lumber that is free of incipient decay and excessive amounts of stains and molds,

2. using designs that keep the wood components dry, and using wood treated with preservatives,

3. using a heartwood from decay resistant species in sections exposed to above-ground decay hazards, and

4. using pressure-treated wood for components in contact with the ground.

## Bacteria

Bacteria causes "wetwood" and "black heartwood" in living trees and a general degradation of lumber. Wetwood is a water-soaked condition that affects the stem centers of living trees; it is most common in poplars, willows, and elms. Black heartwood has characteristics similar to wetwood and causes the center of the stem to turn dark brown or black.

Bacterial growth is sometimes fostered by prolonged storage in contact with soils. This type of bacteria activity produces a softening of the outer wood layers, which results in excessive shrinkage when redried.

## Insects

*Beetles* and *termites* are the most common wood-attacking insects. Several types of beetles, such as bark beetles, attack and destroy wood. Storage of the logs in water or a water spray prevents the parent beetle from boring. Quick drying or early removal of the bark also prevents beetle attack activity. Damage can be prevented by proper cutting practices and dipping or spraying with an appropriate chemical solution.

Termites are perhaps the most destructive organism that attacks wood. The annual damage attributed to termites exceeds losses due to fires. Termites enter structures through wood that is close to the ground and is poorly ventilated or wet. Prevention is achieved by painting and otherwise prohibiting insect entry into areas of unprotected wood through the use of screening, sill plates, and sealing compounds.

## Marine Organisms

Damage by marine boring organisms in the United States and surrounding oceans is principally caused by shipworms, Pholads, Limnoria, and Sphaeroma. These organisms are mostly confined to saltwater or brackish water.

# Wood Preservation

Paints, petroleum-based solutions, and waterborne oxides (salts) are the principle types of wood preservatives. The degree of preservation achieved depends on the type of preservative, the degree of penetration, and the amount of the chemical retained within the wood. Paints are applied on the surface, while the other preservatives are applied under pressure so that they penetrate into the wood (Richardson 1978).

## Petroleum-based Solutions

Coal-tar creosote, petroleum creosote, creosote solutions, and pentachlorophenol solutions are the oil-based preservatives. These preservatives are very effective but are environmentally sensitive. They are commonly used where a high degree of environmental exposure exists and human contact is not a concern. Applications include utility poles, railroad ties, and retaining walls.

## Waterborne Preservatives (Salts)

The typical solutes used in waterborne preservative mixtures are ammoniacal copper arsenate, chromated copper arsenate, and ammoniocal copper zinc arsenate. The

**TABLE 10.4** Minimum Retention Requirements for Different Treatments and Applications of Southern Pine Lumber, Timber, and Plywood, kg/m$^3$*

| | Above Ground | Soil and Fresh-Water Use | Permanent Wood Foundation | Saltwater Use |
|---|---|---|---|---|
| Waterborne preservatives | | | | |
| Ammonical copper arsenate | 4 | 6 | 10 | 41 |
| Ammonical copper zinc arsenate | 4 | 7 | 10 | 41 |
| Chromated copper arsenate | 4 | 7 | 10 | 41 |
| Creosote and oilborne preservatives | | | | |
| Creosote | 131 | 163 | NR[†] | 408 |
| Creosote-petroleum | 131 | 163 | NR[†] | NR[†] |
| Creosote solutions | 131 | 163 | NR[†] | 408 |
| Pentachlorophenol | 7 | 8 | NR[†] | NR[†] |

*Southern Forest Products Association (1990). For complete list of wood products see the American Wood Preservative Association (AWPA) standards.
[†]NR—Not recommended.

advantages of the waterborne preservative over the oil-based are cleanliness and its ability to be painted. The disadvantage of these treatments is their removal by leaching when exposed to moist conditions over long periods of time. These preservatives are also environmentally sensitive and must be applied under carefully controlled conditions. However, once treated, the preserved wood is not environmentally hazardous. Applications include wood structures, such as residential decks and fences, that are exposed to environment and in common contact with people.

### Application Techniques

Preservatives are applied by superficial treatment or by fluid penetration processes. Superficial treatment techniques include coatings applied by painting, spraying, or immersion. Liquid penetration into a porous solid occurs by capillary action and is a function of surface tension, angle of contact, time, temperature, and pressure.

Pressure-treated wood has greater resistance to degradation than surface-treated wood. The preservative is forced into the entire structure of the wood. By thoroughly treating the entire cross section of the wood, decay can be eliminated for an extended period of time. Some vendors of pressure-treated wood provide a life-time warranty when their products are in direct contact with the ground. The key to ensuring long life is the amount of preservative retained in the wood. Table 10.4 gives the minimum retention requirements for different treatments and applications of southern pine lumber, timber, and plywood.

## Engineered Wood Products

Engineered wood includes a wide variety of products manufactured by bonding together wood strands, veneers, lumber, or other forms of wood fibers to produce large and integral units. These products are "engineered" to produce specific and consistent mechanical behavior and thus have consistent design properties. The American Plywood Association (APA) (1980) identifies four categories of engineered wood products.

1. Wood structural-strand panels, including plywood, oriented-strand boards, and composite panels
2. Glued-laminated timber (glulam)
3. Structural composite lumber, including primarily laminated veneer lumber, parallel-strand lumber, and oriented-strand lumber. These products have the dimensions of sawn wood dimensional lumber
4. Wood I-joists

Quality and serviceability of glued-wood products principally depend on

1. gluing properties and preparation of the wood used,
2. type of adhesive, and
3. quality control employed in the gluing process.

In addition to these factors the conditions under which they are used affect the overall performance of a glued product.

Adhesives are either natural or synthetic. Natural adhesives include casein, vegetable protein (soybean being the most common), and blood protein glues. Synthetic adhesives include phenol, urea, resorcinol, polyvinyl, and epoxy resins. Structural panels produced to the standards of the APA are made with phenol formaldehyde adhesives. These adhesives do not have the environmental problems associated with urea formaldehyde adhesives.

## Glued-Laminated Timbers

Laminated timbers are composed of two or more wood layers glued together with the grain of all layers parallel. Softwoods such as Douglas-fir and southern pine are most commonly used for laminated structural timbers. Almost any other species can be used when their mechanical and physical properties are suited for the design requirements (Castle 1980).

Glued-laminated wood (Figure 10.12) is used for structural beams and columns, furniture, sports equipment, and decorative wood finishes, as shown. In such cases glued-laminated wood is preferred over one large-piece member for many reasons, including: the ease of manufacturing large structural members from standard commercial-sized lumber; the opportunity to design large members that vary in cross section along their length, as required by the application; the specialized design to meet architectural appeal; the opportunity to use lower grades of wood within the less stressed areas of the member; and the minimization of checking and other seasoning defects associated with curing large one-piece members.

There are several factors that affect the strength of laminated timbers, such as cross-grain, knots, and the effect of end joining. However, in laminated members it is possible to vary the degree of cross-grain allowed at different points in the depth of the beam based on the needed strengths. The weaker, steeper grain may be used in the interior of the member, and the long sloping grains may be used in the outer heavy tension and compression zones.

Glued-laminated wood is designed and produced to have consistent and uniform properties. Design value properties for glued-laminated timbers can be obtained from the American Institute of Timber Construction.

**FIGURE 10.12** Example of glued-laminated wood. Photo courtesy of American Forest and Paper Association, Washington, DC.

## Plywood

Plywood is a composite of thin sheets of wood, called plies, that are glued together. The grain directions of the adjacent plies are at right angles to each other, as shown in Figure 10.13. Typically, plies range in thickness from 1.6 mm to 7.9 mm (1/16 in. to 5/16 in.). In four- or six-ply panels (even plied panels) the two central plies have the same grain orientation and form a single thick ply. Structural plywood has an odd number of plies, allowing perpendicular orientation of the internal plies. Standard nominal thickness of plywood panels vary from 3.2 mm to 29 mm (1/4 in. to 1-1/8 in.).

To produce plywood, logs are saturated by storing them in ponds, water vats, steam vats, or water sprays. Six hours prior to processing they are moved into a boiling water bath. From the boiling vat they are debarked and sectioned into 8-ft segments. Segments 2.4 m (8 ft) long are rotated in a giant lathe and peeled into a 2.4-m-wide continuous sheet of veneer, or sliced, as shown in Figure 10.14. These segments of veneer are trimmed and combined into continuous rolls. Each roll of veneer is seasoned, dried to a moisture content of 5% or less, and graded. The plies (cut from the

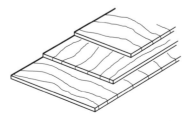

**FIGURE 10.13** Structure of plywood.

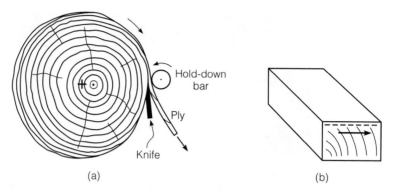

**FIGURE 10.14** Cutting of plies for plywood: (a) rotary cutting and (b) slicing.

veneer rolls) are assembled, glued, and pressed. After drying, the plywood is inspected front and back and stamped with the grade.

Alternating the grain orientation of the plies provides nearly identical properties along the length and width and resistance to dimensional change under varying moisture conditions. The dimensional change for plywood is about one-tenth that of solid lumber under any temperature or moisture condition.

Classification of plywood is based on the type of wood used (soft or hard species), number and grade of plies (veneers), and type of adhesive used. Softwood plywoods are used for structural applications such as subflooring and roofdecks. Hardwood plywoods are used for decorative applications and furniture. These are graded based on whether an interior or exterior glue is used and on the quality of the external sheets of veneer. The grading of the veneer is based on the number of defects and patches. Both exterior sides are graded separately.

## Particle Board and Strand Board

Lumber and paper production claims wood, such as small trees and branches, that is not suitable for lumber products. By gluing together "scraps," boards can be produced that have many practical applications. There are boards produced in 1.2 in. × 2.4 in. (4 ft × 8 ft) sheets in thickness ranging from 12.5 mm to 28.6 mm (1/2 to 1-1/8 in.). Particle board consists of sawdust-sized particles glued together in a dense matrix. Strand board is made from flat chips glued together with a random orientation. Due to the random orientation of the wood fibers, strand board is a dimensionally stable material. Since strand board uses wood scraps that are less expensive than the lumber needed to produce plywood, strand board has replaced plywood in many applications, such as roof sheathing.

As with natural wood products, glued products can be combined in a variety of ways to produce structural elements. For example, I-beams are produced by gluing together two 2 × 4 or 2 × 6 elements separated by a strand board web, as shown in Figure 10.15. Some manufacturers use laminates of plywood instead of the dimensional lumber.

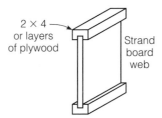

**FIGURE 10.15** Strand board I-beam.

**SUMMARY**

Wood is an extremely flexible building material. In the past, natural wood products were the only option available to engineers. However, modern forestry practices limit the size of natural products that are available. The need to increase the efficiency of wood product use has led to the development of engineered wood products. These products are usually more economical than natural wood, particularly when large dimensions are required. In addition, by careful control of the manufacturing process, engineered woods can be produced that have characteristics superior to natural wood. The characteristics of engineered wood products depend on the wood stock used, the quality of the adhesive, and the manufacturing process. Several factors make wood unique when compared to the other materials used in civil engineering, including anisotropy, moisture sensitivity, creep characteristics, and the existence of defects in wood products. Furthermore, when wood is exposed to the environment, care must be taken to prevent degradation due to fungi, bacteria, and insects.

**QUESTIONS AND PROBLEMS**

**10.1.** What are the two main classes of wood? What is the main use of each class? Name two tree species of each class.

**10.2.** What is the difference between earlywood and latewood? Describe each.

**10.3.** Discuss the anisotropic nature of wood. How does this phenomenon affect the performance of wood?

**10.4.** Briefly describe the chemical composition of wood.

**10.5.** The moisture of content of wood test was performed according to ASTM D4442 procedure and produced the following data:

Mass of specimen in the green condition = 266.7 g
Mass of oven-dry specimen = 152.1 g
Calculate the moisture content of the given wood.

**10.6.** What is the fiber saturation point (FSP)? What is the effect of the FSP on the shrinkage of wood in the different directions? How does this phenomenon affect the properties of lumber?

**10.7.** A stud had dimensions of 38.1 mm × 88.9 mm × 2.438 m and a moisture content of 150% when it was prepared. After seasoning, the moisture content was reduced to 7%. If the tangential, radial, and longitudinal directions of the grains are in the same order of the dimensions indicated above, what are the dimensions of the seasoned stud if the moisture-shrinkage relation follows Figure 10.3?

**10.8.** Wood is cut at sawmills into a variety of products with different sizes and shapes for engineering applications. What are these products?

**10.9.** Construction lumber can be cut from the tree using one of two methods or a combination of them. Name these two methods and show a sketch of each. What is the main advantage of each method?

**10.10.** Why are the actual dimensions of lumber different from the nominal dimensions? Explain.

**10.11.** What are the factors considered in grading lumber? What are the main grades of hardwoods and softwoods?

**10.12.** State five different imperfections that may be found in lumber and briefly define them.

**10.13.** Sketch the typical stress-strain curve for wood. On the graph, show the modulus of elasticity.

**10.14.** Compute the modulus of elasticity of the wood species whose stress-strain relation is shown in Figure 10.7.

**10.15.** List five different tests used to evaluate the mechanical properties of wood.

**10.16.** A wood specimen having a square cross section of 2 in. × 2 in. was tested by applying a load at the middle of the span where the span was 28 in. The deflection under the load was measured at different load levels as follows.

| Load, lb | Deflection, $10^{-3}$ in. |
|---|---|
| 0 | 0 |
| 100 | 27.9 |
| 200 | 55.6 |
| 300 | 83.2 |
| 400 | 111.2 |
| 500 | 140.0 |
| 600 | 166.7 |
| 700 | 194.3 |
| 800 | 222.2 |
| 900 | 250.1 |
| 1000 | 275.4 |
| 1100 | 314.8 |
| 1200 | 359.5 |
| 1300 | 405.0 |
| 1400 | 468.6 (failure) |

**a.** Using a computer spreadsheet program plot the load-deflection relationship.

**b.** Mark the proportional limit on the graph.

**c.** Calculate the modulus of rupture (flexure strength).

**d.** Does the modulus of rupture computed in (c) truly represent the extreme fiber stresses in the specimen? Comment on the assumptions used to compute the modulus of rupture and the actual response of the wood specimen.

**10.17.** A pine-wood specimen was prepared with dimensions of 50 mm × 50 mm × 200 mm and grains parallel to its length. The specimen was subjected to compression parallel to grains, and the load-deformation results are as shown below.

| Load, kN | Deformation, mm |
|----------|-----------------|
| 0        | 0               |
| 8.9      | 0.457           |
| 17.8     | 0.597           |
| 26.7     | 0.724           |
| 35.6     | 0.838           |
| 44.5     | 0.965           |
| 53.4     | 1.118           |
| 62.3     | 1.270           |
| 71.2     | 1.422           |
| 80.1     | 1.588           |
| 89.0     | 1.765           |
| 97.9     | 1.956           |
| 106.8    | 2.159           |
| 111.3    | 2.311           |

**a.** Using a computer spreadsheet program plot the stress-strain relationship.

**b.** Calculate the modulus of elasticity.

**c.** What is the failure stress?

**10.18.** For the purpose of designing wood structures, laboratory-measured strength properties are adjusted for application conditions. State five different application conditions that are used to adjust the strength properties.

**10.19.** What are the four types of organisms that attack wood?

**10.20.** What are the two types of preservatives that can be used to protect wood from decay? How are these preservatives applied?

**10.21.** What are four main types of engineered wood products?

**10.22.** What are the main advantages of engineered wood products over natural-timber members?

**REFERENCES**

American Forest and Paper Association. 1997. *National design specification for wood construction.* American Wood Council. Washington, DC: American Forest and Paper Association.

American Plywood Association. 1980. Special publication. Tacoma, WA: American Plywood Assocation.

Castle, W. 1980. *The Wendell Castle book of wood lamination.* New York: Van Nostrand Reinhold.

Levin, E., ed. 1972. *The international guide to wood selection.* New York: Drake Publishers Inc.

Panshin, A. J. and C. De Zeeuw. 1980. *Textbook of wood technology.* 4th ed. New York: McGraw-Hill.

Richardson, B. A. 1978. *Wood preservation.* New York: Longman.

Southern Pine Products Association. 1990. *Pressure treated southern pine.* Southern Pine Council. Kenner, LA: Southern Pine Products Association.

U.S. Deptartment of Agriculture, Forest Service. 1980. *The encyclopedia of wood.* New York: Sterling Publishing Co.

Van Vlack, L. H. 1982. *Materials for engineering: concepts and applications.* Reading, MA: Addison Wesley.

Van Vlack, L. H. 1989. *Elements of materials science and engineering.* 6th ed. Reading, MA: Addison Wesley.

# 11 ◆ Composites

The need for materials with properties not found in conventional materials combined with advances in technology have resulted in combining two or more materials to form what are called composite materials. These materials usually combine the best properties of their constituents and frequently exhibit qualities that do not even exist in their constituents. Strength, stiffness, specific weight, fracture resistance, corrosion resistance, wear resistance, attractiveness, fatigue life, temperature susceptibility, thermal insulation, thermal conductivity, and acoustical insulation can all be improved by composite materials. Of course, not all these properties are improved in the same composite, but typically a few of these properties are improved. For example, materials needed to build aircrafts and space vehicles must be light, strong, stiff, and exhibit high resistance to abrasion, impact, and corrosion. An example of a composite material that is very useful for civil engineers is fiberglass, which is strong, stiff, and corrosion resistant and can be used to make concrete reinforcing rebars to replace corrosive steel rebars. These combinations of properties are formidable and typically cannot be found in a conventional material.

Composite materials have been used throughout history, with differing levels of sophistication. For example, straw was used to strengthen the mud bricks in ancient civilizations. Swords and armor were constructed with layers of different materials to obtain unique properties. Portland cement concrete, which combines paste and aggregate with different properties to form a strong and durable construction material, has been used for many years. In recent years, fiber-reinforced concrete has been used as a building material that is strong in both tension and compression. The automobile industry has been using composite metals to build lightweight vehicles that are strong and impact resistant. Recently, a new generation of composites has been developed, such as fiber-reinforced and particle-reinforced plastics, that has revolutionized the material industry and opened new horizons for civil and construction engineering applications.

Although several definitions of composites exist, it is generally accepted that a composite is a material that has two or more distinct constituent materials or phases.

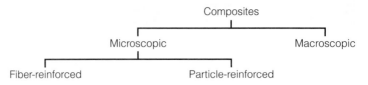

**FIGURE 11.1** Classification scheme of composite materials.

The constituents of a composite typically have significantly different physical properties, and thus the properties of the composite are noticeably different from those of the constituents. This definition eliminates many multiphase materials that do not have distinct properties, such as many alloys with components that are similar.

There are a number of naturally formed composites such as wood, which consists of cellulose fibers and lignin, and bone, which consists of protein collagen and mineral appetite. However, in this chapter we discuss artificially made composites only.

Composite materials can be classified as *microscopic* or *macroscopic,* as shown in Figure 11.1. The distinction between microscopic and macroscopic depends on the type of properties being considered. This distinction seems arbitrary, but normally microscopic composites include *fibers* or *particles* in sizes up to a few hundred microns. On the other hand, macroscopic composites could have constituents of much larger size, such as aggregate particles and rebars in concrete.

## Microscopic Composites

Many microscopic composite materials consist of two constituent phases: a continuous phase, or *matrix,* and the *dispersed phase* or *reinforcing phase,* which is surrounded by the matrix. In most cases, the dispersed phase is harder and stiffer than the matrix. The properties of the composite depend on the properties of both component phases, their relative properties, and the geometry of the dispersed phase, such as the particle shape, size, distribution, and orientation.

As indicated in Figure 11.1, microscopic composites fall into two basic classes: *fiber-reinforced* and *particle-reinforced*. This classification is based on the shape of the dispersed phase. Figure 11.2 shows composites with continuously aligned fibers, random fibers, and random particles. The mechanism of strengthening varies for different classes and for different sizes and orientations of the dispersed shape.

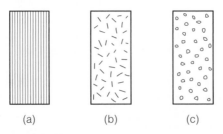

**FIGURE 11.2** Schematic of microscopic composites:
(a) aligned fibers, (b) random fibers, and (c) random particles.

**TABLE 11.1**   Materials and Mechanical Properties of Some Fibers*

| Material | Specific Gravity | Tensile Strength, GPa (psi × 10⁶) | Elastic Modulus, GPa (psi × 10⁶) |
|---|---|---|---|
| Aramid (Kevlar) | 1.4 | 3.5 (0.5) | 130 (19) |
| E-glass | 2.5 | 3.5 (0.5) | 72 (10.5) |
| Graphite | 1.4 | 1.7 (0.25) | 255 (37) |
| Nylon 6,6 | 1.1 | 1.0 (0.14) | 4.8 (0.7) |
| Asbestos | 2.5 | 1.4 (0.2) | 172 (25) |

*Callister (1985).

### Fiber-Reinforced Composites

Fiber-reinforced microscopic composites include fibers dispersed in a matrix such as metal or polymer. Fibers have very high strength-to-diameter ratio, with near crystal-sized diameters. Because of the very small diameter of the fibers they are much stronger than the bulk material. For example, a glass plate fractures at stresses of 10 kPa to 20 kPa, yet glass fibers have strength 3 MPa to 5 MPa or more. Fibers are much stronger than the bulk form because they have fewer internal defects.

Fibers can be classified based on their diameter and character as wiskers, fibers, and wires. Wiskers are very thin single crystals, have extremely large length-to-diameter ratios, have a high degree of crystalline perfection, and, consequently are extremely strong. Fibers have larger diameters than wiskers, while wires are even larger. Wiskers are not commonly used for reinforcement because of their high cost, poor bond with many common matrix materials, and the difficulty of incorporating them into the matrix. Table 11.1 shows some materials used to manufacture fibers and their strength characteristics.

Fibers are manufactured from many materials, such as glass, carbon and graphite, polymer, boron, ceramic, and silicon carbide (Mallick 1993). Because of their low cost and high strength, glass fibers are the most common of all reinforcing fibers for polymer matrix composites. Glass fibers are commercially available in several forms suitable for different applications. Common glass fibers include veils, rovings (continuous fibers), and mats (Figure 11.3). A strand consists of about 30 or 40 fibers twisted together to form a ropelike length.

A common fiber-reinforced composite is fiberglass. Fiberglass is simply a composite consisting of glass fibers, either continuous or discontinuous, contained within a plastic matrix. Typical fiberglass applications include aircraft, automobiles, boats, storage containers, water tanks, sporting equipment, and flooring.

### Particle-Reinforced Composites

Particle-reinforced composites consist of particles dispersed in a matrix phase. The strengthening mechanism of particle-reinforced composites varies with the size of the reinforcing particles. When the size of the particles are about 0.01 micron to 0.1 micron, the matrix bears most of the applied load, whereas the small dispersed particles hinder or impede the motion of dislocations. An example of dispersed-reinforced composite is thoria-dispersed nickel, in which about 3% of thoria ($ThO_2$) is finely dispersed in a nickel alloy to increase its high-temperature strength. On the other hand, when the particles are larger than 1 micron, particles act as fillers to

**FIGURE 11.3** Common glass fibers (veil, roving, and mat).
Photo courtesy of Creative Pultrusions, Inc.

improve the properties of the matrix phase and/or to replace some of its volume, since the filler is typically less expensive. Here, the matrix retains movement in the vicinity of the particle. Thus, the applied load is shared by the matrix and dispersed phases. The stronger the bond between the dispersed particles and the matrix, the larger the reinforcing effect. An example of particle reinforcing is adding fillers to polymers to improve tensile and compressive strengths, abrasion resistance, toughness, dimensional and thermal properties, and other properties.

## Matrix Phase

Typically, the matrix used in most microscopic composites is polymer (plastic) or metal. The matrix binds the dispersed materials (particles or fibers) together, transfers loads to them, and protects them against environmental attack and damage due to handling. Polymers have the advantages of low cost, easy processibility, good chemical resistance, and low specific gravity. The shortcomings of polymers are the low strength, low modulus, low operating temperatures, and low resistance to prolonged exposure to ultraviolet light and some solvents. On the other hand, metals have high strength, high modulus, high toughness and impact resistance, relative insensitivity to temperature changes, and high resistance to high temperatures and other severe environmental conditions. However, metals have high density and high processing temperatures due to their high melting point. Metals also may react with particles and fibers and they are vulnerable to attack by corrosion (Agarwal and Broutman 1990). The metals most commonly used as the matrix phase in composites are aluminum and titanium alloys.

## Fabrication

Fabrication of microscopic composites often combines the production of the material during the fabrication of the composite. The composite is formed by combining the matrix and dispersed material. Several methods have been used to fabricate the composites. The selection of the fabrication process typically is based on the chemical

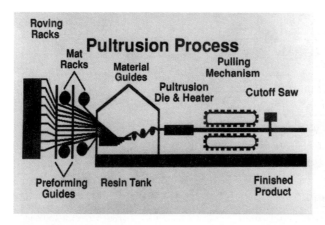

**FIGURE 11.4** Pultrusion scheme used in fabricating structural shape fiber-reinforced composites. Photo courtesy of Creative Pultrusions, Inc.

nature of the matrix and the dispersed phases and the temperature required to form, melt, or cure the matrix. Figure 11.4 illustrates fabrication of structural shape fiber-reinforced composites by using the pultrusion process. Pultrusion is an automated process for manufacturing fiber-reinforced composite materials into continuous, constant cross-section profiles.

## Civil Engineering Applications

Microscopic composites have been used in many civil and construction engineering applications in the last few decades. In fact, composite materials compete with, and in many cases are preferred over, conventional building materials. Composites are used by civil engineers as structural shapes in buildings and other structures and can replace steel and aluminum structural shapes (Figure 11.5). Table 11.2 provides an example of physical properties of fiber-reinforced composite round rods and bars.

**TABLE 11.2** An Example of Physical Properties of Fiber-Reinforced Composite Round Rods and Bars*

| Property | Value |
| --- | --- |
| Tensile strength (ASTM D638) | 830 MPa ($120 \times 10^3$ psi) |
| Tensile modulus of elasticity (ASTM D638) | 45 GPa ($6.5 \times 10^6$ psi) |
| Flexural strength (ASTM D790) | 830 MPa ($120 \times 10^3$ psi) |
| Compressive strength (ASTM D695) | 480 MPa ($70 \times 10^3$ psi) |
| Izod impact strength (ASTM D256) | 2.1 kJ/m (40 ft-lb/in.) |
| Barcol hardness (ASTM D2583) | 50 |
| Water absorption (ASTM D570) | 0.25% max. |
| Specific gravity (ASTM D792) | 2.0 |
| Coefficient of thermal expansion (ASTM D696) | $5.2 \times 10^{-6}$ m/m/°C ($9.4 \times 10^{-6}$ in./in./°F) |

*Creative Pultrusions, Inc. (1997).

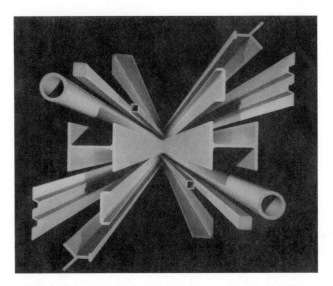

**FIGURE 11.5** Structural shapes made of fiberglass composites.
Photo courtesy of Creative Pultrusions, Inc.

Fiber-reinforced plastic (fiberglass) rebars can also be used for concrete rein-forcement instead of steel rebars. Composites have been used for tanks, industrial flooring, trusses and joists, walkways and platforms, waste treatment plants, handrail-ings, plastic pipes, light poles, door and window panels and frames, and electrical enclosures. Composites can also be used to strengthen and wrap columns and bridge supports that are partially damaged by earthquakes and other environmental factors.

Fiber-reinforced concrete is another composite material that has been used by civil engineers in various structural applications. Different types of fibers, such as separate fibers, chopped-strands, or rovings, can be used to reinforce the concrete. If separate fibers or chopped-strands are used, they are mixed with the fresh concrete in a random order. In such a case, fibers hinder or impede the progression of cracks in concrete. Figure 11.6 shows a scanning electron micrograph of concrete mortar mixed with about 3-mm-long carbon fibers at a volume fraction of 12%. Fiber rovings, on the other hand, are placed in the direction where the tension is applied in the structural member. In this case, fibers carry the tensile stresses. In general, fibers increase the tensile and flexure strength of concrete so that a more efficient structural member can be designed. Table 11.3 shows typical ranges of physical properties of glass fiber-reinforced concrete at 28 days. Research has shown that glass fiber-reinforced con-crete offers two to three times the flexural strength of unreinforced concrete. More-over, the material under increasing load does not fail abruptly, but yields gradually. This gradual yielding occurs because fibers are stronger than the matrix and, there-fore, arrest cracks. Therefore, instead of a worsening of the first crack that occurs in the concrete, more cracks are developed elsewhere, and failure finally occurs when fibers pull out or break (Neal 1977).

**TABLE 11.3** Typical Ranges of Physical Properties at 28 Days of Glass Fiber Reinforced Concrete*

| Property | Value |
|---|---|
| Flexural strength | 21–32 MPa (3.0–4.6 ksi) |
| Tensile strength | 7–11 MPa (1.0–1.6 ksi) |
| Compressive strength | 50–79 MPa (7.2–11.4 ksi) |
| Impact strength | 10–25 kN/m (57–143 in. lb/in.$^2$) |
| Elastic modulus | 10.5–20.5 GPa (1.5–3.0 × 10$^6$ psi) |
| Density | 1.70–2.10 Mg/m$^3$ (105–130 lb/ft$^3$) |

*Neal (1977).

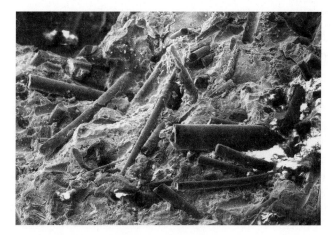

**FIGURE 11.6** Scanning electron micrograph of concrete mortar mixed with carbon fibers.

Entrained air in concrete can also be considered as a component in a microscopic composite material. Entrained air increases the durability of concrete since it releases internal stresses due to freezing of water within the concrete. For the same water-to-cement ratio, however, air bubbles reduce the concrete strength by about 20%. Since entrained air also improves the workability of fresh concrete, the water-to-cement ratio can be reduced to compensate for some of the strength reduction.

## Macroscopic Composites

Macroscopic composites are used in many engineering applications. Because macroscopic composites are relatively large, how the load is carried and how the properties of the composite components are improved vary from one composite to another. Common macroscopic composites used by civil and construction engineers include plain portland cement concrete, steel-reinforced concrete, asphalt concrete, and engineered wood such as glued-laminated timber, and structural strand board.

**FIGURE 11.7** Cross section of portland cement concrete showing cement paste and aggregate particles.

## Plain Portland Cement Concrete

Plain portland cement concrete is a composite material consisting of cement paste and aggregate particles with different physical and mechanical properties as discussed in Chapter 7 (Figure 11.7). Aggregate particles in concrete act as a filler material since it is cheaper than the portland cement. In addition, since cement paste shrinks as it cures, aggregate increases the volume stability of the concrete. When the concrete structure is loaded, both cement paste and aggregate share the load. Both the strength of aggregate particles and the bond between the aggregate and cement paste play an important role in determining the strength of the concrete composite, which is limited by the weaker of the two. The bond between cement paste and aggregate is affected by roughness and absorption of the aggregate particles, as well as other physical and chemical properties of aggregate.

## Reinforced Portland Cement Concrete

Steel-reinforced concrete can be viewed as a composite material consisting of plain concrete and steel rebars, as shown in Figure 11.8. Since concrete has a very small tensile strength, which is typically ignored in designing concrete structures, steel rebars are usually placed in areas within the structure that are subjected to tension. When the concrete structure is loaded, the concrete carries compressive stresses and steel carries tensile stresses. In such cases steel allows concrete to be used as structural members carrying tension. Steel rebars are also used in areas subjected to compression, such as columns, to share the load support. In such cases, steel reduces the required cross-sectional area of the compressive member since the compressive strength of steel is larger than that of concrete. Steel rebars are also used in prestressed concrete where the rebars are prestressed under tension so that the concrete remains under compression even when it is externally loaded. In such cases a smaller cross-sectional area of the concrete member is required. Steel rebars can also be used to control cracking in concrete due to temperature change. For example, concrete pavement is sometimes reinforced by placing longitudinal and transverse steel bars at the midheight of the concrete slab. In this case, when the concrete shrinks due to reduction in ambient temperature, many cracks will develop; these cracks are

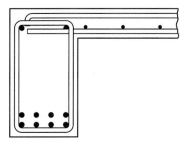

**FIGURE 11.8** An example of steel-reinforced concrete beam and slab details.

uniformly distributed within the pavement section, but each crack will be tight. Typically, tight cracks are not harmful to concrete pavement since they transfer the load from one side of the crack to the other by interlocking. In all applications of steel reinforced concrete, the bond between the rebars and the concrete is important in order to allow the composite to work as one unit. Therefore, bars have a deformed surface to prevent slipping between steel and concrete.

## Asphalt Concrete

Asphalt concrete used in pavements is another composite material. It consists of two materials with distinct properties, as presented in Chapter 9. Asphalt concrete consists of approximately 95% aggregate and 5% asphalt binder by weight. When the traffic loads are applied on the asphalt concrete composite, most of the compressive stresses are supported by the aggregate-to-aggregate contact. The asphalt acts as a binder that prevents slipping of aggregate particles relative to each other. When tensile stresses are applied due to bending of the asphalt concrete layer or due to thermal contraction, the aggregate particles are supported by the asphalt binder. One important property of asphalt is that it gets soft at high temperatures and brittle at low temperatures, whereas aggregate does not change its properties with temperature fluctuation. It is important, therefore, to properly select the asphalt grade that will perform properly within the temperature range of the region in which it is being used. Also, since aggregate represents a major portion of the mixture, it important to use aggregate with proper gradation and other properties. The asphalt binder content must be carefully designed in order to ensure that aggregate particles are fully coated, without excessive lubrication. When the asphalt concrete mixture is appropriately designed and compacted, it should last for a long time without failure.

## Engineered Wood

Engineered wood is manufactured by bonding together wood strands, veneers, or lumber with different grain orientations to produce large and integral units. Since engineered wood consists of components of the same material, it does not qualify as a composite according to our definition. However, engineered wood is presented in this chapter because it follows a strengthening mechanism similar to that of composites. Since wood has anisotripic properties due to the existence of grains, engineered wood produces specific and consistent mechanical behavior and thus has consistent design properties. For example, alternating the grain orientation of the plies of plywood provides nearly identical properties along the length and width and provides resistance to dimensional change under varying moisture conditions. The plywood composite has about one-tenth of the dimensional change of solid lumber under any temperature or moisture condition. As discussed in Chapter 10, engineered wood products include plywood, oriented strand boards, composite panels, glued laminated timber (glulam), laminated veneer lumber, parallel strand lumber, oriented strand lumber, and wood I-joists.

## Properties of Composites

The properties of composite materials are affected by the component properties, volume fractions of components, type and orientation of the dispersed phase, and the bond between the dispersed phase and the matrix. The properties of the composite can be viewed as the weighted average of the properties of the components (Shackelford 1996). Equations can be derived to estimate the composite properties under certain idealized material properties, loading patterns, and geometrical conditions. Assumptions that can be used to simplify the analysis include:

- Each component has linear, elastic, and isotropic properties.
- A perfect bond exists between the dispersed and matrix phases without slipping.
- The composite geometry is idealized and the loading pattern is parallel or perpendicular to reinforcing fibers.

### Loading Parallel to Fibers

When load is applied to aligned fiber-reinforced composite parallel to fibers, as seen in Figure 11.9(a), both matrix and fiber phases will equally deform. Thus, the strains of both phases will be the same (known as isostrain condition) as follows:

$$\epsilon_c = \epsilon_m = \epsilon_f = \epsilon \tag{11.1}$$

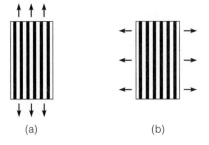

(a)    (b)

**FIGURE 11.9** Patterns of loading continuously aligned fiber-reinforced composites: (a) loading parallel to fibers, and (b) loading perpendicular to fibers.

where

$\epsilon$ = total strain

$\epsilon_c$ = composite

$\epsilon_m$ = matrix strain

$\epsilon_f$ = fiber strain

Also, the force applied to the composite $F_c$ is the sum of the force carried by the matrix $F_m$ and the force carried by the fibers $F_f$.

$$F_c = F_m + F_f \tag{11.2}$$

Thus,

$$\sigma_c A_c = \sigma_m A_m + \sigma_f A_f \tag{11.3}$$

where

$\sigma_i$ = stress of component $i$

$A_i$ = area of component $i$

Replacing $\sigma$ with $E\epsilon$ for each material, Equation 11.3 can be rewritten as:

$$E_c \epsilon A_c = E_m \epsilon A_m + E_f \epsilon A_f \tag{11.4}$$

where $E$ is the modulus of elasticity. Canceling $\epsilon$ and dividing by $A_c$

$$E_c = E_m \frac{A_m}{A_c} + E_f \frac{A_f}{A_c} \tag{11.5}$$

or

$$E_c = \nu_m E_m + \nu_f E_f \tag{11.6}$$

where $\nu$ is the volume fraction of each component and $\nu_m + \nu_f = 1$.

Equation 11.6 shows that the composite's modulus of elasticity is the weighted average of the component moduli.

The share of the load carried by the fibers can be determined as follows:

$$\frac{F_f}{F_c} = \frac{\sigma_f A_f}{\sigma_c A_c} = \frac{E_f \epsilon A_f}{E_c \epsilon A_c} = \frac{E_f}{E_c} \nu_f \tag{11.7}$$

**SAMPLE PROBLEM 11.1**

Calculate the modulus of elasticity of fiberglass under isostrain condition if the fiberglass consists of 70% E-glass fibers and 30% epoxy by volume. Also, calculate the percentage of load carried by the glass fibers. The modulus of elasticity of the glass fibers and the epoxy are 70.5 GPa and 6.9 GPa, respectively.

*Solution:*

From Equation 11.6

$$E_c = (0.3)(6.9) + (0.7)(70.5) = 51.4 \text{ GPa}$$

From Equation 11.7

$$\frac{F_f}{F_c} = \frac{70.5}{51.4}(0.7) = 0.96 = 96\%$$

This example shows that, under the given conditions, 96% of the load is carried by the fibers.

◆

Equation 11.6 can be generalized to cover other composite properties as a function of the properties of the components as:

$$X_c = \nu_m X_m + \nu_f X_f \tag{11.8}$$

where $X$ is a property such as Poisson's ratio, thermal conductivity, electrical conductivity, or diffusivity.

## Loading Perpendicular to Fibers

When load is applied to aligned fiber-reinforced composite perpendicular to fibers [Figure 11.9(b)], both matrix and fiber phases will be subjected to the same stress (isostress condition). In other words,

$$\sigma_c = \sigma_m = \sigma_f = \sigma \tag{11.9}$$

The elongation of the composite in the direction of the applied stress is the sum of the elongations of the matrix and fibers as

$$\Delta L_c = \Delta L_m + \Delta L_f \tag{11.10}$$

Dividing Equation 11.10 by the composite length $L_c$ in the stress direction gives

$$\frac{\Delta L_c}{L_c} = \frac{\Delta L_m}{L_c} = \frac{\Delta L_f}{L_c} \tag{11.11}$$

Assuming that the fibers are uniform in thickness, the cumulative length of each component in the direction of the stress is proportional to its volume fraction. Thus,

$$L_m = \nu_m L_c \tag{11.12}$$

and
$$L_f = \nu_f L_c \tag{11.13}$$

Substituting the values of $L_c$ from Equations 11.12 and 11.13 in Equation 11.11 yields

$$\frac{\Delta L_c}{L_c} = \frac{\nu_m \Delta L_m}{L_m} = \frac{\nu_f \Delta L_f}{L_f} \tag{11.14}$$

Since $\epsilon = \frac{\Delta L}{L}$, Equation 11.14 can be rewritten as

$$\epsilon_c = \nu_m \epsilon_m + \nu_f \epsilon_f \tag{11.15}$$

Replacing $\epsilon$ with $\frac{\sigma}{E}$ gives

$$\frac{\sigma}{E_c} = \nu_m \frac{\sigma}{E_m} + \nu_f \frac{\sigma}{E_f} \tag{11.16}$$

or
$$\frac{1}{E_c} = \frac{\nu_m}{E_m} + \frac{\nu_f}{E_f} \tag{11.17}$$

Equation 11.17 can be rewritten as

$$E_c = \frac{E_m E_f}{\nu_m E_f + \nu_f E_m} \tag{11.18}$$

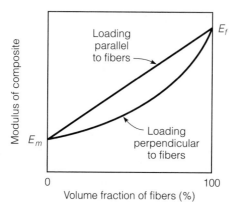

**FIGURE 11.10** Modulus of elasticity of the composite versus fiber volume fraction.

Similar to Equation 11.8, Equation 11.18 can be generalized as

$$X_c = \frac{X_m X_f}{\nu_m X_f + \nu_f X_m} \tag{11.19}$$

where $X$ is a property such as thermal conductivity, electrical conductivity, or diffusivity.

The moduli in Equations 11.6 and 11.18 can be plotted as functions of the volume fraction of the fiber, as shown in Figure 11.10. Clearly, the fibers are more effective in raising the modulus of the composite when loading parallel to fibers than when loading perpendicular to fibers.

## Randomly Oriented Fiber Composites

Unlike continuously aligned fiber composites, the mechanical properties of randomly oriented fiber composites are isotropic. The modulus of elasticity of randomly oriented fiber composites falls between the moduli of loading parallel to fibers and perpendicular to fibers. To estimate the modulus of elasticity of randomly oriented fiber composites, Equation 11.6 can be rewritten as

$$E_c = \nu_m E_m + K \nu_f E_f \tag{11.20}$$

where $K$ is a fiber efficiency parameter (Callister 1985). For fibers randomly and uniformly distributed within three dimensions in the space, $K$ has a value of 0.2.

**SAMPLE PROBLEM 11.2**

A fiberglass composite consists of epoxy matrix reinforced with randomly oriented and uniformly distributed E-glass fibers. The modulus of elasticity of the glass fibers and the epoxy are 65 GPa and 7 GPa, respectively. If the volume percentage of fibers is 30%, and the fiber efficiency is 0.2, calculate the modulus of elasticity of the fiberglass.

*Solution:*

From Equation 11.20

$$E_c = 0.67 \times 7 + 0.2 \times 0.33 \times 65 = 9.0 \text{ GPa}$$

## Particle-Reinforced Composites

The analysis of loading a particle-reinforced composite depends on the specific nature of the dispersed and matrix phases. A rigorous analysis of loading a particle-reinforced composite can become quite complex. Equations 11.6 and 11.18 serve as upper and lower bounds for the particle-reinforced properties.

**SUMMARY**

Combining different materials to produce a composite that has properties superior to the component materials has been practiced since ancient times. In fact, many of the conventional materials currently used in civil engineering are composites, including portland cement concrete, reinforced concrete, asphalt concrete, and engineered woods. Composites are generally classified as either fiber or particle reinforced, depending on the nature of the dispersed phase material. The properties of composites depend on the characteristics of the component materials, the bonding between the dispersed and matrix phases, and the orientation of the dispersed phase.

**QUESTIONS AND PROBLEMS**

**11.1.** What is a composite material? List some composite materials that you use in your daily life.

**11.2.** List five different advantages of composite materials over conventional materials.

**11.3.** Define microscopic composites. What are the two phases of microscopic composites?

**11.4.** What are the two types of microscopic composites? Show the mechanism of strengthening of each type.

**11.5.** Why are fibers much stronger than the bulk material? Give an example of a material that is relatively weak in the bulk form and very strong in the fiber form.

**11.6.** Compare the desired properties of the matrix and the fiber phases of the fiber-reinforced composite.

**11.7.** Name three functions of the matrix phase in fiber-reinforced composites. State the reasons of the need for a strong bond between the fibers and the matrix.

**11.8.** What are the functions of aggregate used in portland cement concrete?

**11.9.** How is the load supported by asphalt concrete in the cases of tension and compression. Under what conditions is the asphalt concrete layer subjected to tension?

**11.10.** Briefly describe why engineered wood is stronger and has better properties than natural wood.

**11.11.** Calculate the modulus of elasticity of carbon-epoxy composite under isostrain condition if the composite consists of 50% carbon fibers and 50% epoxy by volume. Also, calculate the percentage of load carried by the carbon fibers. The modulus of elasticity of the carbon fibers and the epoxy are 350 GPa and 3.5 GPa, respectively.

**11.12.** Repeat problem 11.11 for a 30% carbon fibers by volume.

**11.13.** Repeat problem 11.11 under isostress condition.

**11.14.** A fiberglass composite consists of epoxy matrix reinforced with randomly oriented and uniformly distributed E-glass fibers. The modulus of elasticity of the glass fibers and the epoxy are 70 GPa and 6 GPa, respectively. Calculate the modulus of elasticity of the fiberglass if the volume percentage of fibers is (a) 25, (b) 50, and (c) 75. Plot a graph showing the relationship between the modulus of elasticity of the fiberglass and the percent of fibers. Comment on the effect of the percent of glass fibers on the modulus of elasticity of fiberglass.

**11.15.** A reinforced concrete column is subjected to a 1000 kN axial compressive load. The moduli of elasticity of plain concrete and steel are 25 GPa and 207 GPa, respectively, and the cross-sectional area of steel is 2% of that of the reinforced concrete. Considering the column as a structural member made of a composite material and subjected to load parallel to the steel rebars, calculate the following:

    **a.** the modulus of elasticity of the reinforced concrete

    **b.** the load carried by each of the steel and plain concrete

    **c.** the minimum required cross-sectional area of the column given that the allowable compressive stress of plain concrete is 3 MPa and that the allowable compressive stress of plain concrete will be reached before that of steel.

## REFERENCES

Agarwal, B. D. and L. J. Broutman. 1990. *Analysis and performance of fiber composites.* 2nd ed. New York: John Wiley and Sons.

Callister, W. D., Jr. 1985. *Materials science and engineering, an introduction.* New York: John Wiley and Sons.

Creative Pultrusions, Inc. 1970. *Creative pultrusions design guide.* Alum Bank, PA: Creative Pultrusions, Inc.

Mallick, P. K. 1993. *Fiber-reinforced composites—materials, manufacturing, and design.* 2nd ed. New York: Marcel Dekker.

Neal, W. 1977. "Glass fiber reinforced concrete—a new composite for construction." *The Construction Specifier,* Mar. Washington, DC: Construction Specifications Institute.

Shackelford, J. E. 1996. *Introduction to materials science for engineers.* 4th ed. Upper Saddle River, NJ: Prentice Hall.

# APPENDIX

# Laboratory Manual

The laboratory tests discussed in this appendix can be performed as a part of the civil and construction engineering materials course. More tests are included in this manual than typically can be performed in one semester. The extent of tests provided gives the instructor flexibility to choose the appropriate tests. In order for the students to get the most benefit from the laboratory sessions, tests should be coordinated with the topics covered in the lectures.

This laboratory manual summarizes the main components of each test method. Students are encouraged to read the corresponding ASTM or AASHTO test methods for the detailed laboratory procedures. The ASTM and AASHTO standards are usually available in college libraries.

In many cases, the time available for a laboratory session is not enough to perform the complete test as specified in the ASTM or AASHTO procedure. Therefore, the instructor may limit the number of specimens or eliminate some portions of the test. However, the student should be aware of the complete test procedure and the specification requirements.

In some cases, different experiments are used to obtain the same material properties such as the air content of freshly mixed concrete. In such cases, the instructor may require the specific test used by the state or the test for which equipment is available.

Typically a laboratory report is required by the student after each laboratory session. The format of the report can vary depending on the requirements of the instructor. A suggested format may include the following items:

- Title of the experiment.
- ASTM or AASHTO designation.
- Purpose.
- Significance and use.
- Test materials.
- Main pieces of equipment.
- Summary of test procedure and test conditions.
- Test results and analysis.
- Comments, conclusions, and recommendations. Any deviations from the standard test procedure should be reported and justified.

# Experiment No. 1.
# Introduction to Measuring Devices

## ASTM Designation

There is no ASTM procedure for the main portion of this session. The information at the end of Chapter 1 can be used as a reference. Some discussion on precision and bias can be helpful; this is included in ASTM C670 (Practice for Preparing Precision and Bias Statements for Test Methods for Construction Materials).

## Purpose

To introduce the students to common measuring devices such as dial gauges, LVDTs, strain gauges, proving rings, extensometers, etc. An introduction to precision and bias can also be included.

## Apparatus

The instructor may demonstrate one or more of the following items:

- A few dial gauges with different ranges and sensitivities (Figure 1.23)
- LVDT (Figures 1.25–1.27) and necessary attachments such as power supply, signal conditioner, voltmeter, display device, and calibration device
- Extensometer (Figures 1.24 and 1.28)
- Strain gauge (Figure 1.29) and necessary attachments
- Proving ring (Figure 1.30)
- Load cell (Figure 1.31)

## Calibration

The instructor may require the students to calibrate one or more measuring devices, such as a proving ring or an LVDT. Static loads can be used to calibrate the proving ring to develop a relation between force and the reading of the proving ring. To calibrate the LVDT, a calibration device equipped with a micrometer such as that shown in Figure 1.26 is used. A relation is developed between voltage and displacement of the LVDT.

## Requirements

- Brief description of the demonstrated device(s) including the use, components, theory, sensitivity, etc.
- Calibration table, graph, and equation for each device calibrated

# Experiment No. 2.
# Tension Test of Steel and Aluminum

## ASTM Designation

ASTM E8—Tension Testing of Metallic Materials

## Purpose

- To determine stress-strain relationship
- To determine yield strength
- To determine tensile strength
- To determine elongation and reduction of cross-sectional area
- To determine modulus of elasticity
- To determine rupture strength

## Significance and Use

This test provides information on strength and ductility for metals subjected to a uniaxial tensile stress. This information may be useful in: comparison of materials, alloy developments, quality control, design under certain circumstances, and detecting non-uniformity and imperfections as indicated by the fracture surface.

## Apparatus

- A testing machine capable of applying tensile load at a controlled rate of deformation or load. The testing machine could be either mechanical or closed-loop electrohydraulic. The machine could be equipped with a dial gauge to indicate the load or could be connected to a chart recorder or computer to record load and deformation.
- A gripping device used to transmit the load from the testing machine to the test specimen and to ensure axial stress within the gauge length of the specimen
- An extensometer with an LVDT used to measure the deformation of the specimen
- Caliper to measure the original dimensions of the specimen

## Test Specimens

Either plate-type or rounded specimens can be used as shown in Figure 3.7. Specimen dimensions are specified in ASTM E8.

## Test Procedure

1. Mark the gauge length on the specimen either by slight notches or with ink.
2. Place the specimen in the loading machine. (See Figure 3.8.)
3. Attach the extensometer to the specimen.
4. Set the load reading to zero, then apply load at a rate less than 690 kPa/min (100,000 psi/min). Unless otherwise specified, any convenient speed of testing may be used up to half of the specified yield strength or yield point, or one quarter of the specified tensile strength, whichever is smaller. After the yield strength or yield point has been determined, the strain rate may be increased to a maximum of 0.5 in./in. of the gauge length per minute.
5. Continue applying the load until the specimen breaks.
6. Record load and deformation every 2.2 kN (500 lb) increment for steel and every 890 N (200 lb) increment for aluminum, both before and after the yield point.

## Analysis and Results

- Calculate the stress and strain for each load increment until failure.
- Plot the stress versus strain curve.
- Determine the yield strength using the offset method, extension method (see Figure 1.7), or by observing the sudden increase in deformation.
- Calculate the tensile strength.

$$\sigma = \frac{P_{max}}{A_o}$$

where

$\sigma$ = tensile strength, MPa (psi)

$P_{max}$ = maximum load carried by the specimen during the tension test, N (lb)

$A_o$ = original cross-sectional area of the specimen, mm$^2$ (in.$^2$)

- Calculate the elongation.

$$\text{Percent elongation} = \frac{(L_s - L_o)}{L_o} \times 100$$

where

$L_s$ = gauge length after rupture, mm (in.)

$L_o$ = original gauge length, mm (in.)

For elongation > 3.0%, fit the ends of the fractured specimen together and measure $L_s$ as the distance between two gauge marks. For elongation ≤3.0%, fit the fractured ends together and apply an end load along the axis of the specimen sufficient to close the fractured ends together, then measure $L_s$ as the distance between gauge marks.

*(continued)*

- Calculate the modulus of elasticity.

$$E = \frac{\sigma}{\epsilon}$$

where
  $E$ = modulus of elasticity, MPa (psi)
  $\sigma$ = stress in the proportional limit, MPa (psi)
  $\epsilon$ = corresponding strain, mm/mm (in./in.)

- Calculate the rupture strength.

$$\sigma_r = \frac{P_f}{A_o}$$

where
  $\sigma_r$ = rupture strength, MPa (psi)
  $P_f$ = final load, N (lb)
  $A_o$ = original cross-sectional area, mm$^2$ (in.$^2$)

- Calculate the reduction of cross-sectional area.

$$\text{Percent reduction in cross-sectional area} = \frac{(A_o - A_s)}{A_o} \times 100$$

where
  $A_s$ = cross section after rupture, mm$^2$ (in.$^2$)

To calculate the cross section after rupture, fit the ends of the fractured specimen together and measure the mean diameter or width and thickness at the smallest cross section.

## Replacement of Specimens

The test specimen should be replaced if

- the original specimen had a poorly machined surface.
- the original specimen had wrong dimensions.
- the specimen's properties were changed because of poor machining practice.
- the test procedure was incorrect.
- the fracture was outside the gauge length.
- for elongation determination, the fracture was outside the middle half of the gauge length.

## Report

- Stress-strain relationship
- Yield strength and the method used
- Tensile strength
- Elongation and original gauge length
- Modulus of elasticity
- Rupture strength
- Reduction of cross-sectional area

# Experiment No. 3.
# Torsion Test of Steel and Aluminum

## ASTM Designation

ASTM E143—Shear Modulus at Room Temperature

## Purpose

To determine the shear modulus of structural materials such as steel and aluminum.

## Significance and Use

Shear modulus is a material property useful in calculating compliance of structural materials in torsion provided they follow Hooke's law; that is, the angle of twist is proportional to the applied torque.

## Apparatus

- Torsion testing machine (Figure 3.11)
- Grips to hold the ends of the specimen in the jaws of the testing machine
- Twist gauges to measure the angle of twist

## Test Specimens

Either solid or hollow specimens with circular cross section can be used. Specimens should be chosen from sound, clean material. Slight imperfections near the surface, such as fissures that would have negligible effect in determining Young's modulus, may cause appreciable errors in shear modulus. The gauge length should be at least four diameters. The length between grips should at least be equal to the gauge length plus two to four diameters.

## Test Procedure

1. Measure the diameter (and the wall thickness in case of tubular specimens).
2. Place the specimen in the torsion testing machine with careful alignment and apply torque.
3. Make simultaneous measurements of torque and angle of twist and record the data.
4. Maintain the speed of testing high enough to make creep negligible.

*(continued)*

## Analysis and Results

- Calculate the maximum shear stress ($\tau_{\max}$) and shear strain ($\gamma$) as follows:

$$\tau_{\max} = \frac{Tr}{J}$$

$$\gamma = \frac{\theta r}{L}$$

where

    $T$ = torque
    $r$ = radius
    $J$ = polar moment of inertia of the specimen about its center
    $\theta$ = angle of twist in radians
    $L$ = gauge length

- Plot a graph of shear stress versus shear strain as shown in Figure 3.12.
- Determine the shear modulus $G$ as the slope of the straight portion of the shear stress versus strain relation

$$G = \frac{\tau_{\max}}{\gamma}$$

## Report

- Shear stress versus shear strain graph
- Shear modulus

# Experiment No. 4. Impact Test of Steel

## ASTM Designation

ASTM E23—Test Methods for Notched Bar Impact Testing of Metallic Materials

## Purpose

To determine the energy absorbed in breaking notched steel specimens at different temperatures using the Charpy V Notch test. The energy value is a measure of toughness of the material.

## Significance and Use

This test measures a specimen's change in toughness as the temperature changes.

## Apparatus

Use an impact testing machine of pendulum type of rigid construction and of capacity more than sufficient to break the specimen in one blow (Figure 3.14).

## Test Specimen

Steel specimens prepared according to ASTM E23, as shown in Figure 3.13.

## Test Conditions

The test can be performed at the following four temperatures:

- −40°C (−40°F) (dry ice + isopropyl alcohol)
- −18°C (0°F) (dry ice + 30% isopropyl alcohol + 70% water)
- 4°C (40°F) (ice + water)
- 40°C (104°F)(oven)

*(continued)*

## Test Procedure

1. Prepare the impact testing machine by lifting up the pendulum and adjusting the gauge reading to zero. Since the pendulum is heavy, be careful to handle it safely.
2. Remove the test specimen from the temperature medium using tongs, and immediately position it on the two anvils in the impact testing machine.
3. Release the pendulum without vibration by pushing the specified button. The time between removing the specimen from the temperature medium and the completion of the test should be less than 5 s.
4. Record the energy required to break the specimen by reading the gauge mark.
5. Observe the fracture surface appearance (Figure 3.15).
6. Measure the lateral expansion of the specimen using a caliper or the lateral expansion gauge specified in ASTM E23.

## Report

- Energy required to cause fracture versus temperature plot
- Ductile-to-brittle transition temperature
- Fracture surface appearance (each specimen and temperature)
- Lateral expansion (each specimen and temperature)

# Experiment No. 5.
# Microscopic Inspection of Materials

## ASTM Designation

None.

## Purpose

- To observe the microstructural characteristics of materials.
- To compare the microstructure of different materials.
- To compare metals subjected to different heat treatments.
- To observe grain boundaries of metals.
- To examine fractured surfaces.
- To observe microcracks in concrete.
- To observe fibers in fiber-reinforced concrete.
- To observe the microscopic properties of asphalt concrete.
- To observe grains in wood, etc.

## Apparatus

Scanning electron microscope or optical microscope

## Material

Various materials can be used, such as metals, portland cement concrete, asphalt concrete, or wood.

## Report

Compare and discuss the microstructural characteristics of the materials.

# Experiment No. 6.
# Sieve Analysis of Aggregates

## ASTM Designation

ASTM C136—Sieve Analysis of Fine and Coarse Aggregates

## Purpose

To determine the particle size distribution of fine and coarse aggregate by dry sieving.

## Significance and Use

This test is used to determine the grading of materials that are to be used as aggregates. It ensures that particle size distribution complies with applicable requirements and provides the data necessary to control the material of various aggregate products and mixtures containing aggregates. The data may also be useful in developing relationships concerning porosity and packing.

## Apparatus

- Balances or scales with a minimum accuracy of 0.5 g for coarse aggregate or 0.1 g for fine aggregate
- Sieves
- Mechanical sieve shaker (Figure 5.7 or A.1)
- Oven capable of maintaining a uniform temperature of $110 \pm 5°C$ ($230 \pm 9°F$)
- Sample splitter to reduce the quantity of the material to the size required for sieve analysis (Figure 5.12)

## Test Specimens

Thoroughly mix the aggregate sample and reduce it to an amount suitable for testing using a sample splitter or by quartering. The minimum sample size should be as follows:

| | Minimum Mass, kg |
|---|---|
| Fine aggregate with at least 95% passing 2.36-mm (No. 8) sieve | 0.1 |
| Fine aggregate with at least 85% passing 4.75-mm (No. 4) sieve | 0.5 |
| Coarse aggregate with a nominal maximum size of 9.5 mm (No. 3/8 in.) | 1 |
| Coarse aggregate of a nominal maximum size of 12.5 mm (1/2 in.) | 2 |
| Coarse aggregate of a nominal maximum size of 19.0 mm (3/4 in.) | 5 |
| Coarse aggregate of a nominal maximum size of 25.0 mm (1 in.) | 10 |
| Coarse aggregate of a nominal maximum size of 37.5 mm (1-1/2 in.) | 15 |

**FIGURE A.1** Sieve shaker for small samples of aggregates.

## Test Procedure

1. Dry the aggregate test sample to a constant weight at a temperature of $110 \pm 5°C$ then cool to room temperature.
2. Select suitable sieve sizes to furnish the information required by the specifications covering the material to be tested. Common sieves in millimeters are 37.5, 25, 19, 12.5, 9.5, 4.75, 2.36, 1.18, 0.6, 0.3, 0.15, and 0.075 mm.
3. Nest the sieves in order of decreasing size of opening and place the aggregate sample on the top sieve.
4. Agitate the sieves by hand or by mechanical apparatus for a sufficient period. The criterion for sieving time is that, after completion, not more than 1% of the residue on any individual sieve will pass that sieve during 1 min of continuous hand sieving.
5. Determine the weight of each size increment.
6. The total weight of the material after sieving should be compared to the original weight of the sample placed on the sieves. If the amounts differ by more than 0.3% based on the original dry sample weight, the results should not be used for acceptance purposes.

## Analysis and Results

1. Calculate percentages passing, total percentages retained, or percentages in various size of fractions to the nearest 0.1% on the basis of the total weight of the initial dry sample
2. Plot the grain size distribution on a semi-log graph paper (Figure A.2).
3. Plot the grain size distribution on a 0.45 power graph paper (Figure A.3).
4. Calculate the fineness modulus.

## Report

- Percentage of material retained between consecutive sieves, cumulative percentage of material retained on each sieve, or percentage of material passing each sieve. Report percentages to the nearest whole number, except if percentage passing 0.075 mm (No. 200) sieve is less than 10%, it should be reported to the nearest 0.1%.
- Grain size distribution plot.
- Fineness modulus to the nearest 0.01.

*(continued)*

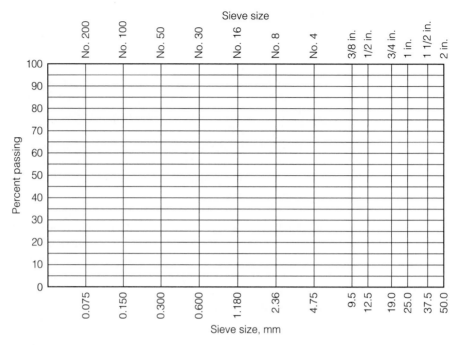

**FIGURE A.2** Semi-log aggregate gradation chart.

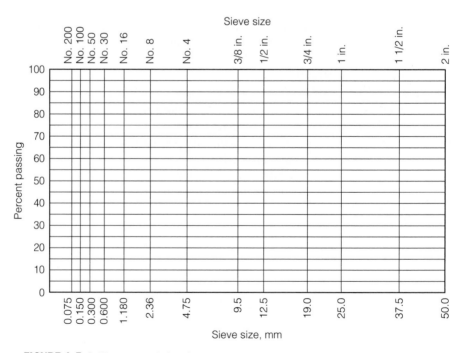

**FIGURE A.3** 0.45 power gradation chart.

# Experiment No. 7.
# Specific Gravity and Absorption of Coarse Aggregate

## ASTM Designation

ASTM C127—Specific Gravity and Absorption of Coarse Aggregate

## Purpose

To determine the specific gravity and absorption of coarse aggregate. The specific gravity may be expressed as bulk specific gravity, bulk specific gravity SSD (saturated-surface dry), or apparent specific gravity.

## Significance and Use

Bulk specific gravity is generally used for the calculation of the volume occupied by the aggregate in various mixtures containing aggregates including portland cement concrete, bituminous concrete, and other mixtures that are proportioned or analyzed on an absolute volume basis. Bulk specific gravity SSD is used if the aggregate is wet. Absorption values are used to calculate the change in the weight of aggregate due to water absorbed in the pore spaces within the constituent particles, as compared to the dry condition.

## Apparatus

- Balance accurate to 0.05% of the sample weight or 0.5 g, whichever is greater
- Wire basket 3.35 mm (No. 6) or finer mesh (Figure A.4)
- Water tank
- 4.75-mm (No. 4) sieve or other sizes as needed

## Test Specimens

- Thoroughly mix the aggregate sample and reduce it to the approximate quantity needed using an aggregate sample splitter or by quartering.
- Reject all materials passing 4.74 mm sieve by dry sieving and thoroughly washing to remove dust or other coatings from the surface.
- The minimum weight of test specimen to be used depends on the nominal maximum size as follows:

**FIGURE A.4** Wire basket and water tank used to determine bulk specific gravity and absorption of coarse aggregate.

| Nominal Maximum Size, mm | Minimum Mass, kg |
|---|---|
| 12.5 | 2 |
| 19.0 | 3 |
| 25.0 | 4 |
| 37.5 | 5 |

*(continued)*

## Test Procedure

1. Immerse the aggregate in water at room temperature for a period of $24 \pm 4$ h.
2. Remove the test specimen from water and roll it in a large absorbent cloth until all visible films of water are removed. Wipe the larger particles individually.
3. Weigh the test sample in saturated-surface dry condition and record it as $B$. Record this weight and all subsequent weights to the nearest 0.5 g or 0.05% of the sample weight, whichever is greater.
4. Place the specimen in the wire basket and determine its weight while it is submerged in water at a temperature of $23 \pm 1.7°C$ and record it as $C$. Take care to remove all entrapped air before weighing it by shaking the container while it is immersed.
5. Dry the test sample to a constant weight at a temperature of $110 \pm 5°C$ and weigh it and record this weight as $A$.

## Analysis and Results

- Bulk specific gravity $= \dfrac{A}{(B - C)}$

  where
  - $A$ = weight of oven-dry sample in air, g
  - $B$ = weight of saturated-surface dry sample in air, g
  - $C$ = weight of saturated sample in water, g

- Bulk specific gravity (SSD) $= \dfrac{B}{(B - C)}$

- Apparent specific gravity $= \dfrac{A}{(A - C)}$

- Absorption, % $= \dfrac{(B - A)}{A} \times 100$

## Report

- Bulk specific gravity
- Bulk specific gravity SSD
- Apparent specific gravity
- Absorption

# Experiment No. 8.
# Specific Gravity and Absorption of Fine Aggregate

## ASTM Designation
ASTM C128—Specific Gravity and Absorption of Fine Aggregate

## Purpose
To determine the specific gravity and absorption of fine aggregate. The specific gravity may be expressed as bulk specific gravity, bulk specific gravity SSD (saturated-surface dry), or apparent specific gravity.

## Significance and Use
Bulk specific gravity is the characteristic generally used for calculating the volume occupied by the aggregate in various mixtures including portland cement concrete, bituminous concrete, and other mixtures that are proportioned or analyzed on an absolute volume basis.

## Apparatus
- Balance or scale with a capacity of 1 kg or more, sensitive to 0.1 g or less, and accurate within 0.1% of the test load
- Pycnometer or other suitable container into which the fine aggregate test sample can be readily introduced. A volumetric flask of 500 cm$^3$ capacity with a pycnometer top is satisfactory for a 500 g test sample of most fine aggregates (see Figure A.5).
- Mold in the form of a frustum of a cone
- Tamper having a mass of $340 \pm 15$ g

**FIGURE A.5** Mold, tamper, and volumetric flask used to determine bulk specific gravity and absorption of fine aggregate.

*(continued)*

## Test Procedure

1. Measure the weight of the pycnometer filled with water to the calibration mark. Record the weight as $B$.
2. Obtain approximately 1 kg of the fine aggregate sample.
3. Dry the aggregate sample in a suitable pan to constant weight at temperature of $110 \pm 5°C$ ($230 \pm 9°F$), allow it to cool; then cover it with water, either by immersion or by the addition of at least 6% moisture to the fine aggregate, and permit to stand for $24 \pm 4$ h.
4. Decant excess water with care to avoid loss of fines, spread the sample on a flat, nonabsorbent surface exposed to a gently moving current of warm air, and stir frequently to cause homogeneous drying. If desired, mechanical aids such as tumbling or stirring may be used to help achieve the saturated-surface-dry condition. Continue this operation until test specimen approaches a free-flowing condition.
5. Hold the mold firmly on a smooth, nonabsorbent surface with the large diameter down. Place a portion of the partially dried fine aggregate loosely in the mold by filling it to overflowing and heaping additional material above the top of the mold by holding it with the cupped fingers of the hand.
6. Lightly tamp the fine aggregate into the mold with 25 light drops of the tamper. Each drop should start about 5 mm above the top of surface of the aggregate. Permit the tamper to fall freely under gravitational attraction on each drop.
7. Remove loose sand from the base and lift the mold vertically. If the surface moisture is still present, the fine aggregate will retain the molded shape. If this is the case, allow the sand to dry and repeat steps 4, 5, and 6 until the fine aggregate slumps slightly indicating that it has reached a surface-dry condition.
8. Weigh $500 \pm 10$ g of SSD sample and record the weight; record as $S$.
9. Partially fill the pycnometer with water and immediately introduce into the pycnometer the SSD aggregate weighed in step 8. Fill the pycnometer with additional water to approximately 90% of the capacity. Roll, invert, and agitate the pycnometer to eliminate all air bubbles. Fill the pycnometer with water to its calibrated capacity.
10. Determine the total weight of the pycnometer, specimen, and water and record it as $C$.
11. Carefully work all of the sample into a drying pan. Place in a $110 \pm 10°C$ oven until it dries to a constant weight. Record this weight as $A$.

## Analysis and Results

- Bulk specific gravity $= \dfrac{A}{(B + S - C)}$

  where

  $A$ = mass of oven dry specimen in air, g
  $B$ = mass of pycnometer filled with water, g
  $S$ = mass of the saturated surface-dry specimen
  $C$ = mass of pycnometer with specimen and water to the calibration mark, g

- Bulk specific gravity (SSD) $= \dfrac{S}{(B + S - C)}$

- Apparent specific gravity $= \dfrac{A}{(B + A - C)}$

- Absorption, % $= \dfrac{(S - A)}{A} \times 100$

## Report

- Bulk specific gravity
- Bulk specific gravity SSD
- Apparent specific gravity
- Absorption

# Experiment No. 9.
# Slump of Freshly Mixed Portland Cement Concrete

## ASTM Designation

ASTM C143—Slump of Portland Cement Concrete

## Purpose

To determine the slump of freshly mixed portland cement concrete both in the laboratory and in the field.

## Significance and Use

This method measures the consistency of freshly mixed portland concrete cement PCC. To some extent, this test indicates how easily concrete can be placed and compacted, or the workability of concrete. This test is used both in the laboratory and in the field for quality control.

## Apparatus

- Mold in the form of lateral surface of frustum with a top diameter of 102 mm (4 in.), bottom diameter of 203 mm (8 in.), and height of 305 mm (12 in.) (Figure 7.3)
- Tamping rod with a length of 0.6 m (24 in.), diameter of 16 mm (5/8 in.), and rounded ends

## Test Procedure

1. Mix concrete either manually or with a mechanical mixer. If a large quantity of mixed concrete exits, obtain a representative sample.
2. Dampen the mold and place it, with its larger base at the bottom, on a flat, moist, nonabsorbent rigid surface.
3. Hold the mold firmly in place by standing on the two foot pieces.
4. Immediately fill the mold in three layers, each approximately one-third of the volume of the mold. Note that one-third of the volume is equivalent to a depth of 67 mm (2-5/8 in.), whereas two-thirds of the volume is equivalent to 155 mm (6-1/8 in.).
5. Rod each layer 25 strokes using the tamping rod. Uniformly distribute the strokes over the cross section of each layer. Rod the second and top layers each throughout its depth so that the strokes penetrate the underlying layer. In filling and rodding the top layer, heap the concrete above the mold before rodding is started. If the rodding operation results in subsidence of concrete below the top edge of the mold, add additional concrete to keep an excess of concrete above the top of the mold at all times.

**FIGURE A.6** Measuring the slump of freshly mixed concrete.

**6.** After the top layer has been rodded, strike off the surface of concrete by means of a screening and rolling motion of the tamping rod.

**7.** Remove the mold immediately from the concrete by raising it up carefully without lateral or torsional motion. The slump test must be completed within 2.5 min after taking the sample.

**8.** Measure the slump by determining the vertical difference between the top of the mold and the displaced original center of the top of the specimen as shown in Figure A.6. If two consecutive tests on a sample of concrete show a falling away or a shearing off of a portion of concrete from the mass of the specimen, the concrete probably lacks the necessary plasticity and cohesiveness for the slump test to be applicable and the test results will not be valid.

## Report

- The slump value to the nearest 6 mm (1/4 in.)

# Experiment No. 10.
# Unit Weight and Yield of Freshly Mixed Concrete

## ASTM Designation

ASTM C138—Unit Weight, Yield, and Air Content (Gravimetric) of Concrete

## Purpose

To determine the unit weight, yield, cement content, and air content of freshly mixed portland cement concrete. Yield is defined as the volume of concrete produced from a mixture of known quantities of the component materials.

## Significance and Use

The unit weight value is used to calculate the volume of portland cement concrete produced from a mixture of known quantity.

## Apparatus

- Measure. Use a rigid metal watertight container with a known volume (Figure A.7). A minimum volume of the measure is required for different nominal maximum sizes of coarse aggregate. For a 25-mm (1 in.) nominal maximum aggregate size, a minimum volume measure of 0.006 m$^3$ (0.2 ft$^3$) is required.
- Balance, tamping rod, internal vibrator (optional), strike-off plate, and mallet

**FIGURE A.7** Measure used to determine unit weight of concrete.

## Test Procedure

1. Place the freshly mixed concrete in the measure in three layers of approximately equal volume.
2. Rod each layer with 25 strokes of the tamping rod when a 0.014 m³ (0.5 ft³) or smaller measure is used, otherwise use 50 strokes per layer. Distribute the strokes uniformly over the cross section of the measure. Rod the bottom layer throughout its depth, but do not forcibly strike the bottom of the measure. For the top two layers, penetrate about 25 mm (1 in.) into the underling layer.
3. Tap the sides of the measure smartly 10 to 15 times with the mallet to release trapped air bubbles.
4. An internal vibrator can be used instead of the tamping rod. In this case, the concrete is placed and vibrated in the measure in two approximately equal layers.
5. When the consolidation is complete, the measure must not contain a substantial excess or deficiency of concrete. A small quantity of concrete may be added to correct a deficiency.
6. After consolidation, strike-off the top surface of the concrete and finish it smoothly with the flat strike-off plate, using great care to leave the measure just level full.
7. After strike-off, clean all excess concrete from the exterior of the measure and determine the net weight of the concrete in the measure.

*(continued)*

## Analysis and Results

- $W = \dfrac{\text{(Net weight of concrete)}}{\text{(Volume of the measure)}}$

- $Y(\text{m}^3) = \dfrac{W_1}{W}$

  $Y(\text{ft}^3) = \dfrac{W_1}{W}$

  $Y(\text{yd}^3) = \dfrac{W_1}{(27W)}$

- $R_y = \dfrac{Y}{Y_d}$

- $N = \dfrac{N_t}{Y}$

- $A = \dfrac{(T - W)}{T} \times 100$

where

$W$ = unit weight of concrete, kg/m$^3$ (lb/ft$^3$)
$Y$ = yield = volume of concrete produced per batch, m$^3$ (yd$^3$)
$Y_f$ = yield = volume of concrete produced per batch, ft$^3$
$W_1$ = total weight of all materials batched, kg (lb)
$R_y$ = relative yield
$Y_d$ = volume of concrete that the batch was designed to produce, m$^3$ (yd$^3$)
$N$ = actual cement content, kg/m$^3$ (lb/yd$^3$)
$N_t$ = weight of cement in the batch, kg (lb)
$A$ = air content (percentage of voids) in the concrete
$T$ = theoretical unit weight of concrete computed on an air-free basis, kg/m$^3$ (lb/ft$^3$)

## Report

- The unit weight, yield, relative yield, actual cement content, and air content

# Experiment No. 11.
# Air Content of Freshly Mixed Concrete by Pressure Method

## ASTM Designation

ASTM C231—Air Content of Freshly Mixed Concrete by Pressure Method

## Purpose

To determine the air content of freshly mixed portland cement concrete by the pressure method.

## Significance and Use

Air content plays an important role in workability of freshly mixed concrete and the strength and durability of hardened concrete. The air content of freshly mixed concrete is needed for the proper proportioning of the concrete mix.

## Apparatus

- Air meter Type B consisting of measuring bowl of a capacity at least $0.006 \text{ m}^3$ $(0.2 \text{ ft}^3)$ and cover assembly fitted with air valves, air bleeder valves, petcocks, and suitable hand pump as shown in Figures 7.5 and A.8. The air meter must be frequently calibrated according to ASTM C231 procedure to ensure proper measurements.
- Miscellaneous items including trowel, tamping rod, mallet, and strike-off bar

## Test Procedure

1. Place a representative sample of the plastic concrete in the measuring bowl in three equal layers.
2. Consolidate each layer of concrete by 25 strokes of the tamping rod evenly distributed over the cross section.
3. After rodding each layer, tap the sides of the measuring bowl 10 to 15 times with the mallet to remove any voids.
4. Strike-off the top surface by sliding the bar across the top rim with a sawing motion.
5. Thoroughly clean the flanges and cover the assembly to obtain a pressure tight seal.
6. Using a rubber syringe, inject water through one petcock until water emerges from the opposite petcock.
7. Jar the meter gently until all air is expelled from the same petcock.

*(continued)*

**FIGURE A.8** Type B meter for measuring air content of concrete.

8. Pump air into the air chamber until the gauge indicator is on the initial line.
9. Open the air valve between the air chamber and the measuring bowl.
10. Tap the sides of the measuring bowl sharply, and lightly tap the pressure gauge.
11. Read the percentage of air content on the dial gauge.
12. Determine the aggregate correction factor according to ASTM C231 and subtract it from the reading obtained in step 11.

## Report

- The air content and the method used (pressure method)

# Experiment No. 12.
# Air Content of Freshly Mixed Concrete by Volumetric Method

## ASTM Designation

ASTM C173—Air Content of Freshly Mixed Concrete by Volumetric Method

## Purpose

To determine the air content of freshly mixed portland cement concrete by the volumetric method.

## Significance and Use

Air content plays an important role in the workability of freshly mixed concrete and the strength and durability of hardened concrete. The air content of freshly mixed concrete is needed for the proper proportioning of the concrete mix.

**FIGURE A.9** Air meter for measuring air content of concrete.

## Apparatus

- Air meter consisting of a bowl and a top section, as shown in Figures 7.6 and A.9
- The bowl has a diameter of 1 to 1.25 the height and a minimum capacity of 0.002 m$^3$ (0.075 ft$^3$). The capacity of the top section is 1.2 times the capacity of the bowl.
- Miscellaneous items including funnel, tamping rod, strike-off bar, measuring cup, syringe, pouring vessel, trowel, scoop, isopropyl alcohol, and mallet

## Calibration

1. The volume of the bowl must be calibrated by accurately weighing the amount of water required to fill it at room temperature and dividing this weight by the unit weight of water at the same temperature.
2. The accuracy of the graduation on the neck of the top section and the volume of the measuring cup must be calibrated according to ASTM C173.

*(continued)*

## Test Procedure

1. Fill the bowl with freshly mixed concrete in three layers of equal depth.
2. Rod each layer 25 times with the tamping rod.
3. After each layer is rodded 10 to 15 times, tap the sides of the measuring bowl with the mallet to release air bubbles.
4. After placement of the third layer of concrete, strike-off the excess concrete with the strike-off bar until the surface is flush with the top of the bowl. Wipe the flanges of the bowl clean.
5. Clamp the top section on the bowl, insert the funnel, and add water until it appears in neck. Remove the funnel and adjust the water level, using the rubber syringe, until the bottom of the meniscus is leveled with the zero mark. Attach and tighten the screw cap.
6. Invert and agitate the unit many times until the concrete settles free from the base.
7. When all the air rises to the top of the apparatus, remove the screw cap. Add, in 1-cup increments using the syringe, sufficient isopropyl alcohol to dispel the foamy mass on the surface of the water. Note that the capacity of the cup is equivalent to 1.0% of the volume of the bowl.
8. Make a direct reading of the liquid in the neck to the bottom of the meniscus to the nearest 0.1%.
9. The percent air content is calculated as the reading in step 8 plus the amount of alcohol used.

## Report

- The air content and the method used (volumetric method)

# Experiment No. 13.
# Making and Curing Concrete Cylinders and Beams

## ASTM Designation

ASTM C31—Making and Curing Concrete Test Specimens

## Purpose

To determine how to make and cure concrete cylindrical and beam specimens.

## Significance and Use

This practice provides standardized requirements for making and curing portland cement concrete test specimens. Specimens can be used to determine strength for mix design, quality control, and quality assurance.

## Apparatus

- Cylindrical molds made of steel or another nonabsorbent and nonreactive material. The standard specimen size used to determine the compressive strength of concrete is 152 mm (6 in.) diameter by 304 mm (12 in.) high for a maximum aggregate size up to 50 mm (2 in.). Smaller-sized specimens such as 102 mm (4 in.) diameter by 203 mm (8 in.) high are sometimes used, but they are not ASTM standards.
- Beam molds made of steel or another nonabsorbent nonreactive material. Several mold dimensions can be used to make beam specimens with a square cross section and a span three times the depth. The standard ASTM inside mold dimensions are 152 × 152 mm (6 × 6 in.) in cross section and a length of not less than 508 mm (20 in.), for a maximum aggregate size up to 50 mm (2 in.).
- Tamping rod with a length of 0.6 m (24 in.), diameter of 16 mm (5/8 in.), and rounded ends
- Moist cabinet or room with not less than 95% relative humidity and 23 ± 1.7°C (73 ± 3°F) temperature or a large container filled with lime saturated water for curing
- Miscellaneous items including vibrator (optional), scoop, and trowel

*(continued)*

## Test Procedure

1. Weigh the required amount of coarse aggregate, fine aggregate, portland cement, and water.
2. Mix the materials in the mixer for 3 to 5 min. If an admixture is used, it should be mixed with water before being added to the other materials.
3. Check slump, air content, and temperature of concrete.
4. For cylindrical specimens place concrete into the mold using a scoop or trowel. Fill the cylinder in three equal layers, and rod each layer 25 times. Tap the outside of the cylinder 10 to 15 times after each layer is rodded. Strike-off the top and smooth the surface. Vibrators can also be used to consolidate the concrete instead of rodding. Vibration is optional if the slump is between 25 mm to 75 mm (1 in. to 3 in.) and is required if the slump is less than 25 mm (1 in.).
5. For beam specimens, grease the sides of the mold and fill the molds with concrete in two layers. Consolidate the concrete by either tamping each layer 60 times until uniformly distributed throughout or by vibrating. After consolidation, finish the surface by striking off the surface and smoothing.
6. Cover the mold with wet cloth to prevent evaporation.
7. Remove the molds after 16 h to 32 h.
8. Cure the specimen in a moist cabinet or room at a relative humidity of not less than 95% and a temperature of $23 \pm 1.7°C$ ($73 \pm 3°F$) or by submersion in lime saturated water at the same temperature.

## Precautions

1. Segregation must be avoided. Over vibration may cause segregation.
2. In placing the final layer the operator should attempt to add an amount of concrete that will exactly fill the mold after compaction. Do not add nonrepresentative concrete to an under-filled mold.
3. Avoid overfilling by more than 6 mm (1/4 in.).

## Report

- Record mix design weights, slump, temperature of the mix, and air content
- Include this information with the report on the strength of the concrete

# Experiment No. 14.
# Capping Cylindrical Concrete Specimens with Sulfur or Capping Compound

## ASTM Designation

ASTM C617—Capping Cylindrical Concrete Specimens

## Purpose

To cap hardened portland cement concrete cylinders and drilled concrete cores with sulfur mortar or other capping compounds to prepare the specimen for compressive strength testing.

## Significance and Use

This procedure provides plane surfaces perpendicular to the specimen axis on the ends of concrete cylinders before performing the compression test.

## Apparatus

**FIGURE A.10**
Alignment device for capping concrete cylinders.

- Alignment device consisting of a frame with guide bars and a cup as shown in Figure A.10. The size of the alignment device should match the specimen size.
- Melting pot used for melting sulfur mortars or capping compound equipped with automatic temperature control. The melting pot should be used either outdoors or under an exhaust hood. Heating over an open flame is dangerous because the mixture may ignite if overheated.

## Capping Procedure

1. Prepare the sulfur mortar or capping compound by heating to about 130°C (265°F). Use a metal thermometer to check the temperature. Make sure to empty any old mortar and use fresh mortar to avoid the loss of strength due to successive heating. The fresh sulfur mortar must be dry when it is placed in the pot because dampness may cause foaming.
2. Warm the capping cup or device slightly before use to slow the rate of hardening and to permit the production of thin caps.
3. Oil the capping cup lightly and stir the molten sulfur mortar or the capping compound immediately prior to pouring in the cup. Make sure the ends of moist-cured specimens are dry enough at the time of capping so there will be no steam or foam pockets.
4. Hold the concrete cylinder with two hands and push it against the guide bars of the capping device. Carefully lower the specimen until it rests in the cup. This step must be completed quickly before the sulfur or capping compound solidifies. The thickness of the cap should be about 3 mm (1/8 in.) and not more than 8 mm (5/16 in.) in any part.

## Experiment No. 15.
## Compressive Strength of Cylindrical Concrete Specimens

### ASTM Designation

ASTM C39—Compressive Strength of Cylindrical Concrete Specimens

### Purpose

To determine the compressive strength of cylindrical PCC specimens, such as molded cylinders and drilled cores.

### Significance and Use

This test provides the compressive strength of concrete, which is used universally as a measure of concrete quality.

### Apparatus

Loading machine with two hardened steel breaking blocks. The upper block is spherically seated, and the bottom block is solid surface (see Figure A.11).

**FIGURE A.11** Concrete cylindrical specimen being tested for compressive strength.

## Test Specimens

- The standard specimen size used to determine the compressive strength of concrete is 152 mm (6 in.) diameter by 304 mm (12 in.) high for a maximum aggregate size up to 50 mm (2 in.). Smaller-sized specimens such as 102 mm (4 in.) diameter by 203 mm (8 in.) high are sometimes used, but they are not ASTM standardized.
- Conduct the compression test on the moist-cured specimens directly after removing them from the curing room. Test specimens must be moist when tested.
- Neither end of compressive test specimen shall depart from perpendicularity by more than 0.5°, approximately 3 mm in 0.3 m (1/8 in. in 12 in.)
- If the ends of the specimen are not plane within 0.05 mm (0.002 in.), they should be capped with sulfur or capping compound. Neoprene caps may be used, but they are not ASTM standards.
- Specimen age, at time of testing, should be 24 hours ± 0.5 hours, 3 days ± 2 hours, 7 days ± 6 hours, 28 days ± 20 hours, or 90 days ± 2 days.

## Test Procedure

1. Measure the diameter of the test specimen to the nearest 0.25 mm (0.01 in.) by averaging two diameters measured at right angles to each other at the middle height of the specimen.
2. Adjust the bearing blocks into position.
3. Clean the faces of the bearing blocks and the specimen.
4. Carefully align the axis of the specimen with the center of the thrust of the spherically seated block.
5. Apply the load continuously and without shock. For screw type machines, use a rate of loading of 1.25 mm/min (0.05 in./min). For hydraulically operated machines, apply the load at a constant rate within the range of 138 kPa/s to 335 kPa/s (20 psi/s to 50 psi/s). During the first half of the anticipated loading phase a higher rate of loading is permitted. No adjustment in the control of the testing machine should be made while the specimen is yielding rapidly, immediately before failure.
6. Continue applying the load until the specimen fails.
7. Record the maximum load carried by the specimen during the test.
8. Note the type of failure and the appearance of concrete (see Figure A.12).

    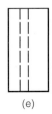

(a)           (b)           (c)           (d)           (e)

**FIGURE A.12** Types of fracture of concrete cylinders (a) cone, (b) cone and split, (c) cone and shear, (d) shear, and (e) columnar. Copyright ASTM. Reprinted with permission.

*(continued)*

## Analysis and Results

- Calculate the compressive strength as

$$f'_c = \frac{P_{max}}{A}$$

where

$f'_c$ = compressive strength, MPa (psi)
$P_{max}$ = maximum applied load, N (lb)
$A$ = cross-sectional area, mm$^2$ (in.$^2$)

## Report

- Specimen identification number
- Diameter (and length, if outside the range of 1.8 to 2.2 times the diameter)
- Cross-sectional area
- Maximum load
- Compressive strength calculated to the nearest 0.07 MPa (10 psi)
- Type of failure, if other than the usual one
- Defects in either specimen or caps
- Age of specimen

# Experiment No. 16.
# Flexural Strength of Concrete

## ASTM Designation

ASTM C78—Flexural Strength of Concrete (Using Simple Beam with Third-Point Loading)

## Purpose

To determine the flexural strength of portland cement concrete by using a simple beam with third-point loading.

## Significance and Use

The flexural strength of concrete is a measure of concrete quality.

## Apparatus

- Loading machine capable of applying loads at a uniform rate
- Loading device capable of applying load configuration as shown in Figure 7.13. Forces applied to the beam shall be perpendicular to the face of the specimen and applied without eccentricity.

## Test Specimens

- The standard ASTM specimen dimensions are 152 mm × 152 mm (6 in. × 6 in.) in cross section and a length of not less than 508 mm (20 in.) for a maximum aggregate size up to 50 mm (2 in.).
- Sides of the specimen should be at right angles to its top and bottom. All surfaces in contact with load-applying and support blocks should be smooth and free of scars, indentations, holes, or inscribed identifications.

## Test Procedure

1. Turn the test specimen on its side, with respect to its position as molded, and center it on the bearing blocks.
2. Center the loading system in relation to the applied force. Bring the load-applying blocks in contact with the surface of the specimen at the third points between the supports.
3. If full contact is not obtained at no load between the specimen and the load-applying blocks and the supports so that there is a 25 mm (1 in.) or larger gap in excess of 0.1 mm (0.004 in.), grind or cap the contact surfaces of the specimen, or shim with leather strips.
4. Apply the load rapidly up to approximately 50% of the breaking load. Thereafter, apply the load continuously at a rate that constantly increases the extreme fiber stress between 860 kPa and 1210 kPa (125 psi and 175 psi)/min until rupture occurs.

*(continued)*

## Analysis and Results

- Take three measurements across each dimension (one at each edge and at the center) to the nearest 1.3 mm (0.05 in.) to determine the average width, average depth, and line of fracture location of the specimens at the section of fracture.
- If the fracture initiates in the tension surface within the middle third of the span length, calculate the modulus of rupture as follows:

$$R = \frac{(PL)}{(bd^2)}$$

where

  $R$ = modulus of rupture, MPa (psi)
  $P$ = maximum load, N (lb)
  $L$ = span length, mm (in.)
  $b$ = average width, mm (in.)
  $d$ = average depth, mm (in.)

- If the fracture occurs in the tension surface outside the middle third of the span length, by not more than 5% of the span length, calculate the modulus of rupture as follows:

$$R = \frac{(3Pa)}{(bd^2)}$$

where

  $a$ = average distance between line of fracture and the nearest support on the tension surface of the beam in millimeters (inches)

- If the fracture occurs in the tension surface outside the middle third of the span length, by more than 5% of the span length, discard the results of the test.

## Report

- Specimen identification number
- Average width
- Average depth
- Span length
- Maximum applied load
- Modulus of rupture to the nearest 0.03 MPa (5 psi)
- Curing history and apparent moisture condition at time of testing
- If specimens were capped, ground, or if leather shims were used
- Defects in specimens
- Age of specimens

# Experiment No. 17.
# Rebound Number of Hardened Concrete

## ASTM Designation

ASTM C805—Rebound Number of Hardened Concrete

## Purpose

To determine the rebound number of hardened portland cement concrete.

## Significance and Use

The rebound number may be used to assess the uniformity and strength of concrete. It can also be used to determine when forms and shoring may be removed.

## Apparatus

Rebound hammer (Figure 7.14)

## Test Procedure

1. Grind and clean the concrete surface by rubbing with the abrasive stone that accompanies the rebound device.
2. Firmly hold the instrument in a position that allows the plunger to strike perpendicularly to the test surface.
3. Gradually increase the pressure on the plunger until the hammer impacts.
4. After impact, record the rebound number to two significant digits.
5. Take 10 readings from each test area.

## Test Conditions

1. No two impact tests shall be closer together than 25 mm (1 in.).
2. Discard the reading if the impact crushes or breaks through a near-surface air void.
3. Discard readings differing from the average of 10 readings by more than 7 units and determine the average of the remaining readings.
4. If more than two readings differ from the average by 7 units or more, discard all readings.
5. The rebound hammer should be periodically serviced and verified using metal anvils, according to manufacturer recommendations.

## Report

- Structure identification, location, and curing condition
- Average rebound number
- Position of the hammer during the test, such as downward, upward, or at a specific angle
- Estimated compressive strength using available correlations such as those obtained from the manufacturer. It should be noted, however, that the rebound hammer test is not intended as an alternative for strength determination of concrete.

# Experiment No. 18.
# Penetration Resistance of Hardened Concrete

## ASTM Designation

ASTM C803—Penetration Resistance of Hardened Concrete

## Purpose

To assess the uniformity of hardened portland cement concrete and indicate its in-place strength.

## Significance and Use

The penetration resistance may be used to assess the uniformity and strength of concrete.

## Apparatus

The penetration resistance (Windsor Probe) apparatus (Figure 7.15) consists of:

- Driver unit
- Probe made of hardened steel alloy
- Measuring instrument
- Positioning device

## Test Procedure

1. Place the positioning device on the concrete surface at the location to be tested. Two positioning devices are available, a single positioning device and a triangular device.
2. Mount a probe in the driver unit.
3. Position the driver and probe in the positioning device.
4. Fire and drive the probe into the concrete. In case of the triangular device, repeat for each position.
5. In the case of single positioning device place the measuring base plate over the probe and measure the length of the probe not embedded in the concrete with the measuring instrument. In the case of triangular device, place the triangular plate on the three probes and position the measuring instrument in the hole at the center of the triangular plate and read the average of the exposed length.

## Report

- Structure identification, location, and curing condition
- The average exposed length of the probe
- Estimated compressive strength using available correlations such as those obtained from the manufacturer. Note, however, that the penetration resistance test is not intended as an alternative for strength determination of concrete.

# Experiment No. 19.
# Testing of Concrete Masonry Units

## ASTM Designation

ASTM C140—Sampling and Testing Concrete Masonry Units

## Purpose

To test concrete masonry units for compressive strength, absorption, moisture content, and density.

## Significance and Use

The test produces compressive strength, absorption, moisture content, and density data for the control and specification of concrete masonry units. These data are important for the safety and proper performance of masonry structures.

## Apparatus

- Testing machine
- Steel-bearing blocks and plates
- Balance

## Test Procedure for Compressive Strength

1. Three representative units are needed for testing within 72 h after delivery to the laboratory, during which time they are stored continuously in air at a temperature of $24 \pm 8°C$ ($75 \pm 15°F$) and a relative humidity of less than 80%.
2. The test is performed on either a full-sized unit or a part of a unit prepared by saw cutting. A part of a unit is used if the capacity or size of the testing machine does not allow testing a full-sized unit.
3. Measure the length, width, and height of the specimen.
4. Cap the bearing surfaces of the unit using either sulfur and granular materials or gypsum plaster.
5. Position the test specimen with its centroid aligned vertically with the center of thrust of the spherically seated steel-bearing block of the testing machine.
6. Apply the load up to one-half the expected maximum load at any convenient rate, after which apply the load at a uniform rate of travel of the moving head so that the test is completed between 1 min and 2 min.
7. Record the maximum compressive load in newtons (pounds) as $P_{max}$.

*(continued)*

## Test Procedure for Absorption

1. Three representative full-sized units are needed for absorption testing.
2. Weigh the specimen immediately after sampling and record the weight as the received weight $(W_r)$.
3. Immerse the specimen in water at a temperature of 15°C to 26°C (60°F to 80°F) for 24 h.
4. Weigh the specimen while it is suspended by a metal wire and completely submerged in water; record the weight as the immersed weight $(W_i)$.
5. Remove the specimen from the water and allow it to drain for 1 min by placing it on a 9.5-mm (3/8 in.) or coarse wire mesh and removing visible surface water with a damp cloth. Weigh the specimen and record the weight as the saturated weight $(W_s)$.
6. Dry the specimen in a ventilated oven at 100°C to 115°C (212°F to 239°F) for not less than 24 h and until two successive weights at intervals of 2 h show a difference of not greater than 0.2%. Record the weight as the oven-dried weight $(W_d)$.

## Analysis and Results

- Gross area compressive strength, MPa (psi) $= \dfrac{P_{max}}{A_g}$

where

$P_{max}$ = maximum compressive load, N (lb)
$A_g$   = gross area, mm$^2$ (in.$^2$) = $L \times W$
$L$    = average length, mm (in.)
$W$   = average width, mm (in.)

- Net area compressive strength, MPa (psi) $= \dfrac{P_{max}}{A_n}$

where

$A_n$  = average net area, mm$^2$ = $\dfrac{V_n}{H}$

$A_n$  = average net area, in.$^2$ = $\dfrac{(V_n \times 1728)}{H}$

$V_n$  = net volume, mm$^3$ = $(W_s - W_i) \times 10^6$

$V_n$  = net volume, ft$^3$ = $\dfrac{(W_s - W_i)}{62.4}$

$H$    = average height, mm (in.)

$W_s$  = saturated weight of unit, kg (lb)

$W_i$  = immersed weight of unit, kg (lb)

- Absorption, kg/m$^3$ $= \dfrac{(W_s - W_d)}{(W_s - W_i)} \times 1000$

- Absorption, lb/ft$^3$ $= \dfrac{(W_s - W_d)}{(W_s - W_i)} \times 62.4$

- Absorption, % $= \dfrac{(W_s - W_d)}{W_d} \times 100$

- Moisture content, % of total absorption $= \dfrac{(W_r - W_d)}{(W_s - W_d)} \times 100$

- Density, kg/m$^3$ $= \dfrac{W_d}{(W_s - W_i)} \times 1000$

- Density, lb/ft$^3$ $= \dfrac{W_d}{(W_s - W_i)} \times 62.4$

where
$W_d$ = oven-dry weight, kg (lb)
$W_r$ = received weight, kg (lb)

## Report

- Gross area and net area compressive strengths to the nearest 70 kPa (10 psi) for each specimen and the average for three specimens
- Absorption, moisture content as a percent of total absorption, and density

# Experiment No. 20.
# Viscosity of Asphalt Binder by Rotational Viscometer

## ASTM Designation

ASTM D4402—Viscosity Determinations of Unfilled Asphalts Using the Brookfield Thermosel Apparatus

## Purpose

To determine the apparent viscosity of asphalt binder from 38°C to 260°C (100°F to 500°F) using the Brookfield Thermosel apparatus.

## Significance and Use

The viscosity is needed to ensure proper handling of the asphalt binder and for quality control and quality assurance. It is also used to determine the mixing and compaction temperatures of asphalt concrete. This test is used for the Superpave performance grading of asphalt binders.

## Apparatus

- Rotational viscometer (see Figure 9.8)
- Spindles
- Thermosel system consisting of thermo-container and sample chamber, SCR controller and probe, and graph-plotting equipment

## Test Procedure

1. Turn on the Thermosel power and set the proportional temperature controller to the desired test temperature.
2. Wait 1-1/2 hours (or until equilibrium temperature is obtained) with the spindle in the chamber (check control lamp).
3. Remove sample holder and add the volume of sample specified for the spindle, approximately 8 mL to 10 mL. Exercise caution to avoid sample overheating and to avoid ignition of samples with a low flash point. Do not overfill the sample container. The sample volume is critical to meet the system calibration standard.
4. Thoroughly stir filled asphalt container to obtain a representative sample.
5. The liquid level should intersect the spindle shaft at a point approximately 3 mm (1/8 in.) above the upper conical body–spindle shaft interface.
6. Using the extracting tool, put the loaded chamber back into the thermo-container.
7. Lower the viscometer and align the thermo-container.
8. Insert the selected spindle into the liquid in chamber, and couple it to the viscometer. Either spindle number 27 or 20 are used for asphalt binders.
9. Allow the asphalt to come to the equilibrium temperature (about 15 min).
10. Start the viscometer at a 20 rpm setting.
11. Record three readings 60 s apart at each test temperature.
12. Multiply the viscosity factor by the rotational viscometer readings to obtain viscosity in centipoises.

## Report

- Viscosity at each test temperature
- Spindle number and rotational speed

# Experiment No. 21.
# Dynamic Shear Rheometer Test of Asphalt Binder

## AASHTO Designation

AASHTO TP5—Determining the Rheological Properties of Asphalt Binder for Specification Purposes Using a Dynamic Shear Rheometer (DSR)

## Purpose

To determine the complex shear modulus ($G^*$) and phase angle ($\delta$) of asphalt binders using the dynamic shear rheometer.

## Significance and Use

The complex shear modulus is an indicator of the stiffness resistance of asphalt binder to deformation under load. The complex shear modulus and phase angle define the resistance to shear deformation of the asphalt binder in the viscoelastic region. This test is used for the Superpave performance grading of asphalt binders.

## Apparatus

- Dynamic shear rheometer (DSR) as shown in Figure 9.9
- Test plates
- Temperature controller
- Loading device
- Control and data acquisition system
- Miscellaneous items such as specimen mold, specimen trimmer, environmental chamber, reference thermal detector, and calibrated temperature detector

## Test Procedure

1. Heat the asphalt binder until fluid enough to pour the required specimens.
2. Carefully clean and dry the surfaces of the test plates so that the specimen uniformly adheres to both plates. Bring the chamber to approximately 45°C so that the plates are preheated prior to the mounting of the test specimen.
3. Place the asphalt binder sample in the DSR using one of the following methods:
   - Remove the removable plate and, while holding the sample container approximately 15 mm above the test plate surface, pour the asphalt binder at the center of the upper test plate continuously until it covers the entire plate, except for an approximate 2-mm-wide strip at the perimeter. Wait several minutes for the specimen to stiffen and then mount the test plate in the rheometer for testing.
   - Pour the hot asphalt into a silicon rubber mold that will form a pellet with a diameter approximately equal to the diameter of the upper test plate and a height approximately equal to 1.5 times the width of the test gap Allow the silicon rubber to cool to room temperature. Remove the specimen from the mold and center the pellet on the lower plate of the DSR.
4. Move the test plates together to squeeze the asphalt mass between the two plates. Move the plates until the gap between the plates equals the testing gap plus 0.05 mm. Trim the specimen by moving a heated trimming tool around the upper and lower plate perimeters while trimming excess asphalt.
5. After trimming is completed, decrease the gap by 0.05 mm. The sample thickness should now equal the desired test gap.
6. Bring the specimen to the test temperature ± 0.1°C. Start the test after the temperature has remained at the desired temperature ± 0.1°C for at least 10 min. The test temperature can be selected from the specifications in Table 9.3.
7. When the temperature has been reached, condition the specimen by applying the required shear strain for 10 cycles at a radial frequency of 10 rad/s. Shear strain values vary from 1% to 12% depending on the stiffness of the binder being tested. High strain values are used for relatively soft binders tested at high temperatures, whereas low strain values are used for hard binders. The rheometer measures the torque required to achieve the set shear strain and maintains this as the maximum torque during the test.
8. The data acquisition system automatically calculates $G^*$ and $\delta$ from test data acquired when properly activated.

## Report

- Identification and description of the material tested
- Test temperature and sample dimensions including thickness
- Stress level
- $G^*$ and $\delta$ values

# Experiment No. 22.
# Penetration Test of Asphalt Cement

## ASTM Designation

ASTM D5—Penetration of Bituminous Materials

## Purpose

To determine the penetration of semi-solid and solid bituminous materials.

## Significance and Use

The penetration test is used as a measure of consistency. High values of penetration indicate soft consistency.

## Apparatus

- Penetration apparatus and needle (Figures 9.12 and A.13)
- Sample container, water bath, transfer dish, timing device for hand-operated penetrometers, and thermometers

**FIGURE A.13** Apparatus for penetration test of asphalt binder.

## Test Procedure

1. Heat the asphalt binder sample until it has become fluid enough to pour.
2. Pour the sample into the sample container and let it cool for at least 1 h.
3. Place the sample together with the transfer dish in the water bath at a temperature of 25°C (77°F) for 1 h to 2 h.
4. Clean and dry the needle with a clean cloth, and insert the needle in the penetrometer. Unless otherwise specified place 50-g mass above the needle, making the total moving load 100 g.
5. Place the sample container in the transfer dish, cover the container completely with water from the constant temperature bath, and place the transfer dish on the stand of the penetrometer.
6. Position the needle by slowly lowering it until its tip just makes contact with the surface of the sample. This is accomplished by bringing the actual needle tip into contact with its image reflected by the surface of the sample from a properly placed source of light.
7. Quickly release the needle holder for the specified period of time (5 s) and adjust the instrument to measure the distance penetrated in tenths of a millimeter.
8. Make at least three determinations at points on the surface of the sample not less than 10 mm from the side of the container and not less than 10 mm apart.

## Report

- The average of the three penetration values to the nearest whole unit

# Experiment No. 23.
# Absolute Viscosity Test of Asphalt

## ASTM Designation
ASTM D2171—Viscosity of Asphalts by Vacuum Capillary Viscometer

## Purpose
To determine the absolute viscosity of asphalt by vacuum capillary viscometer at 60°C (140°F).

## Significance and Use
The viscosity at 60°C (140°F) characterizes flow behavior and may be used for specification requirements for cutbacks and asphalt cements.

## Apparatus
- Viscometers such as Cannon-Manning vacuum viscometer (Figure 9.13) or modified Koppers vacuum viscometer
- Bath with provisions for visibility of the viscometer and the thermometer
- Thermometers, vacuum system, and timing device

## Test Procedure
1. Maintain the bath at a temperature of 60°C (140°F).
2. Select a clean, dry viscometer that will give a flow time greater than 60 s, and preheat to 135°C (275°F).
3. Charge the viscometer by pouring the prepared asphalt sample to the fill line.
4. Place the charged viscometer in an oven or bath maintained at 135°C for a period of 10 min, to allow large air bubbles to escape.
5. Remove the viscometer from the oven or bath, and within 5 min, insert the viscometer in a holder, and position the viscometer vertically in the bath so that the uppermost timing mark is at least 20 mm below the surface of the bath liquid.
6. Establish a 300 mm Hg vacuum in the vacuum system, and connect the vacuum system to the viscometer.
7. After the viscometer has been in the bath for 30 min, start the flow of asphalt in the viscometer by opening the toggle valve or stopcock in the line leading to the vacuum system.
8. Measure to within 0.1 s the time required for the leading edge of the meniscus to pass between successive pairs of timing marks. Report the first flow time that exceeds 60 s between a pair of timing marks, and identify the pair of timing marks.

## Analysis and Results

- Select the calibration factor that corresponds to the pair of timing marks. Calculate and report the viscosity to three significant figures using the following equation:

$$P = Kt$$

where

$P$ = absolute viscosity, Poises
$K$ = selected calibration factor, Poises/s
$t$  = flow time, s

## Report

- Absolute viscosity
- Test temperature and vacuum

# Experiment No. 24.
# Preparation of Asphalt Concrete Specimens Using the Superpave Gyratory Compactor

## ASTM Designation

None.

## Purpose

To prepare asphalt concrete specimens to densities achieved under actual pavement climate and loading conditions.

## Significance and Use

The Superpave gyratory compactor is capable of accommodating large aggregates. Furthermore, this device affords a measure of compaction ability so that potential tender mixture behavior and similar compaction problems could be identified. This method of compaction is used for the Superpave volumetric mix design of asphalt concrete mixture.

**FIGURE A.14**  Superpave gyratory compactor.

## Apparatus

- Reaction frame, rotating base, and motor (Figures 9.18 and A.14)
- Loading system, loading ram, and pressure gauge
- Height measuring and recording system
- Mold and base plate

*(continued)*

## Procedure

1. Specimens must be mixed and compacted under equiviscous temperature conditions of $0.170 \pm 0.02$ Pa · s and $0.280 \pm 0.03$ Pa · s, respectively. Mixing is accomplished by a mechanical mixer.

2. After mixing, loose test specimens are subjected to 2 h of short-term aging in a forced draft oven at a temperature of 135°C. During this period, loose mix specimens are required to be spread into a thickness resulting in 21 kg to 22 kg per cubic meter. The sample is stirred every hour to ensure uniform aging.

3. Place the compaction molds and base plates in an oven at 135°C for at least 30 min to 45 min prior to use.

4. If specimens are to be used for volumetric determinations only, use sufficient mix to arrive at a specimen 150 mm in diameter by approximately 115 mm high. This requires approximately 4500 g of aggregates. If needed, specimens with other heights can also be prepared for further performance testing.

5. Turn on the power to the compactor. Set the vertical pressure to 600 kPa.

6. Set the gyration counter to zero, and set it to stop when the desired number of gyrations is achieved. Three gyrations are of interest: design number of gyrations $(N_d)$, initial number of gyrations $(N_i)$, and maximum number of gyrations $(N_m)$. The design number of gyrations is a function of the climate in which the mix will be placed and the traffic level it will withstand.

7. After the base plate is placed, place a paper disk on top of the plate and charge the mold with the short-term aged mix in a single lift. The top of the uncompacted specimen should be slightly rounded. Place a paper disk on top of the mixture.

8. Place the mold in the compactor and center it under the ram. Lower the ram until it contacts the mixture and the resisting pressure is 600 kPa.

9. Apply the angle of gyration of 1.25° and begin compaction.

10. When maximum number of gyrations $(N_m)$ has been reached, the compactor should automatically cease. After the angle and pressure are released, remove the mold containing the compacted specimen.

11. Print the results of specimen height versus number of gyrations.

12. After a suitable cooling period, extrude the specimen from the mold.

13. Measure the bulk specific gravity of the test specimens according to ASTM D2726 procedure.

## Analysis and Results

- Compute the estimated bulk specific gravity for each desired gyration as follows:

$$\text{Estimated bulk specific gravity} = \frac{\text{Net weight of the specimen}}{\dfrac{\pi d^2 h}{4}}$$

where
  - $d$ = mold diameter (150 mm)
  - $h$ = specimen height corresponding to the desired number of gyrations

- Compute the correction factor $C$ as follows:

$$C = \frac{\text{Measured bulk specific gravity}}{\text{Estimated bulk specific gravity at } N_m}.$$

- Compute the corrected bulk specific gravity by multiplying each estimated bulk specific gravity by the correction factor.
- Compute the percentage of the corrected bulk specific gravity relative to the maximum theoretical specific gravity ($\%G_{mm}$).
- Draw a graph of the logarithm of the number of gyrations versus the $\%G_{mm}$ (densification curve) for each specimen.

## Report

- Mixture ingredients, source, and relevant information
- Densification table as shown in Table 9.8
- Densification curve as shown in Figure 9.23

# Experiment No. 25.
# Preparation of Asphalt Concrete Specimens Using the Marshall Compactor

## ASTM Designation

ASTM D1559—Resistance to Plastic Flow of Bituminous Mixtures Using Marshall Apparatus

## Purpose

To prepare asphaltic concrete specimens using the Marshall hammer.

## Significance and Use

This method is used to design the mix using the Marshall procedure and measure its properties.

## Apparatus

- Either mechanical or manual compaction hammer with 4.5-kg (10 lb) weight and 0.48-m (18 in.) drop height can be used, as shown in Figure 9.19
- Molds with 102 mm (4 in.) inside diameter and 75 mm (3 in.) high, base plates, and collars
- Compaction pedestal, specimen extruder, and miscellaneous items such as mold holder, spatula, pans, and oven

## Procedure

1. Determine the mixing and compaction temperatures so that the kinematic viscosities of the binder are $170 \pm 20$ cSt and $280 \pm 30$ cSt, respectively.
2. Separate all the required sizes of aggregates and oven dry.
3. Weigh 1200 g batches so that gradation would satisfy the midpoint of the specification band. Either your state's specifications or ASTM D3515 can be followed.
4. Place both asphalt binder and aggregate in the oven until they reach the mixing temperature (approximately 150°C or 300°F).
5. Add the asphalt to the aggregate in the specified amount.
6. Using the mechanical mixer, or manually, mix the aggregates and asphalt thoroughly.
7. Place a filter paper inside the mold. Then, place the entire batch of asphaltic concrete in the heated mold, and spade with a heated spatula 15 times around the perimeter and 10 times in the middle.
8. Put in place a collar and a filter paper and put the mold on the pedestal. Clamp the mold with the mold holder, and apply the required number of blows. The typical number of blows on each side is either 50 or 75, depending on the expected traffic volume on the road where the mix is intended to be used. Invert the mold and apply the same number of blows on the other face.
9. After cooling to room temperature, extrude the specimen. Cooling can be accelerated by placing the mold and the specimen in a plastic bag and subjecting them to cold water.

## Report

- Mixture ingredients, source, and relevant information
- Number of blows on each side of the specimen

# Experiment No. 26.
# Bulk Specific Gravity of Compacted Bituminious Mixtures

## ASTM Designation
ASTM D2726—Bulk Specific Gravity of Compacted Bituminous Mixtures

## Purpose
To determine the bulk specific gravity of compacted asphalt mixture specimens.

## Significance and Use
The results of this test are used for voids analysis of the compacted asphalt mix.

## Test Specimens
Laboratory molded bituminous mixtures or cores drilled from bituminous pavements can be used.

## Apparatus
- Balance equipped with suitable suspension and holder to permit weighing the specimen while it is suspended from the balance
- Water bath

## Test Procedure
1. Weigh the specimen in air and record it as $A$.
2. Immerse the specimen in water at $25 \pm 1°C$ ($77 \pm 2°F$) while it is suspended from the balance for 3 min to 5 min and record the immersed weight as $C$.
3. Remove the specimen from water, surface dry by blotting with a damp towel, determine the surface dry weight, and record it as $B$.

## Analysis and Results
- Calculate the bulk specific gravity as

$$\text{Bulk specific gravity} = \frac{A}{(B-C)}$$

where
$A$ = mass of specimen in air, g
$B$ = mass of surface dry specimen, g
$C$ = mass of specimens in water, g

## Report
- Report the value of specific gravity up to three decimal places

# Experiment No. 27.
# Marshall Stability and Flow of Asphalt Concrete

## ASTM Designation

ASTM D1559—Resistance to Plastic Flow of Bituminous Mixtures Using Marshall Apparatus

## Purpose

To determine the Marshall stability and flow values of asphalt concrete.

## Significance and Use

This test method is used in the laboratory mix design of bituminous mixtures according to the Marshall procedure. The test results are also used to characterize asphalt mixtures.

## Apparatus

- Testing machine producing a uniform vertical movement of 50.8 mm/min (2 in./min) as shown in Figure 9.27
- Breaking heads having an inside radius of curvature of 50.8 mm (2 in.)
- Load cell or ring dynamometer, strip chart recorder or flow meter, water bath, and rubber gloves

*(continued)*

## Test Procedure

1. Bring the specimen prepared in experiment number 25 to a temperature of 60°C (140°F) by immersing it in a water bath 30 min to 40 min or by placing it in the oven for 2 h.
2. Remove the specimen from the water bath (or oven) and place it in the lower segment of the breaking head. Place the upper segment of the breaking head on the specimen, and place the complete assembly in position on the testing machine.
3. Prepare the strip chart recorder or place the flowmeter (where used) in position over one of the guide rods, and adjust the flowmeter to zero while holding the sleeve firmly against the upper segment of the breaking head.
4. Apply the load to the specimen by means of the constant rate of movement of 50.8 mm/min (2 in./min) until the maximum load is reached and the load decreases. The elapsed time for the test from removal of the test specimen from the water bath to the maximum load determination should not exceed 30 s.
5. From the chart recorder, record the Marshall stability (maximum load) and the Marshall flow (deformation when the maximum load begins to decrease in units of 0.25 mm or hundredths of an inch). In some machines the maximum load and the flow values are read from the ring dynamometer and the flowmeter, respectively.
6. If the specimen height is other than 63.5 mm (2.5 in.), multiply the stability value by a correction factor (ASTM D1559).

## Report

- Specimen identification and type (laboratory prepared or core)
- Average Marshall stability of at least three specimens, corrected when required, kN (lb)
- Average Marshall flow of at least three specimens

# Experiment No. 28. Creep Test of Asphalt Concrete

## ASTM Designation

None.

## Purpose

To determine creep compliance of compacted asphalt mixture.

## Significance and Use

Creep compliance can be used to predict the rutting potential of the asphalt mixture.

## Apparatus

- Stress controlled loading machine (Figure A.15)
- Temperature controlled chamber, if the test is required to be performed at a temperature other than room temperature
- Axial deformation measuring devices, such as LVDTs
- Data acquisition system, such as computer or chart recorder

**FIGURE A.15** Creep test apparatus.

## Test Specimen

Use an asphalt concrete laboratory molded specimen or core. Several specimen sizes can be used such as 102 mm (4 in.) diameter by 63 mm (2.5 in.) high, or 102 mm diameter by 203 mm (8 in.) high. Specimens with a 152 mm (6 in.) diameter can also be used.

## Test Procedure

1. Place the specimen in a controlled temperature chamber at the test temperature for a minimum of 4 h. Typical test temperatures are 25°C (77°F) and 40°C (104°F).
2. Attach two LVDTs to the specimen in the vertical direction at the opposite sides using LVDT holders.
3. Precondition the specimen by applying a haversine loading without impact varying between 0 and the maximum load, with a frequency of 1 cycle per second for a minimum 30 s. A maximum load of 250 N is used for the 40°C test temperature and 500 N for the 25°C temperature.
4. Apply a constant load the same as the maximum load used in preconditioning to the specimen for 10 min.
5. Monitor the vertical deformation during the entire loading time.
6. Release the load and record the rebound or resilient deformation over a period of 10 min.

## Analysis and Results

- Determine the creep compliance at times of 1 s, 3 s, 10 s, 100 s, 600 s, 601 s, 603 s, 610 s, 700 s, and 1200 s using the following equation. Note that unloading starts a time of 600 s.

$$J(t) = \frac{\epsilon(t)}{\sigma}$$

where

$J(t)$ = creep compliance at time $t$, mm$^2$/N (in.$^2$/lb)

$\epsilon(t)$ = strain at time $t$ = $\dfrac{\text{(Deformation at time } t\text{)}}{\text{Gauge length, mm/mm (in./in.)}}$

$\sigma$ = axial stress = $\dfrac{\text{(Axial load)}}{\text{Cross-sectional area, N/mm}^2 \text{ (psi)}}$

- Determine the permanent strain at time $t$ of 1200 s.

## Report

- Specimen identification, materials, dimensions, and test temperature
- Axial load applied
- Creep compliance at times of 1 s, 3 s, 10 s, 100 s, 600 s, 601 s, 603 s, 610 s, 700 s, and 1200 s
- Permanent strain at 1200 s

# Experiment No. 29.
# Bending and Compression Tests of Wood

## ASTM Designation
ASTM D143—Standard Methods of Testing Wood

## Purpose
To determine modulus of rupture and compressive strength of wood by testing clear specimens.

## Significance and Use
These tests provide data for comparing the mechanical properties of various species and data for the establishment of strength functions. The tests also provide data to determine the influence of such factors as density, locality of growth, change of properties with seasoning or treatment with chemicals, and changes from sapwood to heartwood on the mechanical properties.

## STATIC BENDING TEST
Specimens 50 mm × 50 mm × 760 mm (2 in. × 2 in. × 30 in.) are used for the primary method, and 25 mm × 25 mm × 410 mm (1 in. × 1 in. × 16 in.) for the secondary method. For the loading span and supports, use center loading and a span length of 710 mm (28 in.) for the primary method and 360 mm (14 in.) for the secondary method.

## Apparatus
- Testing machine of the controlled deformation type
- Bearing blocks for applying the load

(continued)

## Test Procedure

1. Place the specimen so that the load will be applied at the center of the span. The load is applied through the bearing block to the tangential surface nearest the pith.
2. Apply the load continuously throughout the test at a rate of motion of 2.5 mm/min (0.10 in./min) for primary method specimens, and at a rate of 1.3 mm/min (0.05 in./min) for secondary method specimens.
3. Record the load-deflection curve up to or beyond the maximum load. Continue recording up to a 150 mm (6 in.) deflection, or until the specimen fails to support a load of 890 N (200 lb) for the primary method specimens, and up to a 76 mm (3 in.) deflection or until the specimen fails to support a load of 220 N (50 lb) for secondary method specimens.
4. Within the proportional limit, take deflection readings to 0.02 mm (0.001 in.). After the proportional limit is reached, the deflection is read by means of the dial gauge until it reaches the limit of its capacity, normally 25 mm (1 in.). Where deflections beyond 25 mm (1 in.) are encountered, the deflections may be read by means of the scale mounted on the moving head.
5. Read the load and deflection of the first failure, the maximum load, and points of sudden change, and show them on the curve sheet.

## Analysis and Results

- Calculate the modulus of rupture as:

$$R = \frac{Mc}{I}$$

where

$R$ = modulus of rupture, MPa (psi)

$M$ = bending moment = $\frac{PL}{4}$

$P$ = maximum load, N (lb)

$L$ = span length, mm (in.)

$c$ = distance from neutral axis to edge of sample = $\frac{1}{2}h$

$I$ = moment of inertia = $\frac{bh^3}{12}$

$b$ = average width, mm (in.)

$h$ = average depth, mm (in.)

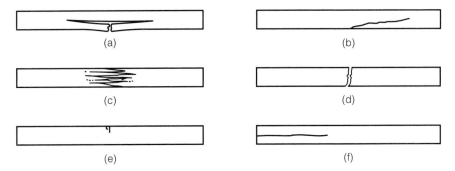

**FIGURE A.16** Types of failure in static bending (a) simple tension (side view), (b) cross-grain tension (side view), (c) splintering tension (view of tension surface), (d) brash tension (view of tension surface), (e) compression (side view), and (f) horizontal shear (side view). Copyright ASTM. Reprinted with permission.

- Static bending (flexural) failures are classified in accordance with the appearance of the fractured surface and the manner in which the failure develops (as shown in Figure A.16). The fracture failure may be roughly divided into *brash* and *fibrous*, the term *brash* indicating abrupt failure and *fibrous* indicating a fracture showing splinters.

## Report

- Specimen identification, dimensions, span length, and other relevant information such as moisture content
- Load-deflection plot
- Modulus of rupture
- Failure condition

*(continued)*

## COMPRESSION PARALLEL TO THE GRAIN TEST

Specimens 50 mm × 50 mm × 200 mm (2 in. × 2 in. × 8 in.) are used for the primary method, and 25 mm × 25 mm × 100 mm (1 in. × 1 in. × 4 in.) for the secondary method. Be careful that the end grain surfaces are parallel to each other and at right angles to the longitudinal axis when preparing the specimens.

### Apparatus

A controlled deformation machine is required. At least one platen of the testing machine is equipped with a spherical bearing to obtain uniform distribution of load over the ends of the specimen.

### Test Procedure

1. Place the specimen perpendicularly on the cross-head of the machine as shown in Figures 10.8(c) and 10.9.
2. Apply the load continuously throughout the test until failure at a rate of motion of 0.003 mm/mm (in./in.) of the nominal specimen length per minute.
3. Measure the deformation over a central gauge length not exceeding 150 mm (6 in.) for primary method specimens, and 50 mm (2 in.) for secondary method specimens. Load-compression readings should be continued until the proportional limit is well passed, as indicated by the curve.

### Analysis and Results

- Plot the load versus deflection diagram.
- Determine the modulus of elasticity as the slope of the straight portion of the stress-strain curve.
- Classify the compression failure in accordance with the appearance of the fractured surface (see Figure 10.10). In case two or more kinds of failures develop, described all fractured surfaces in the order of their occurrence; for example, shearing followed by brooming.

### Report

- Specimen identification, dimensions, and other relevant information such as moisture content
- Load-deflection plot
- Modulus of elasticity
- Failure condition

## COMPRESSION PERPENDICULAR TO GRAIN TEST

Specimens 50 mm × 50 mm × 150 mm (2 in. × 2 in. × 6 in.) are used.

### Apparatus

- Deformation controlled machine
- Metal bearing plate 50 mm (2 in.) wide

### Test Procedure

1. Position the specimen on the crosshead of the machine as illustrated in Figure 10.8(d).
2. Apply the load through a metal bearing plate 50 mm (2 in.) wide, placed across the upper surface of the specimen at equal distances from the ends and at right angles to the length. Measure the actual width of the bearing plate. The specimens are to be placed so that the load will be applied through the bearing plate to a radial surface. The load is to be applied continuously throughout the test at a rate of 0.305 mm/min (0.012 in./min).
3. Take load and deformation readings up to 2.5 mm (0.1 in.) compression, after which discontinue the test. Measure the compression between the loading surfaces.

### Analysis and Results

- Plot the load versus deflection diagram.
- Determine the modulus of elasticity as the slope of the straight portion of the stress-strain curve.

### Report

- Specimen identification, dimensions, and other relevant information such as moisture content
- Load-deflection plot
- Modulus of elasticity

# Experiment No. 30.
# Tensile Properties of Plastics

## ASTM Designation

ASTM D638M—Tensile Properties of Plastics

## Purpose

To determine the tensile properties of unreinforced and reinforced plastic materials, including composites.

## Significance and Use

This test method is designed to produce tensile property data for the control and specification of plastic materials. These data are also useful to characterize the quality of the material and for research and development.

## Apparatus

Testing machine, grips, load indicator, and extension indicator

## Test Specimens

Specimens with an overall length of 150 mm, gauge length of 50 mm, width of narrow section of 10 mm, and thickness of 4 mm are used (see Figure A.17) (Type M-I specimens, ASTM D638M). For isotropic materials, at least five specimens are tested.

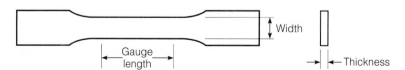

**FIGURE A.17**  Plastic specimen for tension test.

## Test Procedure

1. Measure the width and thickness of specimens with a suitable micrometer to the nearest 0.02 mm at several points along their narrow sections.
2. Condition the test specimens at $23 \pm 2°C$ and $50 \pm 5\%$ relative humidity for not less than 40 h prior to the test.
3. Place the specimen in the grips of the testing machine, taking care to align the long axis of the specimen and the grips with an imaginary line joining the points of attachment of the grips to the machine.
4. Attach the extension indicator. When the modulus is required, the extension indicator must continuously record the distance the specimen is stretched (elongated) within the gauge length as a function of the load through the initial (linear) portion of the load-elongation curve.
5. Set the speed of testing at a rate of travel of the moving head of 5 mm/minute and start the machine.
6. Record load-extension curve of the specimen.
7. Record the load and extension at the yield point (if one exists) and the load and extension at the moment of rupture.

## Analysis and Results

- Tensile strength

$$\sigma = \frac{P_{max}}{A_o}$$

where

$\sigma$ = tensile strength, MPa (psi)

$P_{max}$ = maximum load carried by the specimen during the tension test, N (lb)

$A_o$ = original minimum cross-sectional area of the specimen, $mm^2$ ($in.^2$)

- Percent elongation

If the specimen gives a yield load that is larger than the load at break, calculate the percent elongation at yield. Otherwise, calculate the percent elongation at break. Do this by reading the extension (change in gauge length) at the moment the applicable load is reached. Divide that extension by the original gauge length and multiply by 100.

*(continued)*

- Modulus of elasticity
  Calculate the modulus of elasticity by extending the initial linear portion of the load-extension curve and by dividing the difference in stress of to any segment of section on this straight line by the corresponding difference in strain. Compute all elastic modulus values using the average initial cross-sectional area of the test specimens in the calculations.

## Report

- Complete identification of the material tested
- Method of preparing test specimen, type of test specimen and dimensions, and speed of testing
- Tensile strength, percent elongation, and modulus of elasticity

# INDEX

**Photo and Art Credits,** continued

p. 52: Guy, A.G. *Elements of Physical Metalurgy.* Copyright © 1959 by Addison Wesley Publishing, Reprinted with permission.; p. 92: based on Frank, K.H. and Smith, L.N., *Highway Material Engineering: Steel, Welding and Coatings,* 1990, Publication No. FHWA-HI-90-006, Federal Highway Administration, Washington D.C.; p. 150: based on Kosmatka, S.H. and Panarrese, W.C., *Design and Control of Concrete Mixtures,* 13th edition, 1988; p. 185: ACI Committee 306 report, AC1306R-78 (revised 1983) *Cold-Weather Concreting,* American Concrete Institute; p. 189: Powers, T.C., "Structures and Physical Properties of Hardened Portland Cement Paste," Journal of American Ceramic Society, 41, Jan. 1958; p. 218: Goetz, W.H. and Wood, L.E., "Bituminous Materials and Mixtures," Section 18, Highway Engineering Handbook, 1960, McGraw–Hill, New York, NY: p. 219: The Asphalt Institute, *The Asphalt Handbook,* 1989, MS-4, Lexington, KY; p. 223 a,b: Peterson, J.C., *Rheology of Cold-Recycled Pavement Materials Using Creep Test,* 1984, Vol. 12, No 6, Journal of Testing and Evaluation, ASTM, West Conshohocken, PA; p. 223 c: Peterson, J.C., *Chemical Composition of Asphalt as Related to Asphalt Durability: State of the Art,* in Transportation Research Board, Record 999, Transportation Research Board, National Research Council, Washington D.C. 1984; p. 236: McGennis, R.B., et. al, *Background of Superpave Asphalt Binder Test Methods,* Publication No. FHWA-SA-94-069, Federal Highway Administration, Washington D.C.; p. 254, 257: The Asphalt Institute, *Mix Design Methods for Asphalt Concrete and Other Hot–Mix Types,* 1995, MS–2, Lexington, KY: p. 271, 286: Van Vlack, L.H. *Elements of Materials Science,* 6th edition. Copyright © 1989 by Addison Wesley Publishing Company. Reprinted with permission.; p. 294: Van Vlack, L.H. *Materials for Engineering Concepts and Applications.* Copyright © 1982 by Addison Wesley Publishing Company. Reprinted by permission.